THREAD OF THE SILKWORM

THREAD
of the
SILKWORM

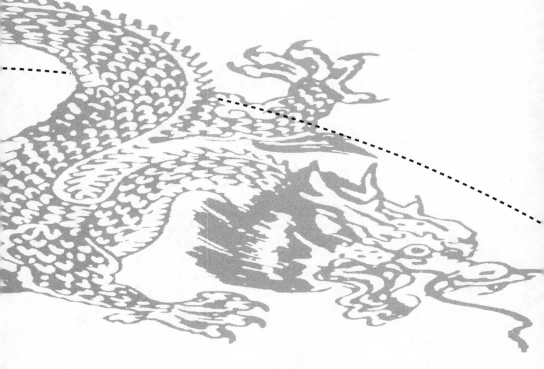

IRIS CHANG

BasicBooks
A Division of HarperCollins*Publishers*

Grateful acknowledgment is made for permission to reprint the following:

From Theodore von Kármán and Lee Edson, *The Wind and Beyond*. Copyright © 1967 by Little, Brown and Company (Inc.). By permission of Little, Brown and Company.

From Milton Viorst, *Hustlers and Heroes*. Copyright © 1967 by Milton Viorst. Reprinted by permission of Wylie, Aitkin & Stone, Inc.

From Edgar Keats, "The Blackboard Jumble," *Saturday Review*. Copyright © 1968. Reprinted by permission.

Designed by Elliott Beard

Library of Congress Cataloging-in-Publication Data
Chang, Iris.
    Thread of the silkworm / Iris Chang.
      p. cm.
    Includes bibliographical references and index.
    ISBN 0-465-08716-7
    1. Tsien, Hsue Shen.   2. Rocketry—China—Biography. 3. Aeronautical engineers—China—Biography.   4. Aeronautical engineers—United States—Biography.
5. Astronautics—China—History.   6. Anti-communist movements—United States—History.   I. Title.
TL781.85.T836C49   1995
629.1'092—dc20
[B]   95-20890
                                              CIP

95 96 97 98 ❖/RRD 9 8 7 6 5 4 3 2 1

Contents

Acknowledgments

A biography such as this one represents not only the solitary work of the author but also, over time, the cumulative help and cooperation of literally hundreds of people. In the writing of this book, I have incurred many debts of gratitude—too many to credit them all. However, here are those who deserve special thanks:

To my husband, Brett, who endured my nocturnal work habits and also my frequent absences when I traveled to China and all over the United States to research this book. His love, good humor, and incredible patience sustained me throughout the development of this biography.

To my editor, Susan Rabiner, who took a gamble on an unknown, untested writer when she signed me on and then spent countless hours nurturing me. Without her guidance, insight, and brilliant editing, this book would simply not be possible. To her assistants, Sarah Stickle and Justin McShea, for their help throughout the project.

To my agent, Laura Blake of Curtis Brown Ltd., who sustained me with her unstinting interest in this book and my career.

To all the participants of the oral history interviews that provided the life and heartbeat of the book, who, alas, are too numerous to list here. To the alumni of Caltech and MIT who took the time to write me letters—some longer than twenty pages—based on a mailing of questionnaires to their homes. Copies of the MIT responses will be donated to the university archives there so that other scholars may make use of them.

To the granting agencies that gave me funds at critical moments of the project. The National Science Foundation provided me with a generous grant to sustain a year of research and writing. An award from the Program on Peace and International Cooperation of the John D. and Catherine T. MacArthur Foundation in Chicago paid for travel to the People's Republic of China and major cities in the United States. A small research grant from the Harry S. Truman Presidential Library made possible the air transportation, room, and board so that I could examine the Truman papers in Independence, Missouri, and the Dwight Eisenhower collection in Abilene, Kansas.

To the archivists, librarians, and historians at various organizations, namely: Archive Photos; Bettmann Archive; Philip Bergen at the Bostonian Society; Judith Goodstein, Bonnie Ludt, Shelley Irwin at the Archives of the California Institute of Technology; Julie Reiz, Kiki Chapman, and especially John Bluth (who proofread the manuscript) at the Jet Propulsion Laboratory; Peggy Hood at the Draper Laboratory; Elizabeth Curran at the Sidney Gamble Foundation; Elizabeth Andrews, Helen Samuels, and Sally Beddow at the Archives of the Massachusetts Institute of Technology; Kara Schneiderman at the MIT museum; Faith Hruby at *Technology Review*; Frank Winter, curator of rocketry of the National Air and Space Museum; the Harry S. Truman Presidential Library; David Haight at the Dwight D. Eisenhower Presidential Library; the history office of the National Aeronautics and Space Administration; John Taylor and Kathy Nicastro at the National Archives; Kenneth John Elwood, assistant district director of the Immigration and Naturalization Service in Los Angeles; Marian Smith, historian for the Immigration and Naturalization Service in Washington, D.C.; David Tulip at NTIS; Jonathan Pollack at RAND; Simon Elliot at the University of California, Los Angeles; Dacey Taub at the University of Southern California; Michael Shulman at Archive Photos; George Cholewczynski and Adam Bernaki at the American Institute of Aeronautics and Astronautics; Kay Schultz at the San Pedro Historical Society; Chi Wang at the Library of Congress; Mitch Sharpe at the U.S. Space and Rocket Center; and R. Cargill Hall, historian for the U.S. Air Force.

To other authors and scholars for their support and assistance: Joseph Bermudez, Frank Boring, Phillip Clark, Barbara Culliton, Fred Dagenais, Hua Di, Hu Guoshu, David Kaplan, Dan Keveles, Arnold Kramish, Dale Maharidge,

Robert McColley, Walter McDougall, Tim Naftali, Robert Reid, Martin Sherwin, George Sutton, Milton Viorst, Frederic D. Wakeman, Jr., and David Wright.

To my dedicated research assistants Yao Shuping, John Sweeney, and Mark McNicholas, who spent weeks finding and translating crucial documents from Chinese into English. To the students from the Applied Learning Program of the University of California at Santa Barbara (UCSB), who worked as my research interns and provided suggestions and much-needed leg work for the project: Connie Crowell, Henrietta Felix, Karin Maloney, Karim Marouf, Susan Moss, Denise Stewart, and Jim Thompson. To Harlan Glatt, who transcribed some of the oral history interviews, and to Marie De Leon, a volunteer from UCSB. To Jenny Liu and Jason Dries-Daffner, who helped provide me with numerous architectural descriptions of buildings in China, the United States, and Europe.

To my parents, for their love, encouragement, and advice, as well as for putting the story in a cultural context. To friends and relatives who have given me all manner of assistance in this project, such as housing, technical expertise, legal advice, and research assistance: Liwen Huang, Paula Kamen, Ron Packowitz, and Carolyn Wu in Chicago; Barbara Chang, Frank and Shirley Chang, Steve Chien, Bryan Davis, and Randy Hein in Los Angeles; Amy Chen in New York; Cheng-Cheng and Carol Chang in Palo Alto; Hui Li in San Diego; Robert Ornstein in Santa Barbara; Mike Ravinitsky in St. Paul, Minnesota; and George Bouza and Tyan Shu-gwei and Ching-Ching Chang in Washington, D.C.

Finally, to those who took the time out of their busy schedules to read over my manuscript, offering insightful comments and valuable criticism. The errors are my responsibility alone.

Introduction

Thread of the Silkworm is the story of Tsien Hsue-shen—a man who hasn't set foot in the United States for almost fifty years and who is known in this country only to a handful of aging scientists. Yet he is considered so important to Chinese space development that newspapers in the People's Republic repeatedly refer to him as the "father of Chinese rocketry," prompting even science fiction author Arthur C. Clarke to name a Chinese spaceship after him in his novel *2010: Odyssey II*.

His life is one of the supreme ironies of the Cold War. Not only was Tsien Hsue-shen (also known as Qian Xuesen) the mastermind and driving force behind the first generation of nuclear missiles and satellites in China (including the infamous Silkworm antiship missile that was later used against the United States during the Persian Gulf War), he had been trained and nurtured for fifteen years in the United States, leaving only because, indirectly entangled with the Chinese role in the Korean war, trumped-up charges of Communism forced his deportation to the People's Republic of China.

Who is Tsien? Born in 1911 as the son of a minor education official in China, he first came to the United States in 1935 on a Boxer Rebellion scholarship. Taken under the wing of Theodore von Kármán, a brilliant aerodynamicist at

Caltech, Tsien helped lay down the foundation of the Jet Propulsion Laboratory. Both during and directly after World War II, Tsien was given clearance to work on classified government projects, despite the fact that he was legally a Chinese national. His work in the fields of fluid dynamics, buckling of structures, and engineering cybernetics made possible the early American entry into the space age.

In 1949, just as China was falling to the Communists, Tsien made the decision to become a U.S. citizen. What he had not counted on, however, was the fact that at this time the United States was entering a period of Cold War hysteria. Many scientists would be caught in its whirlwind.

Tsien would be one of them. During the summer of 1950, a bare year after his return to Caltech as the new Robert Goddard Professor of Jet Propulsion, he was accused of being a former member of the Communist Party—a charge that he vehemently denied. The accusation, however, set off a chain of events that resulted in Tsien being taken into custody and locked in a cell for more than two weeks. Confused, if not panicked, by what was going on, he lost twenty pounds. Upon his release from jail, the Immigration and Naturalization Service started deportation hearings with the expressed purpose of sending Tsien back to China—even though it possessed not one scrap of concrete evidence that Tsien was indeed a Communist.

Despite his own protestations of innocence and further protestations from those who had worked closely with him for many years, Tsien was found guilty and for the next five years lived confined to his house, under the constant surveillance of the FBI. It was a secret cooling-out period. Finally, on September 17, 1955, he was deported to China.

After Tsien was exiled, things began to happen in China at an incredible speed. "From the beginning, 1956 was a year of furious activity," wrote William Ryan and Sam Summerlin in *The China Cloud.* "China's strategic missile program . . . began to take shape during early 1956," wrote Stanford professors John Lewis and Xue Litai. It was "fascinating," Sidney Drell exclaimed in the foreword to their book, *China Builds the Bomb,* how "such a sophisticated technological/military feat could have been accomplished by a poverty-stricken nation with limited industrial and scientific resources—a feat all the more amazing for being accomplished amid the enormous political turmoil of the Great Leap Forward." Missile specialist P. S. Clark observed: "The most important person to return to China was Tsien Hsue-shen. . . . By combining knowledge of the Soviet and American systems—although they were outdated—the Chinese could begin a space program." Ernest Kuh, an electrical engineering professor at the University of California at Berkeley, testified: "Tsien revolutionized the whole of missile science in China—of military science, for that matter. . . . He is the leading scientist and engineer in the country." Zhuang Fenggan, who worked as Tsien's assistant in Beijing and is now the vice president of the China Association for Science and Technology, said: "Tsien started the rocket business from nothing." Without Tsien, Zhuang said, China would have suffered a twenty-year lag in technology. "We wouldn't have the prestige China has in developing a space industry to such an extent today. . . . He was *the* top scientist and most authoritative person."

The story of Tsien Hsue-shen is epic in scale, encompassing some of the greatest technological and political convulsions of this century. It winds from the crumbling of a four-century-old dynasty in China to the terror of Japanese air raids over Shanghai, from the secret American missile tests in a dry river bed in southern California to the deadly concentration camp factories of the V-2 rocket in Germany, from Tsien's imprisonment on a small island in the United States to his conferences with the most powerful members of the Soviet Union and China.

It is the story of one of the most monumental blunders the United States committed during its shameful era of McCarthyism, in which the government's zeal for Communist witchhunting destroyed the careers of some of the best scientists in the country.

It is the story of Tsien's scientific achievement and leadership and how it helped propel not only the United States but his homeland of China into the space age. It is a story of how Tsien was born into a nation of rickshaws and left it a nation of rockets within the span of a single lifetime. It is also the dramatic story of the Chinese struggle to build a nuclear missile and space program amidst decades of internal political chaos.

And finally, it is also the private story of a shy, introspective, brilliant scientist who wanted nothing more in life than to work in peace but was caught up not once, but twice, in the vortex of world politics.

The idea for *Thread of the Silkworm* was not mine but Susan Rabiner's, a senior editor at the Basic Books division of HarperCollins Publishers, who had first heard of Tsien when Judith Goodstein, archivist and adjunct professor of history at Caltech, and Lawrence Badash, history of science professor at the University of California at Santa Barbara, presented their paper "Science in the Haunted Fifties" during a 1990 history of science conference in Seattle. The irony of Tsien's story so intrigued her that she actively sought candidates to write his biography. In 1991, Rabiner offered me the project while I was still in school as a twenty-two-year-old graduate student in the writing program at

Johns Hopkins University. Although I knew absolutely nothing about Tsien's life, his story fascinated me as much it did Susan. I decided to tackle it.

The difficulty of the project became apparent immediately. First of all, Tsien was an expert in aeronautical engineering, an area in which I had no technical training. Being a generalist, he wrote papers on topics that touched on so many fields that not a single expert I spoke with could access the entirety of his work—not even his accomplished former students. Second, much of the material about his life was in Chinese, for which I had only a rudimentary reading knowledge despite oral fluency in the language. Third, the political and secretive nature of his life made it difficult to acquire documents in both the United States and China, for reasons involving national security. And finally, Tsien himself was inaccessible, when a single conversation with him might have forever dispelled some of the shadows surrounding his life. Since his return to China, he has never granted an interview to an American journalist or scholar and has thwarted numerous attempts by Chinese biographers to write his story—giving permission to his secretary to work on a biography only after his death. Tsien once told a Caltech student in the 1950s: "One should never write a book until he is on his deathbed, because he won't live to regret it."

I would like to stress at the outset that this book has certain limitations. For one, much technological detail has been omitted. A redeeming aspect of this omission is that Tsien is primarily remembered not for his scientific achievement in the United States but for his deportation and subsequent leadership in China. There is no question that Tsien was a brilliant, first-rate scholar, but colleagues of his have repeatedly stressed to me that he was not in the same class as Isaac Newton or Albert Einstein or even as his mentor at Caltech, Theodore von Kármán. Tsien may have worked on theoretical problems that proved useful to aerodynamicists in the United States, but he never revolutionized or created his own field. If Tsien had died in 1955 and had never gone to China, his life would not have merited a first-rate biography.

Many believe that as talented as Tsien was as a theoretician, he would be remembered primarily for his leadership in China rather than for his scientific work in the United States. "He didn't see as far—anywhere near as far—as von Kármán did, or Einstein did, or Teller did, or any of those other people who are greats," said Tsien's friend Martin Summerfield, a former aeronautics professor at Princeton University. "He didn't see in the same sense as they did. He would carry out their calculations, be their arms and legs, but not be the mastermind. I think what he took with him was this ability to reproduce. To reproduce what they had produced." Guyford Stever, Tsien's former colleague at MIT, echoed the same belief: "His contribution to the United States was good but not over-whelming. His contribution to the People's Republic of China sounds like it was amazingly great." Holt Ashley, one of Tsien's former students, speculated that his deportation "probably contributed to a relative technical advantage for the People's Republic of China which the United States would not have had, had it not been for Tsien's personally coming back and dedicating his life to some-how getting even with the United States. In retrospect, it's a very unfortunate situation."

In addition, this book makes no pretension to have the final story about Tsien's dealings with the U.S. government during the McCarthy era. There are still numerous Freedom of Information Act (FOIA) requests of mine pending at the FBI that, when finally processed, may shed more light on Tsien's experience with McCarthyism. Legally, the FBI is required to comply with FOIA requests within ten days, but a backlog has prevented the agency from responding for at least two years and sometimes as long as ten years.

Fortunately, there were many other sources available to construct a narrative of Tsien's life. There was no shortage of information about Tsien's twenty-year stay in the United States: a wealth of documents were found in numerous U.S. government and university archives, general publications, and scientific jour-nals. (Portions of his FBI file were tucked away in a U.S. Customs file, and a large Army Intelligence file was located in the National Archives.) The majority of his colleagues and students in the United States are still alive and well, and readily provided oral history interviews about his character, scientific achieve-ment, and problems with the U.S. government. There was also a plethora of news articles about Tsien after his return to China, which was available in English on computer databases or systematically unearthed and translated by my research assistant Yao Shuping, formerly a physicist and official historian for the Chinese Academy of Sciences.

The biggest problem was getting information of a personal nature on Tsien during the years when he helped construct the Chinese missile and space pro-gram. What did he do scientifically? What price did he pay politically? This period of Tsien's life offered nothing more than a gaping black hole for the biog-rapher: indeed, during the 1960s there was barely any news coverage of Tsien in China. Moreover, a sense of secrecy and paranoia among Chinese govern-ment officials prevented other people, such as Milton Viorst, writer for the *New Yorker*, to learn more about him. During my first visit to the People's Republic of China in the summer of 1993, I became keenly aware of this secrecy when I was invited to a dinner in Beijing at which certain colleagues of Tsien's pleaded with me not to write anything that would offend Tsien for fear that they might be punished.

Luckily, I found a small elite of rocket scientists in China who were willing to talk to me. By and large, they had been young engineering graduates of Soviet universities in the 1950s who began their careers right at the time when Tsien first returned to China. A generation younger than Tsien, they were still mentally alert when I talked with them and they remembered with vivid clarity the evolution of the Chinese space program. According to these men, I am the only American to have traveled to China to conduct exclusive interviews with them. I first met with them in Washington, later in Shanghai, and finally in Beijing. I am grateful for their candor and honesty when assessing the flaws and strengths of the Chinese space program, of the political regimes under which they worked, and of Tsien himself. Unfortunately, given the nature of the business of writing about people who are still alive and vulnerable to retaliation in their home countries, many of them will have to remain anonymous in this book.

Contrary to my initial expectations, the vast majority of the people I interviewed in China were eager to talk about Tsien. I was amazed by their kindness and enthusiasm for the project. They invited me into their homes and to their dinner tables and willingly shared with me old photographs, letters, memories. Not one person objected to my use of a tape recorder. The people I interviewed included relatives, friends, former classmates, colleagues, students, and employees of Tsien. Also, there were those who knew Tsien only tangentially but who provided good leads and reference materials: a filmmaker who produced a recent docudrama on the Chinese missile program, news reporters, opera singers, historians, book editors. The interviews took me all over three cities of China: a tower atop a mountain in Hangzhou, a space exhibition in a gleaming Shanghai hotel, an elementary school within a narrow, obscure alley in Beijing, a dinner party within an exclusive government compound.

One of the most important interviews was conducted with Tsien's own son. In 1991, I had the opportunity to talk to Yucon Tsien in Fremont, California, where he was working for a Taiwanese-owned computer company. Yucon was the image of his father when he was forty: he possessed the same small frame, a potato-oval face, round eyes, unblemished skin, a gentle smile, and dark hair parted to one side. We talked in his car and in a nearby park because he did not want his roommate to know that he was Tsien's son.

Yucon was seven years old when he left the United States by ship for China; he would return more than thirty years later. The Cultural Revolution had interrupted his education, forcing him to serve the People's Liberation Army and spend ten years in factories teaching workers how to operate machinery. After the Revolution, he returned to college and received a bachelor's degree in computer science in 1983 from the PLA Defense Science and Technology University in Changsha, a city in the province of Hunan. Three years later, he was admitted to the computer science department of the California Institute of Technology, where he received his master's degree in 1988.

One of the first things Yucon told me during this rare interview was that his father still harbored considerable resentment against the U.S. government for their treatment of him during the 1950s. "It was like having someone as a guest in your house and later kicking him out," Yucon said. "If my father had committed a crime in this country, then my father would have nothing to say. But my father devoted twenty years of his life to service in the United States and contributed to much of this nation's technology, only to be repaid by being driven out of the country."

This is why, Yucon said, that his father refuses to return to the United States—even after being given the Distinguished Alumni Award by Caltech in 1979. He said that Tsien's closest friend at Caltech, aeronautics professor Frank Marble, invited him to return to Pasadena to accept the award during the official ceremony. (Lee DuBridge, then the president of Caltech, even worked with Frank Press, the science advisor to President Carter, to get Tsien's deportation order lifted.) But Tsien did not go back.

Many of Tsien's younger associates in China were completely unaware of his painful history with the INS and were therefore bewildered by Tsien's refusal to return. Some urged him to seize the opportunity to go to the United States and to let bygones be bygones, Yucon recalled.

But Yucon told me there is only one thing in the world that could possibly bring his father back to the United States: an apology from the U.S. government—a gesture of sorts to atone for the decision to lock Tsien away like a common criminal during the 1950s and to make up for the five years of misery he endured before being expelled to China. It doesn't even have to be a presidential apology, Yucon Tsien said, "just an acknowledgment from someone in government that what the United States did forty years ago was wrong."

THREAD OF THE SILKWORM

1

Hangzhou (1911–1914)

His earliest memories, now nearly a century old, may well be of himself as a young boy, no more than three years old and surely not alone, standing at the edge of a lake and looking out over an unobstructed view of hills, pagodas, and temples.

The lake is West Lake. The city is Hangzhou, a beautiful, ancient city. "Above, there's heaven, but below there's . . . Hangzhou" was a popular saying of the time. It has been home to Tsien Hsue-shen's ancestors for more than a thousand years. For the first three years of his life, it will be his home, too.

Three years old is very young to remember anything very much, so it may well be that all Tsien really remembers of the view from the lake is what elders told him of it during the years of his growing up. How could they not want to impart their memories of such a place and time to their young charge? Back then, fishermen drifted across West Lake in wooden skiffs, and lotuses spread themselves in wide green tangles. A famous legend described the spot as a pearl dropped from heaven by a phoenix and a dragon. Each then flew down to its banks and was transformed into one of the two mountains surrounding the city.

In the tenth century, the Wuyue emperor Qian Liu, from whom Tsien was directly descended, deepened and dredged the lake. By the time Marco Polo arrived in 1276, a palace had been built on an island in the center of the lake;

ornately carved boats carried courtesans and musicians arriving to entertain the royalty who now lived at the island palace. Polo himself, the world traveler of his time, declared Hangzhou "the greatest city in the world, where so many pleasures may be found that one fancies himself to be in paradise."

On the east side of the lake stood an ancestral temple. A recent visit to the temple revealed a structure with timber walls painted red and white, its roof covered with gray tile, its entrance guarded by dragon motifs. Also built by Emperor Qian Liu, the temple had survived centuries of war and natural calamities.

From the Phoenix mountain on the north side of the lake, Bao Su Tower rose sharp and small. Standing alongside the family temple, looking across the lake, Tsien would have seen what appeared to be a tiny knife blade of darkness. But if he chose to climb the Phoenix mountain—following a path of stone hemmed in by trees—the blade grew into a massive tower of brick: a gray, ominous structure straining toward the heavens. The sides were patterned with intricate dark squares that resembled windows, and the wide octagonal base of the tower tapered to a small point in the sky.

The original nine-level structure, then named the Heavenly Pagoda, was erected around A.D. 970 by another of Tsien's ancestors, Wu Yanshuang, uncle of the Wuyue emperor Qian Chu. Over the centuries it had been repeatedly destroyed and rebuilt, shrinking the tower to seven levels by the time Tsien was a child. Still, it retained the silhouette of its old grandeur.

Although Tsien was to spend a fraction of his childhood in Hangzhou during the 1910s, the city—or rather, his family's ancient heritage there—was to shape and define his life for years to come. These family legends, as old as they were, instilled in him a sense of pride and reminded him as he grew older—however depressed or tired he might have felt in the moment—that the history of one of China's greatest cities was entwined with the story of his ancestors. If nothing else, it was a reminder that the blood of kings flowed through his veins.

As a family, the Tsiens were scholarly and ambitious, steeped in both Chinese culture and Western ideology. Wealthy and aristocratic, they were nonetheless staunch believers in the principles of education and hard work. They expected their only son, Tsien Hsue-shen, to become a scholar and make a lasting contribution to society.

The father, Tsien Chia-chih (spelled Qian Jiazhi today), was a quiet man, gentle and patient by nature. In his youth, he was tall and slender, a handsome, clean-shaven man moving about in blue cotton robes. Relatives now retain only vague wisps of memory about him, all concentrated in the years when he lived in the British settlement in Shanghai, as a benevolent, elderly gentleman who was a devout Buddhist and who during holidays permitted his great-nieces and -nephews to climb up his knees and play with his long white beard.

Born in Hangzhou in 1882 to a family of prosperous silk merchants, he grew up in an era when the tenets of Western philosophy and methodology were rippling into the Chinese consciousness. As a teenager he studied at Qiushi shuyuan—then a middle school in Hangzhou and later the predecessor of Zhejiang University. Qiushi shuyuan was a cluster of small buildings in the eastern district of the city, only one of which survives today: a whitewashed structure with brown flying rafters and delicately carved wood dragons under the eaves.

One part of the curriculum was traditional, devoted to classical Chinese literature. (One well-remembered professor could recite, entirely by memory, long passages of Cao Xueqin's classic *Dream of the Red Chamber*.) Following China's defeat in the Sino-Japanese War in 1895, activists had demanded additions to the traditional curriculum that would make the country more competitive with foreign powers. Courses in English, biology, and physics were offered, though students could not conduct experiments; they merely watched as the instructors did experiments for them.

In 1902, Chia-chih joined a group of Chinese students traveling to Japan, a country famous at that time for its universities, military academies, and medical schools. Its physical proximity to China—along with cultural similarities—made it a practical alternative to going to Europe or America. In Japan, Chia-chih attended two universities, where he majored in education and philosophy.

When Tsien Chia-chih returned to Hangzhou, he became one of the principals at the Liangzhi Normal School, a teacher's training school. His colleagues consisted primarily of other young men who were in the vanguard of educational reform in China and who, like Chia-chih, had studied in Japan. (One of them was Zhou Shuren, who was to become China's most famous short story writer under the pseudonym Lu Xun.) Tsien Chia-chih taught philosophy and ethics and acted as administrative representative of the school.

There, in his native city of Hangzhou, Chia-chih married Chang Langdran, a woman from a family of silk merchants whose members had obtained powerful political posts in Shanghai and Beijing. On December 11, 1911, his wife gave birth to his first and only child. The very name of the child—Hsue-shen, which means "study to be wise"—reflected the hopes the father had pinned on him.

East of the family temple and parallel to a small stream lay the quiet, genteel road of Fangguyuan. All along the street gates separated the public road from the homes and private grounds of those who lived behind the gates. Behind one such set of gates, beyond three large courtyards, past trees and flowers planted in vases of stone, was a cluster of two-story buildings. Walking toward them, one would arrive at the intricately latticed door of Tsien's childhood home.

Inside, the floors were painted a dark red, and the furniture was made of a scented and expensive carved wood. Delicate scrolls of calligraphy and water-falls hung on the walls, which, like most other homes in Hangzhou, had no glass windows, only square openings that freed the interior to the outside air. In one room, Tsien's parents slept in a bed that was, in the words of one rela-tive, "like a small house." Over the bed arched a blue silk canopy, and along its frame hung silk curtains delicately embroidered with red lotus flowers. Attached to the bed was a rosewood bureau drawer. Tsien had a room of his own.

A day in the Tsien household began, as it did for most other families in Hangzhou, at the crack of dawn. Hu Guoshu, a distinguished Hangzhou histo-rian, described how Tsien and his family might have spent a day in the city in the 1910s.

Upon awakening, Tsien's father most likely put on a *changpao,* a long, flow-ing robe of cotton or silk that was fastened from the neck down with cylindri-cal cloth buttons. It brushed the tips of his cotton, hand-stitched shoes.

All through the town, servants and wives boiled water on earthenware stoves heated with wood, and then poured the water, steaming, into kettles of tea and pots of rice porridge. Breakfast usually consisted of a sweet *fagao* pastry, onion pancakes, or maybe some *baozi,* steamed bread rolls stuffed with meat. From the ceilings of most homes hung bamboo baskets heavy with food; the cracks in the baskets permitted cooling evening winds to act as a natural refrigerant. In Tsien's home, the food was stored in a special cabinet with sides of wire mesh.

Then, perhaps armed with an oilpaper parasol and metal container packed with lunch, Tsien's father would be off to work, either by foot or by rickshaw. A full day of teaching and administrative duties awaited him, and he was unlikely to return until four or five in the afternoon. His son, Hsue-shen, mean-while, would remain at home with his mother.

The feet of Chang Langdran had been bound when she was a child, which broke her foot bones and forced her toes to grow into the balls of her feet. Crippled for life, she was unable to perform any physical labor. But that was the point. Only the wealthiest of men could afford to marry women like Chang Langdran because servants would be required to take care of all the household work. According to the waning memories of Tsien's relatives who knew him in his youth, the Tsien family had three servants: a cook, a maid, and a chauffeur.

Chang Langdran is remembered as an attractive, vivacious woman milling about in a pleated red silk dress. She had a classical Chinese education, rare for women in her generation. In Hangzhou, her family had employed a private tutor to teach her art, calligraphy, history, and literature. As a young woman, she read the Five Classics—the Book of Rites, the Book of History, the Spring and Autumn Annals, the Book of Poetry, and the Book of Changes—that formed the canon of Confucian philosophy. Her admirers remember her as a woman who was quick of wit and smooth of tongue, eloquent in speech and manner. Underneath her poise was a woman bursting with energy. Unburdened by physical labor, she had time to fashion for her only son a cocoon of culture, learning, and gentility.

Beyond the gates of Tsien's childhood home was the rest of Hangzhou, a prosperous city in the 1910s. Geographically, Hangzhou rests some one hundred miles southwest of Shanghai in the Zhejiang province. In the 1910s and 1920s its population was around two hundred thousand. Few other areas in China were blessed with its art, its industry, and its pastoral setting.

Hangzhou was a market city. From its factories came woven silk, cotton, and Longjing tea. In the streets vendors sold silk parasols, sandalwood fans, brocade gowns, and bambooware. The farmers grew rice, millet, sweet potatoes, plums, and watermelon, and around the city were vast groves of mulberry trees and tea plantations. Dinners in Hangzhou were laden with such goods as lumps of meat stewed in soy sauce and ginger, steamed crab, shellfish, and sweet-and-sour carp.

The affluence of both Tsien's family and the city of Hangzhou gave him a sense of security during his formative years. Yet he was living in one of the most unstable times in China's history.

Historically, China has struggled as much as any nation could to keep out foreign influences. As late as 1834 all but one of her ports were closed to foreign trade. But the British in particular were just too eager to gain access to Chinese markets and had too powerful a military force for the Chinese to resist.

After the Opium War of 1839–42, Britain, along with France, Russia, and the United States, began to acquire trading and other rights at a number of Chinese ports. In 1895, with the end of the Sino-Japanese War, the Chinese were forced to sign a humiliating treaty with Japan, which opened Hangzhou, along with three other treaty ports, to foreign trade.

The influx of foreign ideas and technology, while unsettling in the way that all change is unsettling, accelerated the pace of life in Hangzhou. The Tongyi Gong Cotton Mill opened in 1897 and the Zhenjiang Xingye Bank ten years later. Construction began in 1905 on a railway that would connect Hangzhou with Shanghai. The sale of newspapers mushroomed almost overnight, with residents subscribing to the local *Hangzhou Baihua Bao* and *Zhejing Chao,* as well as the regular Shanghai dailies. Hangzhou was in the throes of an industrial revolution.

Then came another revolution. In 1911, after more than two centuries of rule, the Qing (Manchu) dynasty crumbled. Fear and hostility toward the European interventionalists no doubt contributed to anti-Manchu feeling in China; more radical Chinese believed that the nation could be strengthened by overthrowing the Qing and establishing a constitutional government. The very month of Tsien's birth, mutinous troops defeated imperial forces in Nanjing and proclaimed a new government. Sun Yat-sen, leader of the nationalists, returned from exile on Christmas Day to become the first provisional president of the Republic of China.

Although Tsien would have been too young to remember, celebrations broke out across China. In Shanghai, residents tore down the city walls, seen as a relic of medievalism. The pigtailed queue, long considered a symbol of submission to the Manchus, was banned, and barbers were stationed by city gates to seize queues in the street and shear them off. In Hangzhou, local Manchu officials were arrested, while townspeople hoisted paper lanterns and brightly colored sashes and flags onto pavilions across the city to celebrate the first Republican New Year.

It was Sun Yat-sen's dream to have a democratic China with free elections, a senate, and a house of representatives. That dream was shattered when Song Jiaoren, leader of the majority party of the Guomintang (the GMT), was assassinated in 1913—a murder linked to Yuan Shikai, Sun's successor. Then Yuan outlawed the GMT completely and in 1914 dissolved Parliament. Sun was exiled once again, this time to Japan, while Yuan began to take on the trappings of emperor.

Despite his abuse of political power, Yuan pushed for educational reform in China. He wanted compulsory and free primary education for males and started experimental programs in teacher retraining. These reforms coincided not only with the beginning of Tsien's education but with Tsien's father's appointment in the Ministry of Education. It appears that in 1914 Tsien Chia-chih resigned from his post at Hangzhou, packed his belongings, and moved his family to Beijing. The thousand-year connection between the Tsien family and the city of Hangzhou had abruptly come to an end.

2

Beijing (1914–1929)

His family's move to Beijing was the most significant event of Tsien's childhood. He would enter the city as a toddler and leave fifteen years later as a young man.

Beijing was and is the most political of China's cities. For three thousand years it had been the political center of the country and for the past five hundred its capital. In 1908, just a short while before Tsien's arrival, approximately one-third of Beijing's seven hundred thousand people belonged to the military and administrative bureaucracy of the Qing reign. When Tsien arrived, he likely found the streets filled with officials wearing the dignified uniform of China's elite: the blue cotton *changpao* gown, which were as commonplace in Beijing then as the gray pinstriped suits are in Washington, D.C., today.

Fifteen years among Beijing's towers, stupas and gates, breathtaking in their bulk of marble, and among imperial courtyards that stretched unbroken for miles within crenelated walls gave Tsien the opportunity to absorb the thousand-year-old Beijing culture while witnessing firsthand China's turbulent transition into the modern age. This up-close exposure to Beijing's vast halls and palaces, gleaming with lacquered furniture and ceilings of pearl and gold, lavish with jade statues and thrones engraved with dragons, would inevitably impart to Tsien more than a whiff of the arrogance for which Beijing was famous—an arrogance that became infused with his own self-confident personality and would remain a part of him to the end of his days.

Friends and relatives fail to remember exactly where the young Tsien lived in Beijing, recalling only vaguely that it was in a quiet residential area, possibly in the Xuanwumen wai district. If it had been a typical home of an upper-middle-class family living in Beijing at that time, it would have been built in the northern Chinese architectural style, with a courtyard and windows on the south side of the buildings to let in the warmth and light of the sun and a solid wall on the north to shield the home from wind and dust storms. Most families had no plumbing or electricity; lamps were lit by oil, ovens heated by coal. Only the wealthiest of families had running water; the rest bought well water. Every three to five streets had a water house, with two or three rooms and a well in front, run by a certified water master and a dozen or so employees. The water men made daily rounds through the neighborhood, pushing wheelbarrows heavy with two elliptical cases of water. Flush toilets did not exist, and night soil carriers arrived to clean out the latrines. The dung was dried, packed into gunny sacks, and carried by camel out of the city to be sold as fertilizer.

Everything about Beijing—its homes, streets, and ancient landmarks—was designed to accentuate, not diminish, class lines. The city was enclosed within a succession of walls that guarded homes of greater prestige and power as one moved toward the center. First, one moved within the boundaries of the Tartar City, originally reserved for Manchu troops. Closer to the center stood the smaller Imperial City, where the high Manchu officials lived. Finally, at the core of Beijing lay the Forbidden City, the home of the emperor and one of the greatest architectural achievements in the world. The palace grounds were a secluded resort of lakes, statues, and marble bridges once tended by tens of thousands of gardeners and cooks, eunuches and concubines.

Most homes stood in stark contrast with these imperial dwellings. Viewed from a tower, they formed a dark sea of glazed mud tiles, with rooftops curving up like the tips of waves. There was a seamy side to Beijing that Tsien rarely saw. A dirty brook ran through the working-class districts of the city, and during heavy rains it would rise and flow into homes and alleys, drowning children and spreading filth and disease. As many as twenty or thirty laborers slept huddled together on the dirt floors of courtyard tenements and rickshaw garages, and some families were so poor they shared a single pair of trousers.

The most striking symbols of class consciousness and oppression appeared on the streets. The wealthy were carried aloft on sedan chairs and horse carriages; the rest of the population walked. One out of six males in Beijing pulled rickshaws; some of these porters were as young as thirteen, some as old as seventy, with bent backs deformed from a lifetime of labor. They wore wide-legged trousers bound at the ankles with strips of cloth or chicken intestines. The bells of their rickshaws rang constantly, "like a thousand ringing telephones," to warn pedestrians of their approach. With so many human horses at their service, the mandarins of Beijing hardly deigned to walk. They rode in rickshaws in part to express their elevated status and in part to avoid walking on Beijing's unpaved streets, which were thick with dust during hot, dry seasons and swamped by mud during wet ones. "Men in long gowns may not walk," wrote one newspaper columnist. "It's an unwritten law in Beijing." Some, like Tsien's family, employed private rickshaw boys to chauffeur them through the city.

The streets and marketplaces of Beijing must have seemed endlessly entertaining to a small boy of relative privilege like Tsien. Acrobats twisted their bodies into grotesque shapes, while candymakers blew bubbles of toffee into animals and vendors wove reeds into toys. A Beijing native might while away countless hours sipping tea in Chinese opera houses; a boy new to the city would likely be drawn to these streets noisy with puppet performers, fortunetellers, or rickshaw boys, who, like urban taxi drivers, often came to blows during traffic jams.

Tsien could not but observe the special respect Beijing natives reserved for learning. Beijing was a city of bookworms. While some Chinese provinces of the time were 99 percent illiterate, in Beijing one could observe laborers, servants, and rickshaw boys reading newspapers and books. The city's high literacy rate was attributable in part to the triennial palace examinations. Every three years, a fresh wave of scholars descended on the capital to take the last of three tests that would confer upon them the highest educational status in the nation. Only a fraction passed. The rest, either entranced by Beijing or too ashamed to return to their home villages, found work as private tutors or teachers. Thus, for centuries they directly and indirectly enriched the cultural life of the city.

It was in this environment that Tsien was sent to nursery school, one of the first nursery schools established in Beijing. Then, probably around the age of seven, he was enrolled in the Beijing No. 2 Experimental Primary School, organized exclusively for gifted children.

—◌◦—

The school stood in a quiet area of town once reserved for the exclusive use of noblemen and princes. It was housed in a *wangfu*, a luxurious dwelling originally designed for the blood relatives of emperors. This *wangfu* was distinguished by its open-lattice doorways and the lacelike designs on its eaves. The symmetrical layout of the school formed six courtyards, connected by long walkways of expensive rosewood and bamboo. There were gardens of fragrant *haitang* flowers pink with blossom, and undulating white walls and entrances known as moon gates. The setting was a peaceful one, most likely broken only by the laughter of small children and the sound of running feet.

Tsien's father worked in the national government's Ministry of Education, the ministry responsible for inspecting schools, updating textbooks, modernizing private institutions, monitoring foreign-run schools, and starting programs to combat illiteracy. Beijing No. 2 was one of the very first public elementary schools in China, founded on September 19, 1909, as the attached laboratory school to the Jingshi Women's Normal University. Following the trends of educational reform, Jingshi was one of the first universities in China to admit women, and its elementary school accepted girls as well as boys. By 1918, the year that Tsien is remembered to have arrived, Beijing No. 2 was one of several laboratory schools administered by the Beijing Normal University, China's premier institution of modern education.

A diploma from the school practically guaranteed future government employment, and so competition for admission was keen. On the day of the entrance examination, hundreds if not thousands of children arrived at the gates of the school from all districts of Beijing. The wealthy came by rickshaw, the humbler by foot. Freshly scrubbed and wearing their best clothes, children nervously waited in line with their parents. The children were paraded in, one by one, before a panel of judges.

A primitive oral intelligence test was administered. Tsien was probably asked to describe what he saw in drawings, to solve riddles, and to demonstrate an elementary competency in math by counting backwards or making correct change. A physical exam followed. The nearsighted, the weak, and the colorblind were weeded out. In fact, children deemed too short or too tall, too fat or too thin—or even too ugly—were instantly disqualified. "In Tsien's day, the chosen had to be not only intelligent but beautiful and healthy, with the proper height, so they all could stand in a nice neat row and photograph attractively," remembers one teacher, Huo Mianzheng, who was familiar with the history of Beijing No. 2. "There were so many children who applied, the school could afford to be picky."

The teachers themselves were no less rigorously screened. Only the most academically gifted in the nation were permitted to attend Beijing Normal University, and of those who graduated, a handful were invited to stay on as teachers at one of the experimental lab schools. They were a serious, dedicated lot, taking pains to impart the discipline of study through the careful preparation of their own lectures. (It was not uncommon for a teacher to devote an entire evening writing out intricate outlines and sample compositions.) The prestige of their positions resulted in low turnover, and it was not atypical for a teacher at Beijing No. 2 to be hired as a fresh twenty-two-year-old college graduate and retire from the same school forty years later.

As a student at Beijing No. 2, Tsien was spared the rigidity and cruelty that marked much of Chinese education at the time. Beatings were uncommon here; teachers hated to raise their voices at a child, even in anger. In accordance with their philosophy, teachers sought to demonstrate the appropriate code of behavior through their actions, not words. When visiting a teacher in his office, a child was likely to be offered a seat and a cup of tea as if he were an honored adult guest.

But in other ways the school was as strict and formal as all schools in China at this time. Every morning, hair, fingernails, and general cleanliness were checked. There was a strict dress code for both student and teacher. The children wore white cotton uniforms and tiny cloth-soled shoes. The male teachers wore dark jackets and white gowns, and the women white, long-sleeved shirts with stiff collars and white pants. During class lectures, Tsien and his classmates had to sit as rigid as soldiers, shoulders thrown back and hands clasped behind their backs to prevent fidgeting.

In the mornings, students practiced the Chinese ritual of writing. They washed their hands, positioned the paper carefully on desks, and rubbed ink sticks in circles onto wet platforms to form small puddles of the thickest, darkest ink. Daubing a horsehair brush into the ink, they filled row after row of paper with Chinese pictographs. During the first year, when their hands were small and slight of muscle, the calligraphy was large, boxlike, and stiff. By the second year, it began to soften and flow, with lines curving into twists and flourishes. By the third grade, they stopped tracing characters in repetitive rows and started writing on their own. Within the space of six years, a student at the school was expected to memorize more than thirty-five hundred words, enough to permit him or her to read books and newspapers.

In the afternoons, the school turned to exploration. A long-standing tradition at Beijing No. 2 was to encourage children to construct "nature diaries," in which they substituted rose petals, leaves, and pictures for words they had not yet mastered. There were classes in the earth sciences, geography, music, and art. There were also frequent field trips to famous landmarks in Beijing, such as the Great Wall of China, twisting white and serpentine across the northern mountains, the Ming tombs, and the spacious Beihai and Zhongsan gardens.

The years he spent at Beijing No. 2 must have been fond ones for Tsien. A model student, he outshone his peers in course work and was always deferential to his teachers. The instructors, aware of his intellectual precocity, moved him up a grade. His classmates remember him as an exceptionally bright boy whose paper airplanes flew faster and further than all the others. "He folded [them] very precisely and carefully, making them symmetrical and the creases smooth and even, and therefore when the paper airplanes were thrown, they flew stable and far," remembered his friend Zhang Wei years later. "From this little game [one] could see that from the time he was little, when doing something, he always considered it very carefully and thought of a scientific way to achieve [his] purpose."

His school days took on a rhythm of their own. When classes adjourned, the family rickshaw arrived and Tsien climbed in, kneeling on the seat facing the back—his diminutive figure staring back at his classmates as the rickshaw sped away. Then his education continued at home. His father, rather than his mother, appears to have been the dominant influence in the boy's early life. Rather than stuffing the boy's head with facts, he fed his son's curiosity and encouraged the young Hsue-shen to pursue his own interests.

His hobbies flourished under his father's direction. As an amateur taxidermist, Tsien put together his own stuffed bird collection of crows and sparrows. In the summer he caught and studied butterflies, and searched for rocks and fossils. He took lessons for piano, violin, and watercolor painting. His room at home was filled with books on natural science and mathematics. Tsien Chia-chih spared no expense when it came to purchasing books for his son's education. "My father was my first teacher," Tsien said seventy years later for the *People's Daily,* a major Beijing newspaper. "He opened a new world for me in art, music, and literature."

If you catch insects, Tsien's father said, you will begin to understand biology, the study of life. To find a fossil or a fragment of rock is to have a glimpse of geology, a clue to how the earth was formed. To learn to draw, you must ponder the concept of beauty. Hsue-shen loved to draw. He later told his own son Yucon that, had he not become a scientist, he would have become an artist.

As the years loped by, Tsien grew up a Chinese boy in an atmosphere that was not entirely Chinese. It was an atmosphere that encouraged Tsien to question, seek answers, and even challenge authority. Question he did, but the streak of rebellion was not in him. A naturally quiet boy, he spent most of his time at home, shunning sports for study and reading books instead of playing with other little boys in the neighborhood. "No one," a relative said emphatically, "could have asked for a more obedient son."

When he was about ten, Tsien transferred to another school a few blocks away. Beijing No. 2 was coed from grades one to four, but starting with the fifth, classes were segregated by sex. The girls stayed on at Beijing No. 2, but the boys who wanted to continue their education moved onto the Beijing No. 1 Experimental Primary School.

Passage to the upper levels of grade school was not automatic, however. The students in Tsien's class had to take another battery of examinations much more rigorous than the arithmetic and riddle games they had to pass in order to get into Beijing No. 2. Furthermore, they had to compete with applicants from other schools. As many as 1,600 children citywide appeared on examination day for 160 coveted spots. Tsien, however, was an exception. His academic achievements had been so outstanding that the teachers selected him as one of the two or three students who could enter the next level without taking any additional exams. He was, as one teacher put it, one of the top students "academically, physically, and spiritually."

In 1921, together with more than a hundred other boys, Tsien arrived at his new school. The main building was a three-story, gray-brick structure pitched with smooth tan tiles. It stood directly across the street from its parent institution, the Beijing Normal University.

Climbing the staircase of the building and following the circular twists of its red balustrades, one emerged into a long hall flanked on one side by classrooms and on the other by a row of casement windows. The brisk breezes of autumn wafted into the hallway, and from the windows Tsien could see the rickshaw men and pedestrians moving along the street below, and, further away, the dust-colored roof tiles and dark brick buildings of Beijing Normal University.

Entering one of the large, spacious classrooms, he would find himself in a well-equipped room with shiny blackboards, comfortable desks, and chests of books, all brightly lit by the sunlight that poured through the wheel windows on the other side of the building. Two years of hard work and intense competition loomed ahead; then he would face yet another series of tests: tests that would admit him to junior high school, to high school, and eventually to college.

While China was trying to modernize its educational system during the years of Tsien's childhood, his education showed few signs of change. School for Tsien, like school for his father, was marked by memorization, reverence for authority, and an ever-prevalent obsession with taking and passing tests. All had historical precedent and would not easily yield to pressure for change.

China is perhaps most infamous for its test system, whose roots lie in the Han dynasty. Not until the Ming did the tests assume so dominant a role in Chinese culture. On the surface, the system gave the illusion of creating a meritocracy in Chinese politics. In reality, however, it discouraged real insight and boldness of inquiry and mainly benefitted the power elite. The tests were open only to a select group: women were entirely excluded from the process, as were priests, executioners, brothel owners, barbers, men in mourning, and the immediate descendents of actors. But the difficulty and special focus of the tests excluded more people than the entry rules suggest, and only the wealthiest of families could afford to hire tutors to help their sons prepare. The less privileged often spent years toiling over the classics on their own, growing old in their attempts to pass the exams.

Scholars who aspired to recognition and power had to pass three tiers of tests. The first was a preliminary exam given annually to minors by the magistrate of the biggest city in each district. To pass, test takers had to write a series of essays and poems over a period of twelve days. A select few were eligible to take the biennial entrance exam for the district academy. The candidates for this test ranged from teenagers to men in their eighties, all of whom agreed to be shut up in narrow cells for one day and one night to produce more essays and poetry. To guard against cheating, the entire process would then be repeated. Out of some 2,000 test papers, only 20 would be selected, mainly those showing excellence in literary style and beauty of penmanship. The authors would be awarded the "Budding Genius" degree, declared exempt from taxes and corporal punishment, and then admitted to the local district academy, where three years of studies awaited them.

Next came the provincial exam, administered once every three years. The scholars faced stiffer competition: only 100 out of 12,000 candidates would survive this second cut. For three days they were confined in cells barely large enough to sleep in. There they wrote interpretative essays based on the works of Confucius and Mencius and composed a poem that followed a strict literary format. Some devised ingenious methods of cheating—the penalty for those caught was death. The pressure was so intense some went insane, while others fell ill and actually died in the cells. The ones who emerged as winners were, in the words of one historian, "very like the victors in the Olympic games." They were heaped with honors, given powerful official posts in government, and invited to take the third and final test.

The palace examination, held in the imperial palace of Beijing once every three years, was proctored by the emperor himself. The ten top papers were selected, and the best of these was marked in vermillion by the emperor, who proclaimed the author the "Model Scholar of the Empire." The members of this tiny elite of scholars were almost guaranteed positions in the capital, and many became the emperor's personal aides and ministers.

The tests were constantly subject to change, depending on politics and the shifting whims of nationalistic mood. They often took on the characteristics of the emperors themselves. The Qianlong emperor, who was given to luxury, preferred an ostentatious style of writing. "Even calligraphy assumed a fat, rich and satisfied appearance," noted one scholar. In contrast, when the Yongzheng emperor (1723–35), a humanitarian, ruled China, the essays waxed philosophical, with idealistic proclamations to end all war and human suffering.

Sometimes the tests even took the form of subtle espionage. In 1644, the Manchus from China's northern frontiers invaded and conquered the entire country. Preoccupied with uniting and subjugating their empire, the Manchu conquerors used the test system as a way of gathering information from each region of China. Candidates were asked to give short, direct answers to practical problems, such how to suppress rebellion, raise funds for the military, detect government corruption, and promote cooperation between the Manchu and Chinese peoples.

By 1740, the system came under attack by scholars. The population had increased, while the quotas of test candidates remained the same, and even passing the exams did not guarantee a scholar employment. The government had long ignored the problems and complaints of out-of-work scholars, but by the beginning of the twentieth century it was forced to change the test system to address the dominance and technological superiority of the West.

After China's defeat during the Boxer Rebellion in 1901, only a few years before Tsien was born, the Qing government signed a treaty that promised to ban the examinations for five years wherever antiforeign massacres had taken place. In 1901, the Manchu Guangxu emperor abolished the eight-legged essay structure of the ancient examination system, shifting the focus away from calligraphy and classical poetry and toward the more practical problems of government. The examinations themselves were abolished in 1905, and events in China would shake up other aspects of the rigid educational system.

In 1915, Yuan Shikai, who had not only eliminated free elections but proclaimed himself emperor, watched as his administration sank deeply into debt, subsisting primarily on foreign loans. Mass protests against his abuse of power broke out across China. One by one, warlords from each locality declared their independence from the central government. Yuan died in 1916 of uremia, his physical deterioration compounded by shame and grief. Civil war erupted among the various warlords, and natural disaster and famine ravaged the country. As diligent and happy as Tsien was in his studies, he was nevertheless growing up in a nation of chaos.

In the early 1920s a spate of influential visitors arrived in Beijing: the educational innovator John Dewey, the British philosopher Bertrand Russell, the German physicist Albert Einstein, the Indian poet Rabindranath Tagore, and many others. Perhaps the most influential among them was Dewey, who spoke before packed audiences advocating the need to combine education and industry in a democratic setting. Stressing the importance of a child's taking an active rather than a passive role in the classroom, Dewey introduced to China the concept of "play laboratories" equipped with boxes of sand, measuring utensils, and building blocks. Educators in China began to pay more attention to American innovations such as coeducation, IQ tests, psychological testing, and extracurricular activities. These ideas would have reached Tsien both at school—itself a laboratory for educational theory—and at home when his father returned daily from work at the Ministry of Education.

Although the ancient tradition of education in the service of hierarchy crumbled, it did not fully give way to a modern Western system. On one hand, Tsien was encouraged to explore, question, and challenge the body of existing knowledge in the Western empirical tradition. On the other hand, he was not to question the authority of the teacher. Endless hours of memorization and testtaking were as firmly entrenched as ever. As a result, Tsien spent his childhood striving for perfection in a system less than perfect—a system that had become by the 1920s as treacherously shifting as quicksand.

—◌૦——

Two years passed. In 1923 Tsien was admitted to the Beijing Normal University High School, the most prestigious coeducational secondary school in the city and part of the same system of laboratory programs that had nurtured him throughout his childhood. It was right next door to his elementary school and bore the exact same style of architecture, and the teachers tended to be graduates or professors of Beijing Normal University. Going to high school was more or less a continuation of rather than a break from his old routine.

Of course, the high school was much bigger than the Beijing No. 1 Experimental Primary School. It consisted of a series of one-story buildings of classrooms, teacher's offices, and dormitories; a wide dirt oval for track and field; tennis courts; volleyball courts; a new library; well-tended gardens of lilacs; and laboratories stocked with the shiniest new equipment from Germany. A bell fastened to a tree was rung by an old man precisely forty-five minutes after each hour to signify the end of a class period.

"In many ways," recalled one alumnus, "the students at this school were not unlike those at New York City's famous Bronx High School of Science." Most of the students wanted to pursue careers in science, engineering, or medicine. All students took a solid three years of English, mathematics, Chinese, biology, chemistry, history, physics, and a second foreign language. Then they specialized in either the sciences or the humanities for the remaining three years. Some students conducted independent research projects under the guidance of their teachers or enrolled in advanced college-level courses, like calculus and sociology.

Despite its university atmosphere, the high school was run like a military academy. The boys wore gray cotton uniforms, the girls white blouses and black skirts. Dating was strictly forbidden. "There were a few students who fell in love," remembered one graduate. "[But] once the school discovered the romance, one member of the couple was urged to leave the school."

It was also considered taboo to cram the night before an exam. Students were expected to absorb information as it came to them, to truly understand the material and make it a part of their daily life. The teachers helped the pupils master notetaking and mnemonic techniques. Tsien did not cram; indeed, he probably did not have to cram, for he had an excellent memory. But he enjoyed learning and appeared to study all the time. After school, he often sat alone in an empty classroom, poring over books.

He continued to be the perfect student. When asked to describe Tsien years later, his classmates repeatedly used adjectives like "quiet" and "well-behaved." As always, his life was meticulously organized. He played ball with the other boys but went home at exactly the same time each afternoon. When friends visited Tsien at his home, they found his room impeccably tidy.

Once again, he academically outpaced his peers. Wrote alumnus Tan Yinhen: "I heard classmates say that Li Shibo (a veteran teacher of the school, who taught us biology and botany, and who had quite a reputation at the school) often praised Tsien's singleminded devotion and painstaking research in his studies, and admonished us to take him as an example and study hard."

High school gave Tsien three years to explore different career paths. His future began to uncoil before him. He painted animals and waterfalls and flowers. His ecology charts of biological life were so superb they were retained by the school for generations afterwards. He was profoundly sensitive to music and played the violin. He discussed the works of Lu Xun and other famous writers with other classmates and even took up debating as the student representative of his class. As talented as he was, he could have gone into any one of a dozen different fields by the time he graduated. But when Tsien reached his third year, he had made his decision: his future lay in science. For the next three years, he launched into a full course load of advanced chemistry, physics, biology, and mathematics.

If the economy in Beijing had been more stable during the 1920s, Tsien may very well have gone into an entirely different line of work; for example, he had seriously considered a career as an artist. But his high school years coincided with some of the most difficult years of the young Republic of China.

The country was disrupted first by anarchy, then by civil war. Warlords controlled the countryside until 1925, when the Guomintang military leader Chiang Kai-shek led Nationalist armies with the Communist Party to crush the warlords and reunite the country. Three years of fighting ensued, after which the GMT emerged victorious. In 1928 the GMT moved the capital to Nanjing, home of Chiang's power base, while Beijing spiraled into a deep recession.

All except those with the most marketable skills floundered. Beijing was a city in decay in the 1920s: paint flaked off the Forbidden City, construction was abandoned, and the roads were a broken surface of concrete, plumbing, electric wire, and mud. Unemployment soared, and professors, officials, even former generals found themselves pulling rickshaws. (Some college students, driven to desperation by poverty, assumed dual identities: they ducked into alleys, pulled off their gowns, and donned the rough garb of the rickshaw boy.) Teachers, disillusioned with the system, began to turn to socialism. All of this made a lasting impression on the young Hsue-shen. When reminiscing about his high school days, Tsien remembered the courage the headmaster exhibited in pulling the high school together. "You can imagine what a miserable time it was in Beijing in old China," Tsien later recalled. To run a school under such conditions was "an impossible task," he said, and yet the headmaster not only ran it but turned it into one of the best schools in the country. "It was a miracle!"

In 1928 and 1929, when Tsien was a senior, the question of college weighed heavily on his mind. On a class field trip, he visited two famous colleges in town—Qinghua and Beijing University. His schoolmates remembered him examining the laboratories very carefully and making astute comments about each.

That Tsien would go to a top school was beyond doubt. Practically without exception, all the students from his high school would be admitted to college, and half would get into the best ones in the nation. In 1929, the Shanghai newspaper *Shen Bao* published a list of students who had passed the entrance exam of the city's prestigious Jiaotong University, the best engineering college in China. Tsien's score placed him third in the nation in mechanical engineering.

After careful deliberation, he decided to major in railway engineering. It was a relatively new field in China, and a fast-growing one. Vast territories were still untouched by rail in 1929. It was not until very late—1881—that the very first steam engine in China was constructed, pieced together with scrap parts from the West when seven miles of track were laid between the Kaiping mines and a nearby canal. Fifteen years later, in 1896, China had only a meager 370 miles of railway track, compared to the United States's impressive 182,000. But in the ensuing three decades, railroads would emerge as a powerful military and political force, forever scarring the face of China.

The "railroading" of China began at the turn of the century, over the protests of the Qing government. Within their sphere of influence, the Russians began to build a line from Heilongjing province to the port town of Vladivostok; the Japanese lay track from the city of Pusan in Korea to Mukden; the Germans embarked on their projects in Shandong; and the British built railways in the Yangzi valley. Between 1900 and 1905, foreigners built a total of 3,222 miles of railroad track in China. Each country built its track to a different gauge, thereby preventing the linking of rail line. In effect, this modern transportation system began to isolate one region of foreign-dominated China from another, effectively carving the country into separate colonies.

It soon became clear that those who controlled the railways controlled China itself. During the 1900 Boxer uprising, for example, the Qing used Beijing railway lines to speed the movement of their troops while tearing up the track behind them to slow the advance of foreign armies. During the 1911 revolution, the Qing government sped troops from the north down the Beijing-Wuhan railway to crush the rebellion in Wuhan, while mutinous forces rushed from Taiyuan to sever the supply lines. The mutinous forces were ultimately victorious.

When the Nationalists took power, they struggled to centralize the system. Between 1912 and 1920, more than a thousand miles of new railway track were laid in China, much of it funded by Japanese, European, and U.S. bankers. As new tracks were being built, new engineers were being hired, and the demand seemed to rise every year. Here was a chance for a young aspiring student to get into an important industry close to the ground floor. In the fall of 1929, Tsien set aside his high school military uniform for the long gowns of a Chinese scholar and headed for Jiaotong University to become a railway engineer. He bid farewell to Beijing as well.

Shanghai (1929–1934)

Although Shanghai was one of the world's largest cities, with all the ills associated with great urban centers, including widespread crime and social decay, Jiaotong University, located at the city's southwestern outskirts, was in a district more suburban than cosmopolitan, surrounded by quiet residential neighborhoods and small shops. In 1929, the year Tsien arrived, only the occasional rumble of trolleys, the hum of buses, and the screech of automobiles in the distance hinted that the noise of the city proper would eventually make its way to this outermost part. But in 1929 the academic serenity of Jiaotong University was broken only by the occasional pattering of a rickshaw boy or by the calls of vendors selling bowls of wonton soup outside the college gates.

Founded in 1896 as the Imperial Nanyang University, Jiaotong University owed its origins to Sheng Xuanhuai, then the director general of the Imperial Railway administration, who urged the emperor to establish a college of science and engineering. The forced opening of Chinese ports by the better-armed British during the Opium War and the more recent defeat of China by Japanese forces in the first Sino-Japanese War must have reinforced the need to fund

new colleges devoted to advancing the state of Chinese technology. With donations from merchants totaling 8,785 *liang* of silver, the school opened with
three hundred students. In 1897, an American missionary was invited to serve
as the college's first dean of Western studies, marking the beginning of American influence at the school.

For decades, American engineering programs would serve as models for
those offered at Jiaotong University. Many among the faculty had received their
training in the United States, and the curriculum followed that of Cornell University and MIT. Lectures were conducted in English. The school grew swiftly,
and when Tsien arrived on campus, there were 128 teachers, including 33 professors and 800 students belonging to four colleges: electrical, mechanical, and
civil engineering, and railroad management. The railway engineering program
in the mechanical engineering college, Tsien's program, was relatively new:
begun in 1906, it was not offered as a major until 1918. By 1929, several factories in Shanghai donated equipment to the school and recruited its graduates.

For the next three years, Tsien studied the basics: fundamental physics,
chemistry, and mathematics. He would also learn mechanical and electrical
engineering, mechanical design, and even factory training. His last year would
consist of classes in railway engineering and an independent senior research
project largely devoted to the designing of locomotive engines.

The first few days at Jiaotong must have been busy ones: moving in, unpack-
ing, registering for classes. The university was an imposing sight for the young
men arriving to study there. Passing through its gates, they entered a quadrangle surrounded by several brick buildings, all constructed in the Italian Renaissance or Greek styles. The most impressive building was the auditorium: a
three-story symmetrical structure dominated by arches, pilasters, and a clock
tower. The university was in the midst of a building boom, with a half-finished
dormitory and machine factory beyond the quadrangle.

Tsien spent part of his college career in the Zhixin West Dormitory, a three-
story, modern red-brick building with a gray tiled roof. The main section was
flanked by two wings, and in front, atop a concrete column, was the school
emblem: a bronze statue of an anvil, gears, and a furnace within a serrated
circle.

The routine at the university was probably not all that different from the routine at other universities in China. All students awoke early and dressed
quickly, invariably in the scholarly Chinese *changpaos*, which had been washed
and pressed for them by a local laundry service. Breakfast likely included rice

gruel, fried bread, or soup, all served at the campus cafeteria. Mornings were spent in the lecture halls; afternoons, however, were spent in the laboratories, which were housed in the main engineering building: a drab, purely functional Bauhaus structure supported by long vertical columns of precast concrete and enclosing a large courtyard. There, Tsien and his classmates donned trousers and smocks and watched the instructors operate the boilers, locomotive engines, steam turbines, and other pieces of heavy equipment, which generated a cacophony of noises. (Within a few years, Tsien and the other students would have the chance to show off their work to the entire city of Shanghai in an exposition that included, among other things, a demonstration of telephones and a locomotive seating ten to twelve passengers speeding around an oval track on campus.)

As the year progressed, Tsien's peers began to note the grades of this small, reclusive boy from Beijing. "At the time, Tsien was well-known as the best student in the class," remembered one alumnus. "We knew his records. He was number one." At the same time they remembered the boy himself as an odd sort, even by Jiaotong standards. He did not talk much, nor did he study with others. The collegiality of the lecture hall and laboratory held few charms for him, they remember. Tsien, it seemed, wanted only solitude and silence—a blissful atmosphere in which to think and do his own work. He seemed happiest when alone.

Most of his free time was spent in the library, an elegant three-story orange-brick building with palladium windows and gray stone balconies. In front was a white statue of a seated woman, book in lap. Strolling into its cool recesses, Tsien would settle in a back room and absorb, with the greatest concentration, articles in American scientific journals. He would read for hours, breaking his concentration only to jot down in his notebooks intricate mathematical formulae.

He read during lectures as well. While others around him "took down notes religiously, fearful of missing a single word," Tsien sat in the back of the room poring over journal articles, remembered Zhang Xu, one of his classmates. "That was his peculiarity. Other students were mainly concerned about passing their exams. Tsien could easily pass any exam, but he focused his attention on learning more about the subjects he preferred."

In one class, however, he did pay attention. Cheng Shiying, a round-faced, bespectacled mechanical engineering professor who had been educated at MIT, was considered the most brilliant instructor on campus. He lectured in English and wrote the mathematics, line after line, in beautiful tiny print on the blackboards. Everyone, Tsien included, was awed by the meticulous manner in which Cheng gave his lectures and his efficient use of the blackboard, so much so that these two traits would one day become the distinguishing hallmarks of Tsien's own academic style. But most of all, Tsien would pride himself on knowing his material cold and presenting it in the meticulous style Dr. Cheng Shiying had first outlined in his mechanical engineering classes at Jiaotong.

On Sundays Tsien set out alone on his weekly walk through Shanghai.

It must have been exciting to have been a young man in Shanghai in 1929. In the 1920s, Shanghai was to China what New York City was to the United States: a sprawling, convoluted industrial center and, through its seaport, one of China's few links to the outside world in rapidly changing times. Its mills disgorged iron, steel, and textiles. The processing of tea was another addition to its economy. China's most densely populated city, Shanghai was also one of the most rapidly growing: between the year Tsien was born and the year he came to Shanghai to attend college, it had more than doubled in population, ballooning from 1.3 million to 3.1 million people.

The walk from Jiaotong University to the center of Shanghai would have taken Tsien several hours. It was a trip not only of time but of horizons. Surrounding Jiaotong University were the homes of many of Shanghai's wealthiest foreigners, executives of international corporations such as Standard Oil, most of whom lived in large, rambling mansions modeled after English country estates. From the street, Tsien could see people driving by in carriages and the beginning of the trolley track that could take him into the city. If Tsien chose to walk along Nanjing Road or Beijing Road, or another parallel major boulevard, he would walk into an increasingly dense urban setting with temples and red brick housing developments, and attractive Western-style homes and duplexes with private parks.

Soon he would see all the grating contrasts of an international port city. Men strode down the street wearing long Chinese robes and Western-style hats. Chinese women walked through the city unchaperoned, their hair bobbed, wearing flapper skirts and high heels. Ads for Chesterfield cigarettes were plastered, it seemed, on every street corner, while movie posters invited passersby to watch the latest release from America. Buses and trolleys operated alongside rickshaws, mule carts, and even wheelbarrows crammed with people.

By proceeding east, Tsien would find himself in one of the international settlements of the city. To his north was the British settlement, with its classical and Gothic architecture. To the south was the French settlement, where residents lived in neighborhoods reminiscent of the affluent French Place Vendome apartments. The concessions were home not only to foreigners but to wealthy Chinese and Japanese businessmen, many of whom were comfortable conversing in three or more languages. At the confluence of the Suzhou Creek and the Huangpu River on the eastern border of the settlements was the famous Bund: a road winding down the riverside through Shanghai's thriving business district, whose neoclassical buildings of steel, cement, and glass defined the city's skyline.

The Chinese were confined to the poorer sections of the city. Some lived in Old Shanghai, a section of the city once encircled by a three-mile wall. Entire communities were crammed into an intricate labyrinth of meandering streets, side streets, and *longli*: mere cracks between buildings so narrow they barely permitted passage of a bicycle or a man with an umbrella. Others lived in the crowded northern district of Zhabei, a working-class area of factories and homes, or in the slums of the southern Huangpu district. There, opium dens and brothels flourished—some nothing more than rows of sheds in which women who had been sold into prostitution serviced twenty to thirty men a night.

Perhaps nowhere else in China did such extremes of wealth and poverty exist. Working-class Chinese lived in squalid conditions, crammed in huts of mud, straw, reed, and bamboo, while some of the foreign industrialists lived in almost royal splendor in European-style chateaux. Had Tsien ventured into a factory he may very well have seen children working ten to fourteen hours a day plucking silkworm cocoons out of vats of boiling water or factory floors covered with filth in which toddlers wallowed while young mothers worked nearby. Wrote one observer of the factory owner's treatment of child laborers: "He kept them in crowded dormitories, fed them rotten food, and had to see to it that they would not run away."

But most likely Tsien never entered one of these factories, nor did he likely linger in the rougher areas of Shanghai. On one of his walks east he discovered the Lyceum Theatre, an Italian Renaissance–style building marked with keystone arches, fanlight windows, and a crenelated cornice. There, he would spend hours listening to the Shanghai Symphony Orchestra, conducted by Mario Pacci. The pleasure he derived from the music was not only auditory but visual: he would rave, years later, about the "colors" he saw in classical music. Every week he would sit in the Lyceum, shutting out the rest of Shanghai, shutting out perhaps the rest of the world.

In 1930, at the end of summer vacation following his freshman year, Tsien came down with typhus. It started mildly, with a headache and insomnia, then it progressed rapidly into a severe fever and sharp abdominal pains, while rose-colored spots and rashes spread over his chest. By the end of the third week of his infection, Tsien was emaciated and, more likely than not, delirious. So severe was his illness that he was forced to withdraw from school and go home.

Typhus is an infectious disease caused by the bacterium *Salmonella typhi*. It enters the body through the mouth, penetrates the intestinal wall, and poisons the bloodstream in twenty-four to seventy-five hours. It is usually spread through excreta and thrives in impoverished, unsanitary conditions. The crowded, filthy city of Shanghai was a perfect breeding ground for the bacteria.

Fatality rates for typhus were high: left untreated, up to 25 percent of the patients died from the disease. Treatment in 1930 consisted of months of bed rest and plenty of food. Not until 1948 would treatment with antibiotics like chloramphenicol and ampicillin become available.

For the academic year 1930–31, Tsien disappeared entirely from the campus of Jiaotong. Bedridden at his parents' home, he began to take a keen interest in politics. He had time to read books he had bought from vendors in Dong An Lu in Shanghai. These were books about philosophy, politics, and Marxist thought, which reflected the wave of progressive thinking that spread underground across China. It was Tsien's first exposure to Communist thought.

"I read some books on scientific socialism and learned a bit of the background to the government's activities," Tsien would recall later. "My outlook on life rose [to a new level]."

Tsien's college career and "outlook on life" coincided with a period of widespread discontent and activism by students against the central government of China. Student activism was nothing new in Shanghai. During the 1910s and 1920s, foreigners were the target. The students resented their arrogance, of which the famous "No Chinese and Dogs allowed" sign in one of the Shanghai parks was a prime example. The students were enraged when they saw laborers writhing under the brutality of the foreign-run police force, especially when they saw members of this force severely beat coolies for not paying their license fees on time. They were also infuriated by a system of taxation without representation in the government and courts. The 1920s generally was a decade of organized movements against Western imperialists, the most dramatic one ending in bloodshed in May 1925 when the British police fired on a mob in the international settlement.

But near the end of the 1920s and the beginning of the 1930s, things began to change. The brunt of student rage turned away from the foreigners and toward the central government that had failed to protect the Chinese people from these abuses. In the minds of the students, the Guomintang was the new villain.

The GMT earned the students' animosity in three significant ways. The first was its betrayal of the Communist Party. Throughout the 1920s, during Chiang Kai-shek's northern expedition against the warlords, the GMT maintained an alliance with the Communist Party. But on April 12, 1927, with a suddenness that shocked the country, Chiang led a bloody coup against the Communist Party and the unions of Shanghai. Students, workers, and local residents who protested were machine-gunned in the streets. Zhou Enlai, one of the Communist leaders, was fortunate to escape with his life. Executions of Communists continued for the next few weeks, and by the end of 1927, the party was in shambles.

The second way was through the party's rampant graft and corruption. After taking power in the city, GMT officials worked directly with Shanghai drug dealers and split the profits. They extorted money from the wealthy and forced local industrialists to buy short-term government bonds. They arrested the children of the Shanghai elite under trumped-up charges of Communism until handsome "donations" were given. During the late 1920s, Chiang cemented his ties with the foreign industrialists in the city and with the leaders of the Green Gang, the city's most notorious underground crime organization.

The third was the GMT's reaction to the Japanese invasion of Manchuria. On September 18, 1931, Japanese army officers blew up a railway line outside Mukden, a city south of the border of Manchuria. In the confusion, fighting broke out between Chinese and Japanese troops, giving Japan an excuse to attack Mukden and to send forces from Korea to conquer Manchuria. But the GMT, unwilling to engage its troops in battle, withdrew them south of the Great Wall, leaving Manchuria under complete Japanese control.

By the time Tsien returned to campus in the fall of 1931, Jiaotong University had been transformed from a quiet campus into one raging with student activism. Students wore gray uniforms to signify their willingness to fight against Japan and organized anti-Japanese "Save the country" organizations, marches, and boycotts. But the real anger was reserved for their own government.

The government tried to appease the students at first. One strategy was to invite the student leaders to Nanjing, feed and lodge them, and arrange for them to meet with officials from the Ministry of Propaganda. They would have a chance to hear the Generalissimo Chiang himself speak and were sometimes taken to see Sun Yat-sen's tomb. Many of these young and impressionable students left the capital "with a feeling of great accomplishment."

But the feelings did not last, and the movement swung toward violence. On September 28, 1931, students from Shanghai and Nanjing beat up Wang Zhengting, the foreign minister, whom they suspected of being a Japanese traitor. Hundreds of thousands of students across the country boycotted class, and in Shanghai three thousand students decided to confront Chiang directly in Nanjing.

Bursting into the local railway station, they commandeered a locomotive ("Some climbed into freight trains, some found their way into passenger cars, some crowded on top of the train, and suddenly all was chaos," remembered one witness) and sent it steaming toward Nanjing. One classmate of Tsien's remembers that on at least one occasion the GMT barricaded the tracks. But the railway officials in Shanghai, along with the railway workers, cheered the students on.

The movement hit a fever pitch in December, when local GMT officials hired agents to kidnap two student leaders from Beijing who were speaking at a meeting in Shanghai. Before a crowd of spectators, two men pulled up in a dark sedan and forced one of the two speakers into the car. One agent got away with his hostage, but the audience, quick to act, took the second agent as its own hostage. A mob burst into the local GMT *dangbu* and left it a wreckage of broken furniture, then moved into the mayor's office, shutting off the electricity and taking control of the municipal cars. The students, holding the agent hostage, beat him with bamboo rods until he revealed where the student leader was being held. Finally, GMT officials released the boy, who had been locked in a boat on the Huangpu River.

No one remembered Tsien being active in any student rallies, although there is evidence that he too was becoming disenchanted with the central government. After returning from his year of convalescence, Tsien longed to escape the 8:00 A.M. Monday morning Sun Yat-sen commemorative meetings in the auditorium—and, in so doing, avoid listening to the speeches of Li Zhaohuan, the president of the university. As it turned out, Li Jin, Tsien's college classmate and friend from high school, happened to be recruiting people for the brass band, and when Tsien learned that band members were allowed to leave the Monday morning gathering immediately after their performance at the beginning of each meeting, he quickly joined and learned to play the *zhongyin laba*—the euphonium.

It was a dangerous time to be a Jiaotong student, no matter where your loyalties lay. To be politically neutral might bring down on oneself the wrath of the student leaders, who were turning fanatic. Alumni remember that the hardcore radicals among them were going from dorm room to dorm room to guarantee full attendance at anti-GMT meetings. "You *had* to go to these meetings," remembered Tsien's cousin. "If you didn't, they'd beat you with iron sticks."

But at the same time, participation might make you a political target for the GMT. Agents sometimes appeared on campus to make arrests. "We'd see dark vans parked in front of dorms, then hear of an arrest or an expulsion the next day," remembered one student. Ironically, foreign settlements that once drew so much ire from student radicals now served as their refuge. Different laws, confusion, and lack of communication permitted students, when fleeing the police, to move from one concession to the next.

In 1932, China was to see yet more violence. It started with a skirmish in which a group of Japanese Buddhist priests were badly beaten. The Japanese residents in the city held a protest meeting on January 18, and two days later the Japanese consul general presented a petition to the mayor of the Chinese portion of Shanghai demanding an apology.

On January 28, the mayor acquiesced, but poor communication on both sides delayed transmittal of the apology. The Japanese sent two thousand imperial troops into the Zhabei district of the city. When fighting broke out between the troops and the Chinese 19th Route Army, the Japanese navy proclaimed the incident an insult to the Japanese empire and ordered that Zhabei be bombed.

In the early-morning hours of January 29, residents awoke to the rumbling of explosions. As one foreign correspondent recalled:

> I was awakened shortly after four o'clock in the morning of the 29th by a thunder such as I had never heard before and which I did not immediately recognize. I went to the window and then to the roof of the YMCA building where I was staying. Through the rain and the mist I heard the drone of planes, circling around and round over the Chinese city nearby. The droning was periodically punctuated by explosions and bright stabs of light. The Japanese were bombing the crowded Chapei [Zhabei] district of Shanghai.

Shanghai became a city strewn with corpses and the charred ruins of tenement housing. Classes at Jiaotong University were suspended, and Tsien was forced to go home as Chinese and Japanese forces went to war in the streets.

For the next month, chaos ruled Shanghai. Japanese bombers blasted churches, schools, hospitals, cotton mills, industrial plants, and universities. Some six hundred thousand refugees, with their belongings heaped on wheelbarrows and rickshaws, poured into the international settlements, areas even the Japanese did not dare bomb. "Some balanced their bundles on bamboo poles," remembered one observer. "Women carried screaming babies roped to their back."

During the first week of the air war, the Chinese held their own. The GMT aircraft shot down three Japanese bombers, which nonetheless fell directly over Zhabei or the Huangpu River. But a command dispute broke out between the GMT and the Cantonese air force, and the latter withdrew completely from operation. From that moment on, the Japanese air force reigned supreme. Powerful Nakajima biplanes and Mitsubishi attack aircraft bombarded the city, along with several other aircraft fresh from the factories. They destroyed ten planes at Hangzhou and one hundred unassembled planes, most still in their English packing cases at Hungjao Airdrome.

The fighting finally came to a halt on March 3. After a temporary ceasefire, the United States, Great Britain, France, and Italy assisted in negotiations between Japan and China. On May 5, a truce was signed. The Battle of Shanghai was over.

The Japanese attack sent shock waves throughout China and up and down the GMT line of command. The bombing of Shanghai left more than pitted buildings and dead bodies: it seared the national consciousness with the powerful reality that, in the face of modern technology, China was militarily impotent.

There was no denying that, even if the Cantonese air force had remained in place, Chinese aviation was no match for the Japanese. In 1932, the Japanese army and navy had more than 2,000 airplanes. The Chinese had only 270, and fewer than 90 in safe flying condition. Japan had built up an entire aviation industry: companies that were or would be Mitsubishi Heavy Industries Company, Kawasaki Aircraft Engineering Company, Hitachi Aircraft Company, Kawanishi Aircraft Company, Aichi Aircraft Company, Tachikawa Aircraft Company, and Nakajima Aeroplow Company. The Chinese had nothing but imported planes and a few scattered repair factories in Hangzhou, Shanghai, Nanjing, and Wuchang. The Japanese had a system in which military, government, and industrial forces synchronized their activities and hired top engineers from European companies like Junkers and Sopwith to work as consultants. The Chinese had a fragmented air force, in which different warlords each owned a fleet of planes. Wrote Eiichiro Sekigawa, author of *Pictorial History of Japanese*

Military Aviation: "Chinese air power at that time was so insignificant that it could be ignored."

One cannot underestimate the impact of the bombing on Tsien. It was easily the single most dramatic and frightening episode of his life. All too vividly, it demonstrated that technology was the key not only to industrial development but to national security.

Two decades before the battle of Shanghai, foreigners had attempted to extend their power by threading the mainland with railroad track. Back then, the Chinese fought for and won control of the railways. With the Ministry of Transportation firmly in place, hundreds of graduates like Tsien were pouring out of railway engineering schools each year to take their posts in offices across the country.

But the frontiers of technology were ever moving onward, and now they had reached an entirely new arena of competition: the use of aircraft as instruments of terror. The skies, seemingly infinite in their vastness, had become the next battleground of world power.

There was no ritualistic ceremony with cap and gown for the graduates of Jiaotong University, just the sterile formality of taking pictures. As graduation day approached, a tailor came to the dorms to take measurements for Tsien's cap and gown so that he could wear it once as he posed for his photograph. When Tsien's turn came to pose before the camera, the expression on his face was gentle, contemplative, almost sad.

There was plenty to think about, not only with regard to his own future but that of China as well. Through all of Tsien's boyhood memory, never before had China seemed so troubled, beset both by internal corruption and by outside aggressors.

The Japanese had set up a puppet state in Manchuria with Henry Puyi, the last emperor of China, at its head. Japanese aircraft had reduced sections of Shanghai to ruins. The GMT, unable to expel the Japanese, stepped up its efforts to suppress internal criticism of its policies, cracking down on universities and vigorously censoring newspapers, radio broadcasts, books, magazines, and films. Working through the Ministry of Education, the GMT developed a curriculum that included enough compulsory subjects and regular testing to keep most students too busy to think about politics. In predawn raids they arrested radical professors and students and searched classrooms and dorm rooms. Between 1932 and 1934 hundreds of intellectuals across the country were jailed, expelled, and even executed.

Tsien bore witness to many of these tactics of intimidation and pressure. Two of his classmates caught running a Communist cell were expelled. One prominent student leader at Jiaotong University was arrested and thrown in jail for a few weeks. Elsewhere in the city, the GMT massacred a group of left-wing authors and assassinated the editor of the *Shen Bao*, Shanghai's largest daily newspaper.

Tsien confided his worries about the future to Luo Peilin, a mechanical engineering major he befriended during his last two years at Jiaotong University. They lived in the same dorm, with Tsien on the first floor and Luo on the third. Luo was one the few students who had a phonograph—an RCA "His Master's Voice" model that was cranked by hand. Together the two boys would sit in Luo's room and listen to the recordings of Enrico Caruso and Ernestine Schumann-Heink, violinist Fritz Kreisler, pianist Ignace Paderewski, and conductor Leopold Stokowski. Sometimes Tsien dropped by with records of his own. When he graduated as the top student at Jiaotong, with a grade point average of 89.10 out of a possible hundred, he won some prize money from the Phi Tau Phi honor society and used it to buy an album of Aleksandr Glazunov's concert waltzes from a music shop on Nanjing Road.

Tsien declared to Luo that in order to achieve any real progress in China one would need not only scientific talent but power—political power. "He wanted to use revolutionary methods to change the state of China," Luo remembered. "Tsien was dissatisfied—very dissatisfied with the GMT's activities. He thought they were moving against intellectualism. He told me that just studying would not give us hope to correct the current system. You had to use political action."

"For me, that was a rather new concept," Luo said. "I was very much against the current government but I didn't have such concepts. He changed my entire attitude."

Tsien's next move suggests that such concepts had changed *his* "entire attitude" as well. For the past four years he had diligently trained for a career in railway engineering. Having graduated first in his class from the top engineering school in the country, he was now virtually assured a position designing railway engines for the Ministry of Transportation with a starting salary of 60 yuan a month, more than enough to live comfortably.

But his thoughts were apparently elsewhere—out of railway engineering and because of that out of China. The future lay in aviation, but there were no graduate programs in aeronautical engineering in China. The best programs were in Britain, Germany, and the United States. He could go to England, perhaps to Cambridge, to study with G. I. Taylor, one of the giants of aerodynamical theory. Or he could go to the United States, where the Wright brothers had begun it all.

There was even a program he knew of that offered to a handful of hopefuls a ticket out of China and into the best graduate schools in the United States. It was the Boxer Rebellion Indemnity Scholarship fund.

Tsien began to lay his plans.

Boxer Rebellion Scholar
(1934–1935)

Over the twenty-five years of the program, the Boxer Scholarship had become one of the most prestigious awards in China. As one recipient put it, winning this scholarship was more impressive "than [winning] a Marshall, Rhodes, and Fullbright scholarship put together" in the United States. This was so, despite the fact that the Boxer Rebellion Indemnity Scholarship program, as it was officially named, had been conceived in violence and born in an atmosphere of mutual suspicion between China and the United States.

The Boxers were a motley collection of some of the most disenfranchised members of Chinese society, including army deserters, prostitutes, and criminals. In a year of terrible floods in the northwestern province of Shandong and intensified foreign expansion in China by many of the same foreign powers that had earlier sought trading rights and concessions at Chinese ports during the Opium War, gangs of Chinese, calling themselves the Boxers United in Righteousness, began to attack foreign missionaries in Shandong.

The movement gained momentum as the Boxers recruited peasants from a famine-stricken countryside. Practicing secret martial arts rituals, they convinced themselves and each other that they had become impervious to bullets and swords. Wearing turbans and costumes of yellow, red, and black, they drifted into Beijing, Tianjin, Shanxi, Hebei, and Henan, and in 1900 the Boxers laid siege to these cities, massacring scores of Europeans and Americans as well as scores of Chinese practicing Western religions.

At this point, the Manchu government committed a monumental blunder: Cixi, the empress dowager of the Manchu empire, praised the Boxers and gave them official backing. Throughout the long summer of 1900, the Boxers were indiscriminate in their violence: engineers, merchants, and missionaries alike were slaughtered. In response to the killing, the Manchu government promised to give some forty foreign men, women, and children protection but put them to death once they arrived in Taiyuan.

Finally, on August 4, 1900, a column of twenty thousand troops arrived from the United States, Britain, Japan, Russia, and France and quickly crushed the Boxers. The peace treaty signed in 1901 contained severe terms for the Chinese. The Manchus were forced to pay 450 million taels in indemnities—almost twice the Qing national income. The total indemnities paid out to foreigners, factoring in interest payments over thirty-nine years, amounted to almost one billion taels, or U.S.$982 million. The U.S. share of the indemnities was set at $25 million, which with interest would total $46 million after thirty-nine years.

When the Theodore Roosevelt administration learned that the indemnity was nearly twice the amount of actual American claims against China for damages, it decided to return the surplus to China by establishing a scholarship fund to send Chinese students to the United States. The Chinese protested, wanting instead to use the surplus for railway, mining, or banking ventures. But the United States saw the scholarship program as a way to return the money, cultivate an influential body of American-educated leaders in China, and, as one educator commented, bring about "the intellectual and spiritual domination of its leaders."

After four years of complex negotiations between U.S. and Chinese diplomats, the Boxer Rebellion Indemnity scholarship program was set up in 1909. The Chinese would administer their end of the program—which included selecting, training, and transporting to the United States each year's group of scholars—through Qinghua University in Beijing. Once in the United States, the students would be under the auspices of the Chinese Institute, which opened in New York City specifically to manage the finances of the scholarship recipients while they lived in the United States.

The program turned out to be a success. The fund helped educate some of the most influential figures in Chinese education and politics. The first wave of scholarship recipients included Bing Zhi (Cornell University Ph.D., class of 1918), founder of the first biological institute in China; Zhu Kezhen (Harvard University Ph.D., class of 1918), later the president of Zhejiang University; and Hu Shi, later the president of Beijing University and ambassador to the U.S.

Initially each Chinese province was guaranteed a certain number of slots annually proportional to the size of the indemnity from that province. High school students took the competitive exam in their senior year; the winners were enrolled at Qinghua University for their first two years of college and then journeyed to the United States for their junior and senior years.

But after the Japanese attack on Manchuria and Shanghai, the government restructured the program to confront the technological needs of a nation in crisis. All matters of regionalism were tossed aside. The revised exam was open only to the top four graduates in math and science of each university in China and to college graduates who had worked at least two years in their scientific fields. The new requirements were born out of an urgency to find and support the most talented students in the country, those who could rapidly master the available scientific and technological knowledge in the United States and return to build up China's defense industry.

This is how it came to pass that a scholarship program born out of acts of hatred for all Westerners resulted in the best Chinese students being sent to the West to develop the scientific and technological skills that would allow China to combat its oldest and probably still most feared enemy—Japan.

In August 1934 Tsien journeyed to Central University in Nanjing to take the competitive examination for the Boxer scholarship. Along with his diploma, he was required to bring with him his senior thesis, physical exam records, two photographs in addition to a vitae with his picture on it, and the test fee of five yuan. Presenting these to the proctor, he took his place alongside eighty other young men, all no doubt as eager as he to distinguish themselves. Only twenty scholarships would be awarded. For Tsien this exam would be not only the most important he had thus far taken but his last taken as a Chinese student.

The tests began at 8:00 A.M. and continued until 5:00 P.M., with only a short break for lunch. The bulk of the questions, 80 percent, were on scientific topics—physics, calculus, thermodynamics, mechanical engineering, and aeronautics. The remainder tested the applicant's knowledge of the GMT and fluency in Mandarin Chinese, English, and either German or French.

In October twenty winners were announced. One can only imagine Tsien's joy and relief when he learned that he was one of them. As it turned out, he was the only one who would be studying aeronautics.

Decades later, one of Tsien's fellow Boxer scholars, Kai-loo Huang, described what happened next. Each winner was immediately assigned an advisor in China, who decided which school the student would attend, under which expert he would study, in what field he would specialize. Before going to the United States, each scholarship recipient had to spend a year touring China to learn more about the practical national needs in his particular field. (Huang, an economics major, recalls studying the factory and labor situation throughout China. A literature major within the group was forced to study Chinese theater floodlights!)

Tsien's advisor was Shi-Cho Wang, a Qinghua University professor of aeronautics and MIT alumnus who urged Tsien to go to MIT to pursue a doctorate and to spend a year examining the facilities of the Chinese aviation industry.

Frankly, there wasn't much to see. The facilities at Nanjing consisted of a few hangars filled with scattered propellers, engines, and five or six planes, mainly imported Junkers and Corsair planes. Tsien probably visited the one major Chinese aviation expert in Nanjing: Colonel Chien Tang-zho, another MIT alumnus now in charge of aeronautics for the Chinese air force. Tsien no doubt also visited the repair factory on the rural outskirts of Shanghai, which consisted not of hangars but small structures of wood, cement, and mud. Inside American foremen directed the restoration of Curtiss Hawk biplanes. His contemporaries recall one small two-seater Chinese trainer airplane in construction: a skeleton consisting of steel tubing, fabric, and wood. "There wasn't much construction, only repairs," another Boxer scholar remembered. "There couldn't have been more than a hundred airplanes in all of China in 1933."

Two places Tsien did visit were an airplane factory in the city of Nanchang and the Central Aviation Academy, founded in 1932 at Shien Chiao near Hangzhou, the city of Tsien's childhood. When Tsien arrived at Shien Chiao, the school was run by a mission of U.S. military pilots under the direction of U.S. Army Colonel John H. Jouett (Ret.). Its five instructors trained and graduated fifty pilots each year. The facilities, all new, were comprised of cadet barracks that housed some 130 young men, a three-story administration building, a large, steel-framed hangar, a freshly dug well, a radio station, a woodworking shop, an engine overhaul shop, an airplane repair shop, and a clinic. The school also possessed some thirty Curtiss Hawk biplanes from the United

States—an impressive number in those days. Other Boxer scholars guessed that Tsien spent about six months at Shien Chiao.

The tour was a race against time. All throughout 1934, as Tsien scrutinized airfields, repair factories, and schools across the country, matters within China grew worse. The GMT was intensifying its efforts to wipe out the Chinese Communist Party, and by the middle of 1934, they had tightened their blockade around the Communists in the Jiangxi region. On October 16, 1934, the Communists began their famous Long March—a six-thousand-mile retreat out of the Jiangxi Soviet region and toward the northwestern province of Shanxi, during which more than 72,000 Communists out of the 80,000 who started the journey died.

Meanwhile, the Japanese embarked on still another series of military aggressions. This time they planned to expand the territory ceded to them during the Tanggu Truce of 1933 and take over all of Hebei province. Once again students rioted across the country, protesting the irony of having Chinese forces pitted against Chinese while the Japanese enjoyed the spoils.

Time and resources appear to have been so tight that Kai-loo Huang did not remember any official meetings, farewell banquets, or memorable speeches held in 1935 to bring all the scholarship winners together in one group. (In fact, most of them met for the first time on the steamship bound for the United States.) Before their departure most Boxer scholars submitted reports to their advisors that summarized the highlights of their tours. Once in the United States, each student was guaranteed a tuition waiver and a one-hundred-dollar monthly allowance from the China Institute in New York for a maximum of three years, after which he was expected to return to China and accept employment arranged by his advisor.

Late one afternoon in August 1935, Tsien boarded the steamship *President Jackson* in Shanghai with a small group of other Boxer scholars. A picture of them taken on the *Jackson* presents the ultimate image of propriety: a group of dignified young men standing in neat rows with smooth, clean-shaven cheeks, Western suits and ties, and closely cropped black hair. As the *Jackson* slipped from the dock, Tsien watched friends and family of the Boxer scholars recede into the distance. One could more readily intuit the thoughts of Tsien's mother and father—their enormous pride in their son's achievements mixed with the sadness they felt at his departure—than Tsien's. He was finally going to America, a country none of his ancestors had ever seen, and to a strange but famous school called MIT. He knew he had won.

MIT (1935–1936)

Although he could not have known it at the time, his stay at MIT would be a short and somewhat unhappy one. Part of the problem was that MIT in 1935, even as a top-flight technical school, gave short shrift to aeronautical engineering. But another part of the problem was more personal: Tsien and MIT were simply not a match.

At MIT, aeronautics had long been more a sport than a topic of serious study, as students on skates held glider races from the frozen Charles River. The glimmerings of a formal program came in 1913, when A. A. Merrill, a former airplane designer, organized a series of informal talks on aviation. Possibly because of the success of the lectures, MIT allocated thirty-five hundred dollars to set up a laboratory, built a four-foot wind tunnel, and, the following year, offered a graduate course in aeronautical engineering—the first of its kind in the United States. The first student to get a master's degree in the program was the Chinese scholar H. K. Chow. He was followed by several others from his homeland, most of whom obtained prominent positions in government and academia when they returned home.

The program expanded rapidly during World War I. Faced with an acute shortage of aircraft, the military used the MIT wind tunnel facilities to test new airplanes. Army and navy pilots arrived on campus to receive advanced training. But with the end of the Great War, government support quickly dried up; the government cancelled hundreds of its contracts with airplane manufacturers, in the process dealing a heavy blow to the fledgling MIT program.

Fortunately, in 1926 the Guggenheim family put the program on a firmer financial footing by announcing its intention to fund seven four-year bachelor's degree programs in aeronautical engineering. MIT was one of the schools chosen to receive a half-million-dollar grant, with which it erected the Guggenheim Building and equipped it with a new library and a brand new seven-and-one-half-inch wind tunnel. After 1926 a rigorous shop environment pervaded the department. Under three active professors and two instructors, students mastered not only the theory of aerodynamics but actual airplane design. They learned to weld with acetylene torches and bend metal with hammers, tongs, and anvil. At least one student built an actual biplane of wood and wire.

Tsien's own mentor at Qinghua University, Shi-cho Wang, had graduated from the MIT program in 1928. "We were all told by our advisors in China, 'Of course you should go to MIT,'" remembered Tunghua Lin. "It was considered *the* school of engineering in the United States."

By the time Tsien arrived in 1935, the program included several pioneers of the aircraft industry, among them Jerome Hunsaker, head of the department of mechanical engineering in charge of the course in aeronautical engineering. He was famous for his work in airship construction and supervision of the design of the NC-4 flying boat, the first airplane to cross the Atlantic Ocean. There was also Carl Rosby, renowned as the father of American meteorology. The alumni, too, were becoming leaders of an emerging aviation industry. A few would become household names, such as General James Doolittle, who led the first bombing raids over Tokyo; J. S. McDonnell, the founder of McDonnell Aircraft; and Donald Douglas, founder of Douglas Aircraft.

True to his performance since childhood, Tsien's grades at MIT were stellar. According to one story, Tsien took a class in which the professor gave so difficult an exam that the majority of the class failed it. After some discussion, a body of students decided to confront the professor about the unfairness of the exam. "When they arrived at the professor's office," wrote MIT alumnus Webster Roberts, "they found Tsien Hsue-shen's exam posted on the door. It was done in ink. It was complete. It was perfect and there were no scratch-outs or erasures! (They did not confront the professor.)"

But when it came to designing something with his bare hands, as much a part of the MIT education as test taking or theory development, Tsien seemed almost helpless. One of Tsien's closest friends at MIT was a fellow graduate student named William Sangster. The two boys sometimes ate together at the home of a retired Scottish couple from whom Sangster rented a room ("a welcome change to Walker [Memorial] food," he recalled years later, referring to the dining hall where students ate). After dinner Sangster would bang out tunes on the piano while Tsien played along with his euphonium. "When [Tsien] was amused," Sangster remembered, "he would give off a little smile."

One day in the workshop, Tsien came to Sangster's drafting table with a look of concern on his normally "inscrutable" face. Tsien wanted to know how to connect the fuel tank, which was behind the firewall, to the engine, which was in front of the firewall. Sangster could not believe the question. Just make a small hole in the firewall and connect the engine and tank with a copper pipe, Sangster told him, thinking his question "rather stupid." Sangster wrote, "He [Tsien] seemed to have difficulty accepting the fact that it was permissible to make any holes in the firewall."

It may have been that to a mind like Tsien's, breaching the integrity of a device devised to provide a firestop was not something to be lightly contemplated. But there is another possibility to be considered. Tsien's problems with shop work may simply reflect the environment in which he grew up, one that relegated manual labor and physical activity to servants. "There is a general disdain for handiwork," wrote one medical researcher who observed the habits of the Chinese scientific elite years later. "One is reminded of the Mandarin officials who carefully kept their three-inch fingernails protected in bamboo or silver cases. The length of the fingernails indicated how little the hands were used—the ultimate sign of authority." Whether disdain or total unfamiliarity influenced his actions, Tsien clearly found the laboratory work daunting, the shop work tiresome.

But worse lay ahead, when he struggled to come up with results for his master's thesis on the turbulent boundary layer. The boundary layer is the thin layer of flowing gas or liquid that is in contact with the surface of certain objects. In the case of a moving airplane, the wing is sheathed with a thin layer of air particles. These air particles create minute levels of friction that affect the overall flow pattern of the air across the wing. This boundary layer was discovered in 1904 by Ludwig Prandtl, a brilliant German aerodynamicist who found that one had to treat the flow pattern of the boundary layer separately from that of the rest of the air flow. When studying the behavior of an airfoil (anything that cuts the air), he introduced to the world the concept of "drag" in airplane design. If the boundary layer is separated from the surface of an airplane wing, it can slow the plane, possibly to that most dreaded point where it may fail.

There are two kinds of boundary layer flow: laminar and turbulent. The first type will pass smoothly over the surface of a wing, while the second will churn between layers of air with different velocities. Laminar flow can become turbulent, like smoke rising from the tip of a cigarette: it first moves upward in straight filaments, but after some distance in the air, it wavers and curls in a random roiling motion. In general, the flow in a boundary layer is laminar at the leading edge of an airfoil (the upstream portion) and turbulent in the trailing edge (the downstream section). Tsien decided to study the phenomenon experimentally.

During the summer of 1936, Tsien teamed up with W. H. Peters, a second-year master's student in mechanical engineering from Girard, Pennsylvania. Together, they worked on the turbulent boundary layer project under the supervision of instructor Joseph Bicknell, assistant professor Heinrich Peters, and professor Richard Smith. The two boys spent most of their time in the wind tunnel room on the first floor of the Guggenheim Building. The tunnel they used was a rectangular contraption eight feet long, fifty-four inches high, and thirty-three inches wide, with three walls of plywood and a fourth wall of steel plate highly polished to reduce friction. An electric fan powered by a two-horsepower motor sucked the air through the tunnel to simulate the force of wind over the surface of a wing. Brass plugs threaded with Pitot tubes were pushed through the steel plate and linked with manometers, which would measure the pressure of the wind in the test section of the tunnel.

One problem that Tsien and Peters did not foresee was the severe turbulence caused by the equipment itself. To study the boundary layer, Tsien needed air that would move smoothly parallel to the steel plate, but the loud whine of the fan and the shape of the tunnel caused it to tremble and travel in rotational eddies. The pressure in the test section of the wind tunnel fluctuated wildly. Each day, from eight in the morning to ten at night, Tsien tried to smooth out the flow of air.

The work was frustrating and tedious. The wind tunnel was on the ground and Tsien was constanting bending over or crouching to adjust the Pitot tubes, fan, and manometers. He stretched cheesecloth over the mouth of the tunnel in an attempt to eliminate the spin in the air. He placed a honeycomb of sheet metal tubes three inches in diameter inside the tunnel to straighten out the disturbance in the flow. But in the end, he had nothing to show for it.

"We didn't get meaningful results," remembered W. H. Peters, who later threw away his own master's thesis. "The turbulence problem defeated the whole thing." Tsien's own conclusions echoed Peters's words. After thirty-six pages of formulae and charts, followed by snapshots of the wind tunnel facilities, Tsien wrote in his thesis: "The results obtained from the existing tunnel are definitely limited by the present equipment. . . . No definite conclusions can be based upon results from this tunnel until the recommended changes have been made."

If Tsien felt any disappointment, he hid it well. His partner remembered him as a "shy, very nice person" who was good to work with and who did not like to talk. So quiet was Tsien that Peters had no idea that Tsien was planning to leave MIT.

There is a great deal of speculation about Tsien's departure from MIT in 1936, a bare year after his arrival. A number of stories abound about his motives, all of them different. We do know that his time at MIT was marked by isolation. One of the faculty members remembered that Tsien was "kind of lost" in the department. Shy and uncommunicative by nature, further impeded by his broken English, which made it difficult for his classmates to understand him, Tsien essentially kept to himself. For entertainment he returned to his lifelong passion and, perhaps, his best friend—classical music. There were twenty performances given that 1935–36 season by the Boston Symphony Orchestra under Serge Koussevitzky, and Tsien missed not a single one. "It was my only form of entertainment," he later recalled.

But clearly other factors, academic factors, went into his decision. Some say he was forced out of MIT. One version of this story comes from Tsien's close friend Andrew Fejer. Tsien, Fejer said, told Fejer that he had recently gone to speak with Jerome Hunsaker, head of the program, to express his dissatisfaction with the empirical nature of the aeronautics program and that Hunsaker had responded, "Look, if you don't like it here, you better go back to China."

Another version claims that Tsien wanted to stay on for a doctorate, but that one professor, Shatswell Ober, told him to go out and get some practical experience in the aircraft industry first. "He was essentially told, no, don't stay here at MIT, go elsewhere," recalled MIT alumnus and professor Judson Baron, who heard the story secondhand. "I suspect that Tsien preferred to go to [another] graduate school than do that [go into industry]."

Tsien may have been unwilling—or simply unable—to find such work because of anti-Asian prejudice in the business. According to his former secretary Wang

Shouyun, Tsien's study of engineering at MIT would have "required" him to work in a factory when "American aviation factories did not welcome Chinese." His contemporaries concur: "In those pre–World War II days, particularly during the Great Depression, it was taken for granted that only Caucasian Americans could get jobs in well-established U.S. corporations," wrote Hisayuki Kurihara, an acquaintance and classmate of Tsien's. "Asians, with or without citizenship, seeking degrees in colleges and universities, could not expect to get suitable jobs in U.S. corporations, but pursued studies so that they may get some opportunities back in their home countries."

In addition, the 1930s were a difficult time for all young aeronautical engineers, and in all likelihood, even if Tsien were Caucasian he would still not have found a job easily. The Great Depression forced aircraft companies to lay off workers. Worse, the Neutrality Act of 1935, which put restrictions on the sales of munitions to belligerent countries, further limited aircraft production. Tsien's Caucasian classmates sent mass mailings of letters requesting employment to aircraft companies while poring through those copies of *Aero Digest* or *Aviation Week* to find addresses of new companies that might be hiring. "The response was almost always the same," one MIT alumnus remembered. "No job."

Perhaps had times been different—had Tsien been able to get a job in industry, or had the MIT faculty, recognizing the extreme difficulty of Tsien's getting the practical training he probably needed, helped place him in a job or even made a sincere effort to facilitate his search—he might have stuck it out at MIT. But it would have been the wrong decision. It is clear that MIT and Tsien were on opposite ends of the spectrum in personality and scientific approach. Tsien wanted a theoretical education. The aeronautical engineering department at MIT prided itself on producing engineers who could go out and actually build something once they graduated. Tsien grew up in the libraries of China. Most of his American classmates grew up in makeshift home laboratories, tinkering with automobile parts, bicycles, radios, and model airplanes in barns, basements, or garages. Tsien himself once commented to his friends that Americans were "born with monkey wrenches in their hands." Tsien had hoped to study under a great mathematical scientist at MIT, but his professors, according to MIT alumnus Bob Summers, were "not so much scientists as they were adventurers and explorers." As one commentator on the history of science once noted, "The spirit of Edison, not Einstein, still governed their image of the scientist."

But Tsien must surely have left MIT with a heavy heart. If he packed his bags and returned to China now, his options would be severely limited. Armed with a master's degree, he could obtain an administrative position in China, but that was not what he wanted. In essence returning to China now meant giving up his dream of bringing back with him the technical expertise that would help China defend itself against repeated acts of Japanese aggression and perhaps even free itself of the corrupt and ineffective Kuomingtang government that was now in power.

On a more personal level, of course, going back to China with his graduate work unfinished would expose him to the disappointment sure to be in the eyes of his father and his former teachers. Most of all, he would have to find some way to explain to his own satisfaction why he had been unable to make some accommodation at MIT.

But something else may have also gone through his mind as he pondered his future. If he returned to China now his formal education would, in essence, come to an abrupt end, for there simply were no research facilities there that could match those in the United States. For someone as devoted to the learning process as Tsien, this may have been the most painful price to contemplate. Whether or not the Chinese students who came on scholarship to American graduate schools gave much thought to what they would find once they got there, it appears that most were overwhelmed by the resources devoted to pure research they found here, particularly in those fields with potential military and industrial applications. "After we had seen what was available in the American universities," commented another Chinese student pursuing graduate studies in the United States—in fact, at MIT—"we didn't want to go home. Most Chinese students tried to stay in the U.S. as long as possible." It is implausible that Tsien was not as affected by these considerations as his compatriots.

Every consideration, it seemed, led Tsien to one decision—to try to find another graduate school in the United States willing to take him on for his Ph.D. The only program comparable to MIT's in facilities and reputation was a much lesser-known school on the other side of the country—the California Institute of Technology (Caltech). As important, Theodore von Kármán, director of its Guggenheim Aeronautical Laboratory, had earned a reputation instilling the Caltech program with a German tradition of theoretical rigor. The emphasis there was on pen and paper, not on hands-on experience.

To Tsien's good fortune, Kármán had once worked as a consultant at Tsinghua University, the institution that had given Tsien a Boxer Indemnity scholarship. Perhaps Kármán would take him in as student. He could write, or perhaps even call, and ask about the possibilities, but that was not Tsien's way. Rather, it appears that he went straight to Caltech to meet Kármán in person.

Theodore von Kármán

Theodore von Kármán was a giant in aeronautical circles, a legend at Caltech. Kármán, a man of mystery and contradiction: so intellectually sharp that he could, in one moment, jot down on a cocktail napkin the solutions to a complex mathematical problem that had stymied other minds for weeks, and yet so preoccupied the other professors had to hire a chauffeur to take him to and from the university because left on his own he frequently hit another professor's car when backing out of his parking space. Kármán meant so many different things to different people: a meek, docile son to his mother, an outrageous flirt to young women, a father figure to his graduate students, a respected advisor to generals. At Caltech, Kármán was a mischievous, rambunctious, irrepressible character whose short stature belied a powerful physical presence, marked by iron-gray eyes, dark bushy eyebrows, and shock of tousled dark hair. He was a man adored by most, revered by all, and understood by very few.

Throughout his life Kármán would remember fondly his hometown of Budapest, where he was born in 1881. "Horse-drawn droshkies carried silk-gowned women and their Hussar counts in red uniforms and furred hats through the ancient war-scarred hills of Buda," wrote Kármán in his autobiography. But, he added, "such sights hid deeper social currents."

The city was divided in two by the Danube River. On the west bank was Buda, a beautiful, Gothic city with winding, narrow cobblestone streets and the remnants of medieval castles. From its hills rose the magnificent palace of Franz Joseph I, the emperor of Austria and the king of Hungary. On the east bank was the town of Pest, flat in contrast to Buda and thriving with commercial activity. Banks and brokerage firms engaged in the trade of grain, beef, fruit, and wine within Pest's ring of boulevards while industrial factories for iron, metal, textiles, and leathers grew on the outskirts of the city.

During Kármán's youth the city would grow faster than any other on the European continent, numbering almost a million people and sporting its first subway. This was a period of economic prosperity and cultural renaissance for Budapest. The city boasted an opera house, art museums, well-groomed parks and squares with their cavalries of equestrian statues, and a massive, six-story parliament building swelling with a giant dome. Writers, artists, and actors gathered in cafes along the banks of the Danube, "the meeting houses," wrote one Hungarian journalist, "for the intellectuals and those opposed to oppression."

Contributing to this renaissance was an upwardly mobile population of Jews. During the latter part of the nineteenth century, social reforms weakened the feudal structure that had shackled Jews and other ethnic minorities to virtual slavery in the countryside of Hungary. Serfdom was abolished in 1848, and the Nationality Act of 1868 granted civil rights to the non-Magyar population, which included Jews. This legislation permitted many Jewish peasants to migrate to the cities; an ambitious man might journey to a nearby small town to learn some form of trade, thereby giving his son an education and a chance to seek his fortune in Budapest.* By the end of the century, the Jews had formed a prosperous, intellectually distinguished community in Budapest.

Kármán was born into that community. His father was the leading pedagogue in Hungary—the secretary general of the Austro-Hungarian ministry of education—and in that role he founded a model gymnasium (a European high school) for gifted children. His mother, a woman of great culture and refinement, was descended from a long line of scholars (most prominent among them Yehuda Loew Ben Bezalel, a famous sixteenth-century mathematician

---

*By 1910 the Jews, who represented only 5 percent of the population in Hungary, made up 80 percent of its financiers, 59.9 percent of its doctors, 53 percent of its businessmen, and 50.6 percent of its lawyers.

who invented a mechanical robot known as the "Golem" of Prague). Young Kármán grew up within the sheltered confines of his parents' large apartments in
the Jozsefvaros district of Buda and enjoyed the trappings of privilege typical of
his class.

At the age of nine he was enrolled by his father in his father's gymnasium,
where he learned advanced mathematics from graduate students and later won
the Eostvos prize, given annually to the best math student in Hungary. (The
school, known as the "nursery for the elite," later turned out among its graduates internationally famous scientists such as Edward Teller, John von Neumann, and Leo Szilard.) He also learned to master all the charming manners
of a young Budapest gentleman: dancing, gallantry, wittiness, fencing. By the
time he was sixteen, he could have pursued any one of a dozen different
careers, but his dream was to enter a prestigious foreign university and devote
his life to science.

But these plans were thwarted by his father. Maurice von Kármán was a nervous, inflexible man who sought to control every aspect of his children's development. Distrusting the local elementary schools, he had hired a tutor for his
children; distrusting even the available reading material, he penned their
primers, textbooks, and fairy tales himself. When six-year-old Theodore
demonstrated an inordinate ability to multiply five- to six-digit numbers in his
head, his father panicked, fearing his son might turn out to be what was then
described as an "idiot savant." Determined to crush his son's talents as a human
calculator, he took Theodore's math books away, ordering him to read geography, literature, and history instead, until Theodore's gift for calculation disappeared and left him a slow multiplier for the rest of his life. By the time
Theodore graduated from high school, his father had suffered a nervous breakdown, brought on by years of vicious feuding with his colleagues, and he was
forced to enter a sanitarium. To save money Theodore enrolled in the nearby
Royal Joseph University of Polytechnics and Economics, and to please his
father he majored in engineering, which Maurice considered more practical
than pure mathematics.

The next few years were fairly comfortable for young Theodore. Away from
his demanding father, he learned to use his considerable mathematical skills to
solve practical engineering problems. While still an undergraduate at the Royal
Joseph, he developed a mathematical method to eliminate the clatter of engine
valves and wrote an insightful engineering thesis. After graduation he served a
year of compulsory service in the Austro-Hungarian artillery. Then he returned
to the Royal Joseph, whose faculty had been so impressed by Kármán that they

invited him to serve as assistant professor of hydraulics. There he wrote a paper on the buckling of structures. His insights provided engineers with a valuable tool that would be taken into account in the construction of bridges, aircraft, and buildings.

In 1906 the paper was published and gave him his first taste of international recognition. The critical acclaim stirred his old ambition of a career in theoretical science. He faced a difficult choice: Should he stay in Budapest and live out the safe, respectable life of an engineering professor? Or should he pursue the buckling problem as a graduate student at an internationally famous university?

Once again, his father made the decision. Embittered by his own illness and feeling unappreciated by his homeland, he urged Kármán to venture forth and make a name for himself. He applied for a fellowship from the Hungarian Academy of Sciences to pursue the buckling problem, and then applied to the graduate program of mechanics at the University of Göttingen under the world-famous professor Ludwig Prandtl. Göttingen accepted Kármán's proposal and promised to support him for two years. And so in 1906 Kármán left the cozy nest of Budapest engineering to seek his fortune in Germany.

Kármán arrived in Göttingen in October. It was an elegant city in the Leine River Valley with narrow cobbled lanes twisting past brick Gothic church towers and medieval homes. Within its walls was a town of some twenty thousand souls and the world-famous University of Göttingen, founded in 1737. Within Göttingen's spired and gargoyled buildings emerged such literary talents as Henry Longfellow, George Bancroft, and the brothers Jakob and Wilhelm Grimm, who wrote their fairy tales in the university library. From the old, cathedral-style structures of Göttingen came some of the greatest minds in mathematics: Carl Gauss, who had developed the concept of complex numbers and the fundamental theory of algebra; and Bernhard Riemann, who invented function theory and whose work in the geometry of space lay down the foundation for Einstein's theory of relativity.

Kármán came to Göttingen just when it was poised on the brink of a new golden age. During the short space of a few decades, Göttingen would generate some of the most significant scientific developments of the twentieth century. David Hilbert and Felix Klein were there producing groundbreaking work on geometry and cultivating the graduate student Richard Courant, who would later be renowned for developing a calculus of variations. Four men who spent some time at Göttingen would later win Nobel Prizes from the 1910s to the 1930s: Max von Laue, for his work on the X-ray diffraction of crystals, which opened the field of solid state physics; Werner Heisenberg, for formulating quantum theory in terms of matrices; James Franck, for his research on the excitability of electrons; and Paul Dirac, for his work on quantum mechanics. During this same period a young American called J. Robert Oppenheimer would study under Max Born, a Göttingen professor studying the behavior of subatomic particles, and would return to the United States to build the atomic bomb.

Paradoxically, Göttingen was also governed by a rigid caste system in which only the top people enjoyed prestige and security. The system evolved during the nineteenth century in Germany and was exported internationally as a model for the modern research university. At the base of this slippery pyramid were the graduate students, who spent about six years completing a doctoral dissertation of original research. After graduation those who aspired to become professors remained at the university for *Habilitation,* a postdoctoral fellowship that would last another six to ten years. At this rung of the hierarchy, the scholar was required to assist the professor in his research, give lectures to undergraduate students, and complete a crucial second piece of scholarship. If his work proved satisfactory, he was invited to ascend to the third level as a *privatdozent,* analogous to the position of assistant professor. There he worked until he was promoted to *ausserordentlicher Professor*, associate professor, where he stayed unless the chair of full professor became vacant.

Typically, a department would have several postdoctorates in their twenties, a few *privatdozents* in their thirties, one or two associate professors in their late forties, and a single full professor in his fifties or sixties, the last commanding tremendous respect in German society. At prestigious universities it usually took a lifetime of teaching and scientific achievement to reach the top. Competition was brutal. "It was not uncommon," wrote Daniel Fallon, author of *The German University,* "for this waiting period to last twenty years or more, during which many impoverished scholars lost heart and left academic life, or died."

Kármán's earliest years at Göttingen were to be the most miserable of his life. The young man from Hungary was psychologically unprepared for the frigid political climate of the campus. During his first meeting with Prandtl, the bearded, bespectacled professor handed Kármán a menu card of about fifteen subjects from which he had to choose his thesis topic. When Kármán told him that he wanted to pursue his undergraduate interest in the buckling of inelastic columns, Prandtl seemed completely uninterested in giving him any direction for the project. As the months dragged on, Kármán felt lonely and frustrated. The hierarchical nature of the school extended even to its under-graduates, Kármán noted with disgust. They formed dueling and drinking societies that excluded Jews and Catholics and engaged in a vicious, Lord-of-the-Flies sort of pecking order. "There was scarcely any form of social contact," Kármán remembered, "except of the kind that existed between army officers and enlisted men."

Twice he almost quit. In 1907 he fled to the Charlottenberg Technical College, a famous school near Berlin, and enrolled there. But the equipment was shoddy, and he could not get along with the professors there any better. He returned to Göttingen, where, fortunately for Kármán, the German armament manufacturer Krupp decided to supply Prandtl with a large hydraulic press. This made it possible for Kármán to continue his buckling research and to complete his doctorate. Then, in 1908, Kármán's stipend ran out. Instead of trying to find a way to extend his stay at Göttingen, Kármán went to Paris, where for two blissful weeks he roamed like a Bohemian artist about the city, listening to lectures given by Madame Curie, hanging out at cafés, and going to parties at night.

Then came a chance encounter that would change his life. Amid all the confusion and aimlessless of Kármán's departure from Göttingen, there stood a single, vivid day—and one that he would never forget. In the early dawn hours of March 21, 1908, Kármán was walking home from one of his all-night parties. Exhausted, he decided to stop at a coffee house on the Boulevard St. Michel. There he bumped into Margrit Veszi, the daughter of one of his friends, who was working as a newspaper reporter. She asked Kármán to drive her to the Issy-les-Moulineux, an army parade ground on the Left Bank, where Henry Farman, an English aviator, would attempt the first two-kilometer flight in Europe. At first Kármán refused, claiming that he was not interested in watching a "box kite of sticks, wood and paper." But in the end he capitulated and drove her to the airfield.

Against the backdrop of dawn, Kármán watched as the Voisin plane was pushed out of a hangar from the south end of the field. Henry Farman maneuvered himself through the tangle of wires that held the plane together and, with the propeller awhirl and the engine clattering, flew in circles over the strip of airfield. The audience was stunned. They had read of the Wright brothers' invention in the newspapers, but, not having seen it with their own eyes, they were skeptical of it having occurred at all. For many Europeans this was the first conclusive evidence that air flight was possible.

Kármán was so impressed by the flight that he began to investigate the field on his own. He met with airplane manufacturers in Paris and talked about the possibility of designing lighter and more powerful engines. Nothing came of these discussions, but Kármán was eager to pursue a topic within the field of aeronautics. Consequently, when Prandtl wrote to him, inviting him to take a laboratory assistantship in Göttingen's new airship wind tunnel, which was being built under contract by the Zeppelin Company, Kármán jumped at the opportunity. He returned to Germany in the fall and soon wrote a paper predicting the future economic significance of aviation.

For the next four years, Kármán served as Prandtl's *privatdozent*. During this time, the young Hungarian conducted some of his most significant aerodynamical research, including his 1911 discovery through mathematical analysis of the existence of a source of aerodynamical drag that occurs when the airstream breaks away from the airfoil and spirals off the sides in two parallel streams of vortices. A phenomenon now known as the "Kármán Vortex Street," it was to be used for decades to explain the oscillations of submarines, radio towers, power lines, and even, much later, to explain the collapse of the Tacoma Narrows Bridge in Washington State.

But despite his growing reputation, Kármán saw no way to get a promotion at Göttingen. He began to see the school as part of a flawed system, rife with nepotism and exploitation. The disparities in income and prestige between the full professor and his underlings were simply too great. For instance, it was common practice for the state to collect mandatory fees from students for attending "private" lectures; these fees were later turned over to the professor. Since there seemed to be little distinction between these so-called private lectures and public lectures, some professors became immensely wealthy. Meanwhile, *privatdozents* worked long hours without job security for whatever fees students were willing to pay. Kármán joked bitterly about a time-honored tradition at Göttingen: the fastest route to academic stardom was to marry the professor's daughter.

Fuming about his situation, Kármán made a decision that almost ruined his career. He impulsively accepted a full professorship at a backwater college in Selmeczbanya, Hungary. It was funded by the Ministry of Finance rather than the Ministry of Education, because its goal was primarily to turn out experts in gold mining. Kármán cared nothing about gold mining, but he did care about the full professorship, and in the fall of 1912 he moved to Selmeczbanya to fill the chair of applied mechanics. The research facilities were inadequate, and the students' biggest gripe was that he replaced a professor who had used the same test questions year after year. Kármán asked for a one-year leave of absence to finish up his research at Göttingen. In fact, he would never return to Selmeczbanya.

Back at Göttingen, the eminent Felix Klein summoned Kármán to his office. Klein scolded him for accepting the position at the mining academy and promised that he would receive the next important chair to became vacant at a good university. But when a slot at the University of Munich did open up, it was granted to the son-in-law of a retiring faculty member instead. Kármán received the next best position: a professorship of aeronautics and mechanics at the *Technishe Hochschule*, literally, the Technical University, at Aachen, Germany, and even the directorship of the Aachen Aerodynamics Institute.

In Aachen, Kármán set about redesigning its wind tunnel, hiring new staff, working out an innovative theory on airfoil design, and collaborating with Hugo Junkers, the professor and industrialist responsible for setting up a pipeline between the university and German aircraft industries. He was only thirty-one years old, but finally he was doing work he truly enjoyed.

On July 28, 1914, a little more than a year after Kármán moved to Aachen, Austria declared war on Serbia. Kármán, who held a commission in the Austro-Hungarian army, was called to active duty as first lieutenant. The war interrupted his research for the next four years, but during those years he mastered the art of overseeing a large budget, supervising engineering work in the military, and establishing rapport with powerful generals in Hungary, Austria, and Germany.

During the four years of the war, Kármán did not receive a pass to go home. Sadly, during that period, his father had died. Finally, in 1918, Kármán was able to return to Hungary to comfort his aging mother. Later that year, in November, Austrian Emperor Charles abdicated the throne, and the Hungarian parliament declared the country a free republic. The new government gave rise to a Socialist regime under Count Michael Karolyi and later fell under the Communist rule of Béla Kun. The Communists invited Kármán to stay in Hungary as the new undersecretary of education. He agreed and introduced into the Hungarian university curriculum new subjects such as atomic physics and modern biology.

His political career lasted about two months. In August 1919, Romanian armies invaded Hungary, the Bela Kun government collapsed, and Budapest was sacked. Kármán hid in the home of a friend until it was safe for him to leave Hungary. Kármán returned to the Technical University in Aachen. Miraculously, his position was still open, but the institute itself had been shut down for five years and had fallen into neglect. Ingeniously, Kármán persuaded Belgian troops to restore the laboratory in exchange for free lessons in aeronautics. Years later, Kármán would say that political experience "saved me for all my life from having any belief in Communism—I saw it in operation and that was sufficient."

Gradually, by 1921, his life returned to normal. His mother and his younger sister Josephine (whom everyone called "Pipö") left Hungary and moved in with him in his new home in Vaals, Holland, which was near the university in Aachen. There, they hosted lively parties and managed Kármán's domestic and social affairs, leaving him free to do research.

The next few years were supremely happy ones for Kármán. He enhanced his reputation by publishing several important papers on aerodynamical drag, boundary layer, surface friction of fluids, and chaotic motion. He attracted lucrative contracts from airplane manufacturers like Hugo Junkers and the von Zeppelin Company. On campus he was a popular lecturer, illustrating scientific concepts with vivid images like comparing a vortex to soapy water swirling down the bathtub drain or inertia to two monkeys dangling from the ends of a pulley-suspended rope. He was able to work with students from all over Europe, for not only was he fluent in Hungarian but in Italian, French, German, English, and "very little Yiddish—enough to tell jokes." He soon became not only their scientific advisor but their confidant and friend.

By the mid-1920s, however, events beyond Kármán's control began to intrude upon his life. Hyperinflation drained research funds for universities to a mere trickle. The rate of exchange in Germany sank from 400 marks to the dollar in 1922 to 4.2 trillion marks to the dollar by the end of 1923. Banks hired bookkeepers good with zeros and paid out cash withdrawals by weight. The Technical University at Aachen, dependent as it was on private and state endowments, suffered a drastic plunge in enrollment. During the academic year 1921–22 there were ninety-four students enrolled in the math and physics department at Aachen. By 1924–25 that number had dropped to twenty-nine.

Then in 1922, Kármán's mentor at Göttingen, Ludwig Prandtl, received an offer to chair the department of applied mechanics at the Munich Polytechnic Institute. It came as a surprise to no one that Prandtl suggested Kármán succeed him as the director of the Institute for Applied Mechanics at Göttingen. This recommendation was quite an honor, for Prandtl was generally conceded to be the single most important figure in European aeronautics. But Kármán's appointment was resisted by other members of the Göttingen faculty because "of the anti-Semitic composition of the natural science faculty." There were already four Jewish professors in the natural sciences—Max Born, James Franck, Edmund Landau, and Richard Courant—and the faculty would not tolerate another Jewish appointment. Significantly, Prandtl did not defend his former student but instead stayed neutral. The appointment was denied Kármán.

For four years, Kármán continued his duties at Aachen. Then, in 1926, he suddenly received a telegram from Robert Millikan, chairman of the board at the California Institute of Technology: "What is the first boat you can take to come to Pasadena?" Was this rather cryptic note in fact an offer to become a professor at the second most technological university in the United States? To say the least, Kármán was intrigued.

In 1921, when Robert Millikan, the first American-born scientist to win the Nobel Prize, agreed to become chairman of the board at Caltech, his initial efforts were directed toward building up Caltech's physics department. By 1925, Millikan was ready to turn his attention to his next goal: developing a first-rate department of aeronautics. He requested a half-million-dollar grant from the Daniel Guggenheim Fund for the Promotion of Aeronautics and even before securing it purposefully set about looking for a world-famous expert to fill the position of director. Millikan and the board of trustees considered the British aerodynamicist Geoffrey I. Taylor and even Ludwig Prandtl, but they decided to offer the position to Theodore von Kármán. Millikan had met him in Holland in 1924 at the First International Congress of Applied Mechanics.

Professor Paul Epstein wrote to Kármán inviting him to spend the fall of 1926 consulting for Caltech. When no reply was received, Millikan assumed that he had approved of the offer; in reality, Kármán had not gotten the letter because he was on vacation in Belgium. He received first the telegram asking Kármán the earliest possible date he could sail for the United States.

Kármán replied with a cable expressing interest in visiting the campus but requesting clarification. He received it, and later that year, he and his sister Pipö journeyed to New York, where they met with Harry Guggenheim in his enormous mansion in Long Island. Then they traveled to California by car, arriving at Pasadena, then a small town, to tour Caltech and examine its aeronautical facilities. The department had on staff only five professors; they were divided in teaching philosophy by the theoretical approach of mathematician Harry Bateman and the empirical method of inventor Albert Merrill. The department was beginning to award bachelor's degrees and master of science degrees in aeronautics, and it had started a cooperative teaching venture with the Douglas Aircraft Company of Santa Monica in which professional engineers would teach students the basics of aircraft structures. The program was young and filled with great promise.

Kármán agonized over the offer. His sister, and especially his mother, protested the move vehemently; in addition to the uprooting, his mother feared the ocean voyage. But several dark events in Germany prodded him to act. He was disturbed by the rise of anti-Semitism in his country, which infected the German academic community. Kármán witnessed it himself in Aachen when one of his best students arrived in class sporting a swastika in his buttonhole. Shortly after that, a local fraternity harassed a Hungarian graduate student, calling him the "Hungarian Semite."

There were also signs that the German government was secretly rearming itself. A friend of his died in a mysterious airplane crash that seemed military-related. When the Ministry of Transportation warned Kármán not to ask any questions when German aircraft firms used the Aachen wind tunnel for high-speed tests, he truly sensed the prelude to yet another war. "The much discussed illegal army of Germany," wrote Kármán years later, "was a reality." Suddenly, he had "a terrible foreboding for the future."

Kármán also remembered that in Göttingen, Prandtl made a promise to him, then betrayed him. When Kármán realized that the highest reaches of German academic society would be forever barred to him because he was a Hungarian Jew, he knew he had no choice. He would make a new home in a country that lagged far behind Germany in the field of aeronautics at the time, but that would give him opportunities.

"Those were anxious days," Kármán remembered. "The thought of leaving Germany still tormented me, but the advance of the Nazis toward domination of Germany was too great." To Germany's everlasting disgrace and loss, and to the United States's pride and benefit, Kármán decided to leave his homeland and take his rightful place in the burgeoning American scientific community.

In December 1929, Kármán and his family turned their backs on Europe and stepped on a ocean liner bound for New York. He would not return for another fifteen years.

Upon their arrival, Kármán and his family searched for a suitable home. They found it in a house in the southern end of Pasadena only two miles from Caltech. It was a two-story Spanish villa-style home with a tiled courtyard and roof, set within a rambling garden and lawn near a street lined with pepper trees.

Before long, the interior resembled a museum. The Kármáns decorated the living room with Persian rugs, Chinese and Japanese floor screens, oriental chairs upholstered in silk, and an international collection of dolls, wall scrolls, vases, and other exotic ornaments. The downstairs room was converted into a study for Kármán's use, and a nearby cottage became his library of aeronautical journals, magazines, and books. It would also become the site of numerous parties and meetings that people at Caltech would remember for decades to come.

At the university, Kármán encountered none of the coldness and formality that marked the German universities at which he had spent most of his life. There was a refreshing lack of tradition at Caltech and an excitement that was linked to the mushrooming young aircraft community of Los Angeles. Douglas Aircraft was born in the back room of a barbershop in Santa Monica, and Lockheed was formed by two brothers who flew their homemade airplanes in the little coastal town of Santa Barbara. There were also other major firms in the area—Consolidated-Vultee, North American, Boeing, Hughes, and Curtiss-Wright—attracted by the fair weather that permitted the yearlong testing of airplanes. Local aircraft engineers from these companies used the ten-foot wind tunnel of the Guggenheim Aeronautical Lab (GALCIT), subjecting hardwood prototypes of their aircraft to different wind velocities and altitudes and consulting with professors on methods of improving the designs of their planes. In return, aircraft companies hired graduates of the GALCIT program, placing them in prominent engineering positions. Within just a few years, Kármán would help transform Caltech into a nerve center of aeronautical activity in southern California.

Kármán also made a number of significant new contributions to aeronautics. He designed a simple, effective fillet for the famous Douglas DC-3 that eliminated severe wind buffeting. He helped aircraft industries complete the transition from wood and fabric planes to those with sheet metal skins by calculating how to put stiffeners along the surface to prevent the metal from warping under pressure. He developed a fundamental law of turbulence and skin friction, beating out his mentor, Prandtl, who had spent years on the same problem.

His contributions were not restricted to research. Kármán was becoming one of Caltech's most popular professors. Despite his broken English, he was a natural lecturer, slicing the air with his hands as he talked and charming his audience with his wit. (When one French scientist questioned the name applied to his theory, Kármán Vortex Street, Kármán quipped, "You prefer Rue de la Kármán?") At times, he walked a class through a complex equation so deftly that, when he had finished, the audience broke into applause as if they were watching a sporting event. His students idolized him. So powerful was Kármán's charisma that many of his graduate students were emotionally dependent on him for approval, jockeying for position whenever he arrived on the scene.

At the age of fifty-five, Kármán was living a somewhat paradoxical life. He was a man who awed executives and world-famous scientists, and yet in many ways he resembled a small child. At parties he could be seen pumping on a swing set or dressed as Santa Claus during Christmas festivities. Every time he traveled, friends and assistants had to pick up the stream of hats and papers he left behind. He was naturally impulsive and curious, and completely unable to resist playing with any device or gadget within reach. Perhaps he was able to retain his child-like nature because he still lived with his mother and sister, who removed from his life the mundane, tedious aspects of adulthood. They fed him, made sure he was properly dressed, and wielded tremendous control over his social life.

Kármán never married. His maiden sister was his closest confidant and friend and took on many of the duties normally reserved for a wife. Their relationship can be compared to that of the Cambridge mathematician G. H. Hardy, who remained single, and his sister, with whom he shared a deep and chaste intellectual bond. It is not clear why Kármán chose not to marry: some suspect that his mother and sister had such a stranglehold on his social life that he never found the opportunity. However, his bachelorhood may have been a key factor in his success, for, without a wife or children, he had more time and freedom to devote to his scientific and administrative activities.

Because of his status as a single man, Kármán gained the reputation on campus of being a womanizer. "We *never* permitted Kármán to dance with our wives," one student remembered years later. "He had a way of looking at a woman as if she was the only thing that existed for him." He loved to be photographed with attractive young girls and spiced his speeches with jokes and sexual innuendos that were repeated gleefully for generations afterward at Caltech. Once, when asked by aviatrix Amy Johnson about the spin of aircraft, he answered: "Young lady, a spin is like a love affair. You don't notice how you get into it, and it is very hard to get out of." Another time, when asked why he chose not to marry, he grinned lasciviously and replied, "I never found the need to."

Nevertheless, Kármán harbored deep paternal instincts. He treated his graduate students as surrogate sons, inviting them to his home regularly, listening to their personal problems, and even hosting and paying for their weddings. "We were like his children, playing in his big house," one student remembered fondly. Those who struggled with difficult theoretical problems might find on their desks the next morning a beautiful set of equations in Kármán's handwriting along with a note: "I think this is what you want!" He also adored the children of his students, his fondness for them perhaps reflecting his wistfulness over not having any of his own. Kármán insisted that they call him "Grandpa," and wrote to them regularly and sent them gifts of toys and pianos.

With these displays of affection, Kármán fast became one of the most beloved professors ever to walk the halls of Caltech. By 1936 he was certain Pasadena was his true home. On July 24 of that year, he stood in the U.S. District Court of Los Angeles and took the oath of citizenship.

Day by day, his life followed a familiar schedule. He woke at 6:00 each morning. His sister and mother had ready for him a delicious Hungarian breakfast, which he frequently enjoyed in the company of guests. Then he attacked a thick pile of correspondence, dictated letters, and fielded phone calls. Afterward, he would either go to Caltech for a day of teaching and research, or stay at home to pursue his own work. At school Kármán taught courses on theoretical aerodynamics—focusing on the dynamics of real fluids and "perfect" fluids—and on the elasticity of airplane structures. He also worked with students on advanced problems in theoretical aerodynamics and presided over seminars where they presented their results.

In the evening, he lit a cigar, poured himself a glass of Jack Daniels whiskey or maybe some Slivovitz, his favorite brand of Hungarian plum wine, and then enjoyed a fine Hungarian dinner. A stream of visitors would call on him: writers, movie stars, priests, colonels, businessmen. They were greeted in the living room by his mother, sitting with a shawl over her knees, while his sister bustled about, serving refreshments.

Frequently, Kármán invited his students to his home for informal seminars, several hours of stimulating mathematical talk. Sometimes, in the middle of one of these gatherings, Kármán disappeared into the study with one or two students at a time to discuss their research. Occasionally he retreated into the study alone to pursue his own private thoughts.

His life was rich. He was finally a full professor at a top, world-class university. He lived in the warm paradise of southern California, surrounded by bright, adoring students. An ocean separated him from the ominous rumblings in Europe. His family took care of all his personal and domestic needs. There was almost nothing more he could ask for.

Then one day, a young Chinese man named Tsien Hsue-shen sought him to inquire about graduate studies at Caltech.

7

Caltech (1936)

The meeting was brief, but Tsien made a striking impression before Kármán's shrewd and appraising eye. As Kármán later wrote in his memoirs about Tsien:

> One day in 1936 he came to see me for advice on further graduate studies. This was our first meeting. I looked up to observe a slight short young man, with a serious look, who answered my questions with unusual precision. I was immediately impressed with the keenness and quickness of mind, and I suggested that he enroll at Cal Tech for advanced study.

Tsien accepted Kármán's offer immediately and wrote a letter to Luo Peilin, one of his closest friends in China, to share with him the good news. Though Luo no longer has the original letter, he recalls that Tsien was ecstatic. With Kármán's promise secured, all that was left to Tsien were the more mundane tasks of registering for classes, exploring his new environment, and looking for a place to live.

He had time to explore the city in which he would spend the next few years. Pasadena must have been a welcome contrast to Boston and its winters. The city lay at the foot of the San Gabriel Mountains, in a valley of hills and orange groves. Mediterranean-style palaces of white stucco and red tile stood in the west end of town, while the scent of rose and wisteria floated from their gardens through streets lined with palm trees. Formerly a small rural town to which tuberculosis patients migrated in hopes of recovery, it had become in 1936 a quaint and cultured city of some eighty thousand residents and a retirement community favored by some of the most privileged families in the country. "Pasadena is ten miles from Los Angeles as the Rolls Royces fly," wrote one commentator during the 1930s. "It is one of the prettiest towns in America, and possibly the richest."

Of course, Tsien and the rest of the student population lived in relative poverty. The campus of Caltech stood east of the thriving civic center of Pasadena, surrounded by tiny woodframe houses posted with Rooms for Rent signs to attract student boarders. In September, Tsien moved into one such house at 344 South Catalina Avenue and took his meals at another boarding house at 290 South Michigan. The second house had a big kitchen and a gas stove, on which Tsien and three other Chinese students prepared meals of rice and vegetables. Each morning Tsien would eat breakfast there and then proceed to class impeccably dressed in jacket, suit, and tie.

A few paces away was a small rectangle of campus, just minutes east of the thriving civic center of Pasadena. The buildings at Caltech—eighteen in all— were a curious but beautiful blend of Mediterranean and Spanish architecture fast becoming known to architects worldwide as "Californian." Entering the campus, Tsien might well have imagined himself in some Florentine villa or Spanish city plaza were it not for the other students walking past him with their textbooks and slide rules.

In 1936 Caltech was on a tight budget. The Great Depression curtailed much-needed construction and forced the lawns to be landscaped in ice plant—a cost-cutting move that financed four research fellowships. The National Youth Administration gave Caltech twenty-two hundred dollars a month to help needy students pay their way through school, while the Boxer Rebellion Indemnity Fund supported Tsien and a few other Chinese students by paying their annual three-hundred-dollar tuition. Despite the limited funds, from this 1930s tight-knit community of some 780 students and 80 professors would emerge many of the major scientific discoveries of the century.

There was something almost miraculous about Caltech's growth from a humble vocational school in a warehouse into a world-class university in just four decades. It began in 1891 as Throop University, a little college that offered machine and tool training to students of elementary school age and up. No one suspected then that its location would be key to its future success. The school lay at the foot of Mount Wilson, a mountain that attracted George Ellery Hale, founder of observational astrophysics. He rode by mule up its winding, dusty path to study the stars and to build at its peak in 1902 the world's finest telescope. It was Hale who enticed Arthur Noyes, the nation's preeminent physical chemist, and Robert Millikan, physicist and future Nobel laureate, to join him in Pasadena. Together the three worked at recruiting top scientists to their institution, renamed in 1920 as the California Institute of Technology.

During the 1930s Caltech attracted famous scientists across the globe and cultivated new ones among its own students. Albert Einstein wintered at Caltech between 1931 and 1932, pedaling about campus on his bicycle and attending "every luncheon, every dinner, every movie opening, every marriage and two-thirds of the divorces" before leaving for Princeton. In roughly the same period, a young physics graduate student named Carl Anderson discovered the positron, the first empirical evidence of the existence of antimatter. Thomas Hunt Morgan, chair of the biology department, won the Nobel Prize for medicine for his genetic studies of fruit fly chromosomes. Later, a young thirty-five-year-old scientist named Linus Pauling was appointed the new chair of the chemistry department, while another young researcher, Charles Richter, worked on a earthquake scale that would be forever linked with his name.

It was in this environment—in the vaulted ceilings and cool terra-cotta classrooms at Caltech—that Tsien took his first courses as a graduate student. In addition to his class work, he conducted his own study of aeronautics. Decades later he told newspaper reporters that during his first year at Caltech he collected all the material he could on aeronautics from different countries and read it through systematically—a task that averaged ten hours a day. During his first semester he rarely mingled with other students, who considered him a rather mysterious figure on campus. But they did notice that in class, Tsien tended to ask astute, complex questions that pleased his instructors and baffled his peers. Word of Tsien's brain power spread not only among the student population but among the Caltech faculty. As Kármán wrote:

I remember that Professor Paul S. Epstein of the Physics Department, a great theoretician, once said to me: "Your student H. S. Tsien is in one of my classes. He is brilliant."

"Ja, he is good," I replied.

"Tell me," Epstein said with a twinkle in his eye. "Do you think he has Jewish blood?"

Tsien had found his intellectual home in the German approach to aeronautics that Kármán brought to Caltech, an approach that rigorously applied the laws of fundamental mathematics and physics to basic engineering problems—just the approach Tsien hungered for. Interestingly, this theoretical and highly mathematical method of problem solving is what also characterized physicists' moves into astronomy. Calling themselves particle physicists and cosmologists, depending upon whether their immediate focus was on elementary particles or the large structures of time and the cosmos, they brought to the study of the cosmos the same belief that through mathematics and physics one could come to define the forces at work in the universe. Both the physicists and Kármán and Tsien proved correct in this belief.

For Tsien, it was the beginning of one of the most important intellectual relationships in his life, and the start of a career that could only be described as stellar. The young Chinese scholar was so respectful toward Kármán that he always referred to him as "my revered teacher," which Kármán understood to be the greatest compliment one person could give to another in China. Apparently, it was also the beginning of an important relationship for Kármán. In Kármán's autobiography, *The Wind and Beyond,* Tsien was the only student to whom he devoted an entire chapter. As Kármán wrote:

> He worked with me on many mathematical problems. I found him to be quite imaginative, with a mathematical aptitude which he combined successfully with a great ability to visualize accurately the physical picture of natural phenomena. Even as a young student he helped clear up some of my own ideas on several difficult topics. These are gifts which I had not often encountered and Tsien and I became close colleagues.

Tsien's arrival at Caltech just as Kármán and others were about to inaugurate the dawn of theoretical aerodynamics meant that Tsien would be there at its creation. Better, he and Kármán would together start to answer some of the most basic questions about flight. In time they would solve problems in compressible fluids and the buckling of structures, and derive a famous pressure correction formula that would be used in the design of subsonic aircraft.

The two of them were a marriage. Kármán had the genius of physical insight—the ability to visualize aerodynamical problems and to pluck out their key elements. Tsien, meanwhile, had the tenacity and the gift of applied mathematics necessary to work out the details on paper. The division of labor seemed well defined. If Kármán saw before him, in a flash, the entire grand sweep and structure of a theory, it was often Tsien who painstakingly erected it with line after line of mathematical formula. If the spontaneous, gregarious Kármán saw mathematics primarily as a tool, as the means to an end, the more bookish Tsien saw it as an end in itself; relishing its elegance and grace of form.

Tsien's friend Martin Summerfield recalled just how close intellectually the relationship became:

> He was Kármán's right-hand man. He would carry out all kinds of projects and thoughts that Kármán would have and carry them out with alacrity, and by working night and day he would deliver the manuscript or calculations very quickly, but also very brilliantly. He became a close assistant—Kármán's arms and legs— working out formulas that Kármán had masterminded. He had the brilliance and he had the speed. It was unusual to find someone like him.

Strangely, the two worked so well together partly because their intellectual styles were so very different. Kármán was not uncomfortable working through a problem in front of a group of students or colleagues; indeed, he appears to have thrived on the give and take of a group dynamic and even the pressure of coming up with a solution under the scrutiny of others. "Kármán could look at a problem, turn it inside out, and solve it, right there!" observed Caltech professor Frank Marble.

In contrast, Marble notes, "Tsien wouldn't wrestle with a problem in front of you. He did not have the open, give-and-take personality of intellectuals like Kármán. He would take a problem home, think it through, and at night, the answer would come to him."

That the answer came to Tsien was no miracle. During his first year as a graduate student he worked almost continuously, from eight in the morning to eleven in the evening. Through long, often sleepless nights, Tsien, armed with paper, pencil, and before the arrival of the modern computer, a slide rule, would labor on those problems Kármán seemed to toss off so effortlessly in the hubbub of a crowd. Invariably, Tsien did his best work in complete solitude. While Kármán was at his best before an audience of fellow academics, Tsien seemed to have been born with no stronger desire than to be left alone, to be allowed to do his thinking. Through all the years of schooling in China, through MIT, and into his years as a graduate student at Caltech, the dominant memory everyone has of him was of someone happiest when left to his own thoughts. He appears to have had no need to talk through his ideas with others. While

never unfriendly, he also seemed to have had a minimal need for companion-
ship or even the validation of colleagues. To a striking degree, few of his schoolmates ever remember having had a personal conversation with him. Talk, if it
occurred at all, centered around work or politics or school affairs. Given the
choice, Tsien seemed perfectly capable of filling endless hours working out in
his own mind the intellectual problems of the moment. (There are few anecdotes
about Tsien's work during this period of his life, probably because he spent so
much of his time in isolation.)

By all accounts, the results were invariably superb. "Tsien definitely made
von Kármán more productive because . . . [with his] superb mathematical
technique [he] could work out . . . Kármán's ideas very rapidly," remembers
Caltech professor Hans Liepmann, who arrived as an aeronautics research fellow at GALCIT in 1939.

In the days before computers, translating general theory into actual formula
by what was known as numerical analysis involved a long and tedious process,
to say the least. Tsien was willing to plot the functions, point by point. Other
scientists of his rank and reputation considered the work intimidating or
beneath them. One Chinese graduate student, asked to do this type of work,
actually put off obtaining his degree rather than devote the time and energy to
the work. As Liepmann recalled:

> One student left because he didn't want to do any numerical work. His supervi-
> sor told him that after he had done the theory, he should work out a numerical
> example, and he considered this below his dignity and left.
>
> Thirty-eight years later, when I was the director of the institute here, I got a let-
> ter from China with a numerical example. I went to the supervisor, who by that
> time was retired, and said 'What do you think?" and then we decided, why not,
> and let him have his degree. . . . The only thing that worried me later was that
> maybe he didn't do it himself!

Tsien was by no means the only graduate student working closely with Kármán. In 1936, by joining Kármán's intimate club of young theoreticians, he was
thrown together with other bright young graduate students, some of whom
would later assume powerful positions in the U.S. military, in industry, or in
aeronautics departments through the country. They included the outgoing and
popular Homer Joseph Stewart, who would later become a professor at Caltech
and a key figure in the American space program; the bespectacled William
Sears, who eventually became head of Cornell University's applied mathemat-

ics program, and William Duncan Rannie, a shy, narrow-faced student from Canada who would later, like Stewart, become part of the Caltech aeronautics faculty. These three, along with others, would meet at Kármán's home typically at least once a week for evening discussions on aeronautics and mathematics. "Kármán was a lucky man," Stewart remembered. "Whenever he came up with a great idea he had a energetic young graduate student at his elbow to pursue it."

It is interesting to note that, despite this wealth of talent, "Tsien was the greatest collaborator von Kármán ever had," Frank Marble observed. "I've worked with both of them, and when you saw them together you would see . . . creation."

8

The Suicide Squad
(1937–1943)

In the spring of 1937, Tsien was given his first office, sharing it with a fellow graduate student named Apollo Milton Olin Smith, known to his friends as "Amo." Smith recalled that Tsien was "not talkative," "really arrogant," and oblivious to everything but his work. But one thing about Smith did pique Tsien's interest: his membership in a small group on campus that conducted experiments with small rocket motors.

The group was started by Frank Malina, a graduate student in aeronautical engineering who arrived at Caltech two years before Tsien did. Born in 1912 in the small town of Brenham, Texas, Malina—a thin, tall, soft-spoken young man with dark hair and eyes—had dreamed of the possibility of space flight ever since he read Jules Verne's *From the Earth to the Moon* when he was twelve years old. While some other professors scoffed at the idea of their graduate students building rockets, Kármán was impressed and agreed to be Malina's sponsor.

Essentially, the experiments were a labor of love, planned at night and conducted on weekends between course work, research, and outside jobs the young men had to take to pay for spare junk parts. They had pooled their savings and drove all over Los Angeles to find secondhand equipment for their

rocket tests. In the fall of 1936, the group—which consisted of Caltech graduate students and local rocket enthusiasts—conducted their first primitive
experiments in rocketry in the Arroyo Seco, a dry river bed canyon a few miles
from Caltech near the massive concrete arches and supports of Devil's Gate
Dam and stretching to the foot of the San Gabriel Mountains.

The work was physically exhausting and downright dangerous. In late October 1936, the group worked until three o'clock in the morning to pick up
equipment and prepare a small rocket motor for the test. After catching three
hours of sleep, they drove at dawn to the Arroyo Seco and mounted a gleaming duralum model rocket motor with its end pointing skyward from the spring
of a test stand. They connected one tube that fed the combustion chamber
gaseous oxygen; another tube provided the chamber with methyl alcohol liquid propellant fuel. Before attempting to ignite the fuel-oxygen mixture, the
group dug trenches to hide in and piled sandbags around the apparatus.

Upon ignition, the flame of the blast would jump from the nickel-steel nozzle of the motor, forcing the body of the motor down on the spring. A speck of
diamond attached beneath the motor would scratch against a parallel glass
cylinder thrust recorder, etching a line by which the force of thrust could be
measured. During the tests, there were moments of excitement that the boys
would relish for years afterward. In October 1936, the oxygen hose broke and
caught on fire, prompting them to flee across the canyon. (They later regretted
not capturing the whole episode on film.) Later, in January 1937, the rocket
motor ran for a record forty-four seconds, causing them to break out into
applause.

Malina and Smith often discussed their results in Smith's office, and whenever they did, Smith noticed that Tsien would listen intently. "He was curious
when he overheard Malina and me talking," he said. One day in class, Tsien
came up to Malina and asked him questions about a copy of an article Malina
had written for a magazine on the subject of rockets.

After an animated discussion of the possibilities of space travel, Malina said
to Tsien, "If you're interested in rockets, why don't you join our little group?"
Tsien eagerly agreed. Shortly afterward, on May 22, 1937, Malina wrote home
to his parents: "A Chinese graduate student interested in some theoretical problems of rocketry has also joined our group. We have now five . . . Parsons, Smith,
Tsien, Arnold, and myself."

During the 1930s, few viewed the field of rocketry as a worthwhile scientific
endeavor. They had good reason not to do so. The technology was barely in its

infancy, pioneered by three scientists working alone in relative isolation in three different countries.

The progenitor of the field was Konstantin Tsiolkovsky, a deaf and introspective Russian research scientist who in the 1890s was one of the first men to address mathematical problems of rocket flight in space in his work, "Exploration of Cosmic Space by Means of Reaction Devices." Unfortunately, his findings were largely ignored by the Russian Academy of Sciences and the world scientific community until 1921.

Then came Dr. Robert Goddard, a professor of physics at Clark University in Massachusetts, who had been fascinated since adolescence by the concept of space flight. Upon graduation from college, he recorded in his notebooks ideas that would take shape half a century later: staged rockets, liquid- and solid-propellent rockets, manned exploration of the moon. In his home laboratory, Goddard conducted experiments on small solid-propellent rocket motors just over the hill from the Arroyo Seco in 1918 and published his findings in his 1919 classic paper, "A Method of Reaching Extreme Altitudes." The press sensationalized his findings, giving rise to rumors that Goddard was planning to shoot himself to the moon. Embittered by the experience, the reclusive scientist withdrew from public view and harbored a lifelong obsession with secrecy.

The third pioneer was Dr. Hermann Oberth, a German scientist and writer. In 1922, at the age of twenty-eight, he sought a Ph.D. at Heidelberg with a dissertation that included a design for a long-range liquid-propellent rocket and a mathematical explanation of how rockets could escape the pull of gravity. The university rejected it. Undaunted, Oberth published it as a book entitled *Die Rakete zu den Planetenräumen* (The rocket into interplanetary space). The book was scarcely noticed at first, except by a professor who denounced it, claiming that a rocket cannot work in a vacuum because its exhaust has nothing to push against.

The turning point came when Oberth's book attracted the attention of Max Valier, an indefatigable popular writer in Munich. In 1924 Valier turned his energies to a crusade for Oberth's ideas, writing popular books and articles based on Oberth's thesis and going on endless lecture tours in Germany and Austria. As a result, both Oberth's and Valier's books sold out. With Valier's help, Oberth's book gave rise to a popular fad in rocketry—a fad dominated mainly by science fiction fans and eccentric playboys. A spate of popular and technical books on rockets appeared in Germany, followed by the founding of the world's first journal and society devoted to space exploration.

Yet in 1926, Valier was disappointed with what he considered the lack of public response. It prompted him to launch a national campaign to raise money for a gradual transition from rocket planes to space ships. He commissioned artists to draw sensational pictures of his technological fantasies, causing Oberth to break with him because he felt some of the pictures were not technologically accurate. Nevertheless, the campaign intrigued Fritz Van Opel, a playboy who was the darling of the German media and heir of the Opel car manufacturing fortune. Before long, in 1928, Opel was careening around racetracks in rocket cars before packed audiences of thousands.

By this time, the rocket craze was hitting its peak in Europe. In 1929 German film director Fritz Lang released *Frau im Mond*, a three-hour epic movie about the construction and flight of a rocket to the moon. This generated even more excitement. There were also additional stunts conducted by the Valier-Opel team: rocket ice sleds, rocket rail cars, rocket gliders—even rocket bicycles, until a chamber with liquid-propellent fuel exploded in 1930 and left Valier dead with a splinter in his chest. His death drew forth outcries in Germany to ban private rocket research, and the fad began to peter out.

Rocketry did not arouse serious interest at Caltech until 1935, when William Bollay, another graduate student of Kármán's, delivered a talk about Eugen Sänger's rocket motor experiments in Vienna and the possibility of future development of rocket-powered aircraft. This served as a catalyst for bringing together the group that Tsien later joined. Intrigued by Bollay's talk, Frank Malina examined the scientific literature on rocketry and asked Clark Millikan, the son of Robert Millikan and a professor of aeronautics at Caltech, if he would sponsor a doctoral dissertation on the flight characteristics of sounding rockets. But Millikan, who shared the academic community's tendency to associate rocketry with Hollywood films and thrill seekers, rebuffed his proposal, even suggesting that Malina quit school and join the aircraft industry instead. Fortunately for Malina, Kármán was impressed with his idea and agreed to be his advisor.

Meanwhile, two other young men equally interested in rocketry appeared on campus. John Parsons, a self-taught chemist, and Edward Forman, a skilled mechanic, lacked the official academic credentials but had experimented with small black-powder rockets in Pasadena for years. When they read of Bollay's comments on rocket planes in the *Pasadena Post*, the two young men boldly came to Caltech with hopes of acquiring technical resources and assistance to build liquid-propellent rocket motors. Malina agreed to work with them to further the common goal of rocketry. After a series of meetings, the three young men sketched out a plan to develop sounding rockets that could penetrate the upper levels of the atmosphere, which they hoped would have practical use for meteorologists.

In early 1936, Malina, Bollay, Parson, and Forman laid out two initial goals: to solve the theoretical problems of a sounding rocket related to reaction propulsion and flight performance and also to build a small rocket motor for physical experiments. Their team was soon joined by A. M. O. Smith, another graduate student at Caltech who had a National Youth Administration job working as Kármán's assistant. The diversity of the backgrounds in the team provided a unique range of talents, blending the mathematical skills of Malina, Bollay, and Smith with the risk-taking, experimental "hands-on" approach of Parsons and Forman. While Malina, Smith, and Bollay worked on mathematical calculations for rocket performance, Parsons and Forman built the rocket chamber with pieces of metal found in junkyards and secondhand shops. Unlike Tsiolkovsky, Goddard, and Oberth, three solitary scientists who toiled away in isolation on their projects, the Pasadena scientists were functioning as part of a team to merge theory and experimentation. They represented in skeletal form the systematic division of labor that was to so heavily characterize the later development of aeronautics in the United States and throughout the world.

The biggest difference, of course, between the "big science" of today and the early efforts of the Caltech group in the 1930s was money. Since there were no government grants in rocket research at the time, the Caltech student group had to raise the funds themselves. Malina was paid 80 cents an hour to work in the wind tunnel at Caltech. Smith received a small stipend for drawing illustrations for technical papers and textbooks. In June 1936 Malina was on the verge of giving up on the rocketry project because they lacked the $120 needed to purchase two instruments. They hoped Irving Krick, a Caltech professor of meteorology and a commercial weather forecaster, might convince his wealthy friends to advance money for their rocket studies, but that turned out to be nothing more than a pipe dream. Desperate for cash, Malina and Parsons even tried to write an antiwar science fiction novel about a group of evil rocket scientists under a foreign dictator that they hoped to sell to a Hollywood studio.

An unexpected source of revenue came in June 1937 from Weld Arnold, a graduate student in meteorology. Arnold had been so inspired by a lecture Malina had given on campus that he promised to raise one thousand dollars for the rocket project. A few days later, Arnold astounded everyone by presenting them with the first installment—one hundred dollars in one- and five-dollar bills wrapped in old newspaper. Decades after the incident, Smith said that

Weld never told them where he got the money, nor did the team press him too closely for details. Malina, however, could not resist showing the bills to Clark Millikan, the aeronautics professor who had disapproved of the project from the very beginning. "Clark, how do we open up a fund for the rocket research project at Caltech?" he asked Millikan proudly. (Millikan, to his delight, was "flabbergasted.") The thousand dollars paid for the rocket motors and freed Malina from the routine of working on the novel in Parson's kitchen, which Parson also used as warehouse for accumulating tetranitromethane—a clear, toxic liquid with a pungent odor and high explosive potential.

With this fresh injection of funds, the project began to pick up speed. This appears to be about the time when Tsien joined the group, and not surprisingly, he took on the job of resident mathematician. On May 29, 1937, he finished a report bearing the long and impressive title of "The Effect of Angle of Divergence of Nozzle on the Thrust of a Rocket Motor; Ideal Cycle of a Rocket Motor; Ideal Efficiency and Ideal Thrust; Calculation of Chamber Temperature with Disassociation." It was a theoretical diagram of an ideal rocket motor with a fixed-volume chamber and exhaust nozzle. He concluded that the flame bursting out of the end of a rocket should be narrow in circumference to provide a focused thrust in space; a wider flame was more likely to send the rocket out of control. Tsien's paper joined a body of collected work that the rocket group called its bible. It consisted of reports of previous experiments and analyses of rocket motors, liquid fuels, rocket planes, and rocket shells that laid the foundation of the group's research.

In June 1937 the group was officially recognized by Caltech as the Guggenheim Aeronautical Laboratory of the California Institute of Technology (GALCIT) Rocket Research Project. As one of its first official acts, the group secured permission from Kármán to use the facilities of the Guggenheim Aeronautical Laboratory for its future experiments, eliminating the need to lug heavy tanks and equipment around the Arroyo Seco. The following month the rocketeers moved their apparatus to the basement. Gaining access to the ceiling through the open space of the staircase, they fastened four wires and attached to them a square cagelike metal framework on which they positioned a small eight-inch rocket pointed horizontally. This served as a fifty-foot pendulum. The distance of its swing when the rocket was fired would give them a measure of the thrust of the rocket.

No sooner had they moved in than disaster struck. First, Malina and Smith accidentally sprayed a cylinder of nitrogen tetroxide on the pristine green lawn of the Gates Chemistry Building, leaving behind a large brown patch that greatly annoyed the gardener. (The cylinder was left over from World War I and so corroded that Malina and Smith could not shut the valve.) Then, during one experiment, the nitrogen tetroxide and alcohol fuel combination misfired, causing the rocket to froth red and spew forth fumes that oxidized all the shiny exposed metal surfaces—such as steel scales—in the laboratory. The smell was terrible ("If there were any rats in the building, they have probably left," Malina commented), and the corrosive fumes coated the scales and equipment so prized by the aeronautics department with a thin layer of bright red rust. Frantically, the rocketeers tried to scrub the rust away with oily rags, but much of the equipment was ruined.

"The event was considerably horrifying," remembers Martin Summerfield, the bespectacled, Brooklyn-born physics student who was then Malina's roommate. "It was very serious at the time. They could have been banished, they could have been closed down completely. At that time it was underfunded, so why should they keep it? Yet von Kármán must have persuaded the Caltech administrators that it was a valuable thing to do, even though they suffered all of this corrosion."

But they were kicked out of the lab permanently. "Out!" Kármán yelled as the students hastened to move the apparatus outdoors. Placing the equipment atop a concrete loading platform, open to wind and rain, at the east end of the Guggenheim Laboratory, they suspended the pendulum from beams protruding from the roof of the building. Occasionally, a strong gust would send the oxygen, fuel tanks, and dial gauges trembling, forcing them to wait until the wind died down.

Nevertheless, the outdoor porch served as a suitable laboratory for the rest of the year. The campus soon rang out with the sound of their experiments. "I remember looking out of my window and seeing the boys out there on the ramp exploding," one professor remembered. The other students soon called them "the Suicide Squad."

In late 1937, the team was under considerable pressure to put together a theoretical paper that Malina would present at the sixth annual meeting of the Institute of Aeronautical Sciences (IAS) in New York City in January of the following year. (Clark Millikan may have been the pressure source; he was IAS president in 1938.) The paper, entitled "Flight Analysis of the Sounding Rocket," authored by Smith and Malina, investigated the mathematics of a simple model rocket, which they defined as a wingless shell of revolution in vertical flight through a vacuum. In four pages they laid down some basic equations for optimum rocket motion. On January 20, 1938, Kármán surprised Malina by giving him two hundred dollars to pay for travel expenses to New York. Malina boarded the train for the East Coast the following day, and en route carefully prepared for the talk he was scheduled to give on January 26, during the conference's aerodynamical session.

The paper was extremely well received in New York. It was the first one on rocketry ever to be presented at an IAS conference, and it won the Caltech rocket group not only respect from the aeronautical community but nationwide publicity. The Associated Press and *Time* magazine picked up the story and described the group's scheme to send a rocket far beyond the atmospheric reaches obtained by sounding balloons. "This analysis," the AP quoted Malina saying, "definitely shows that, if a rocket motor of high efficiency can be constructed, far greater altitudes can be reached than is possible by any other known means." Soon, headlines like "Scientists Plan to Shoot Rocket 967 Miles in Air" appeared in newspapers across the country.

When Malina returned to campus, he found the rocket group excited and nervous about the publicity. Stories about their IAS paper had surfaced in newspapers all over California, including an editorial in the *Los Angeles Times*. While looking over the conference paper together, Tsien and Malina thought they detected an error and panicked. It took them some time to calm down and convince themselves that they were correct after all. "Needless to say," Malina wrote, "there were a few worried moments."

The students soon relished their new roles as rocket celebrities on campus. Even Tsien, it seems, couldn't resist the glare of the spotlight. On February 3, 1938, the student newspaper, the *California Tech*, carried a front-page article on the group that contained a fairly detailed description of results that Malina did not want published yet.

> The rocket has emerged from the realm of fiction. In the next three months, Frank J. Malina, A. M. O. Smith, and Hsue-shen Tsien, Caltech graduate students in aeronautics, will have more reliable information about rocket motors than the whole world has been able to learn by all its previous attempts.

The article described a second rocket motor being developed in the Caltech machine shop. Unlike the motor dangling from the roof of the Guggenheim, this one had steel packets, a copper nozzle, and a carbon lining to protect the chamber from disintegrating under the tremendous heat of combustion. (Carbon, after all, has an extremely high melting point—6,233°F.) They decided to use a form of carbon called graphite, a soft, black material with a metallic sheen often used in pencils. The plan, according to the article, was to inject a carbon-hydrogen fuel mixture into one end of the combustion chamber, and oxygen through the other, and then ignite the combination with electricity. By lining the exhaust nozzle and combustion chamber with carbon, the group hoped to solve the problem of designing a chamber to survive the high temperatures associated with combustion in a rocket that never left the earth's atmosphere or one that was destined for Mars.

The article contained some of Tsien's dreams for future experimentation:

One object of this experimentation is to learn some of the characteristics of the earth 600 to 900 miles above the surface. The proposed rockets will be composed of three separate parts. A great deal of energy is consumed in rising through the lower layers of dense air, and if possible will be launched from a high mountain. Once above the dense air, the rocket will drop its dead weight and proceed upwards with a decreased fuel consumption. Finally at a predetermined time, a second section will be dropped and the rocket will "coast" to a higher altitude.

"We were even more surprised when we learned that Tsien, who usually says very little, had spilled the beans," Malina wrote to his parents. "Nothing serious in the slip, however."

The next few months were a time of intense teamwork between Tsien and Malina. If they weren't tinkering with the rocket apparatus, they were working out theoretical calculations of rocket flight. So absorbed were they in their work that sometimes they forgot to stop for lunch. By April 1938 they had completed a paper on rocket flight and submitted it to Clark Millikan for review. The team also tested the rocket apparatus by filling it with water to see if the pipe joints leaked.

By the end of spring and beginning of summer, their work began to pay off. In May the team obtained its best results yet. The rocket motor ran for a full minute and made such a loud popping noise that it quickly drew a large audience at Caltech. An AP reporter began to follow Malina around campus and eventually wrote a lengthy story that was circulated nationally. Soon, Tsien, Malina, Parsons, and Forman were providing information for science writers from national magazines like *Popular Mechanics* as well as the local Pasadena and Los Angeles newspapers.

The response was immediate and sensational: one publication carried five-column sketches of rocket ships with passengers blasting off from the Los

Angeles Civic Center, while a Hollywood radio station offered to broadcast the sound of the rocket motor. One enterprising stunt man from New York even volunteered to be shot up one thousand feet by rocket during county fairs and to float down from the sky by parachute. "Such fuss!" exclaimed Malina. "The reporters seem to have better imaginations than we do."

Though pleased by the attention, the group worried that the publicity might backfire. Robert Goddard was almost "blasted away" by the media during the 1920s, Malina observed. If somebody had called Goddard a crackpot during those years, the Yankee scientist would have considered "the donor of the compliment a very mild human being." Theodore von Kármán urged the students to keep a low profile and to focus on their research.

In mid- to late 1938, Tsien began to see the rewards of his efforts in both aerodynamics and rocketry. In May his first paper with Theodore von Kármán, "Boundary Layer in Compressible Fluids," was published. It was a study of the behavior of boundary layers over objects moving at high speeds, such as rockets and missiles. Tsien and Kármán estimated the ratio between wave resistance and frictional drag of a rocket and also calculated a mathematical relationship between drag and heat transfer. Then, in October, another of his papers, "Supersonic Flow Over an Inclined Body of Revolution," appeared in the *Journal of the Aeronautical Sciences*. Here, Tsien investigated the lift of a pointed projectile at supersonic speeds, and found that the lift at any fixed Mach number was directly proportional to the angle of attack of the body. Finally, in December, Tsien and Malina's paper, "Flight Analysis of a Sounding Rocket with Special Reference to Propulsion by Successive Impulses," appeared in the same journal. In this paper they proposed a reloading type of solid-fuel black-powder motor that would propel a rocket with a series of explosions rather than with a steady flow of continuously burning liquid fuel. Theoretically, this motor would push rockets to much greater heights than that currently reached by sounding balloons.

By December Tsien focused his efforts on finishing his dissertation, although he took time off to aid Malina on other scientific ventures. For instance, when Malina worked on a paper that he planned to submit to a scientific contest held in Paris, Tsien was unstinting in his help. (The paper was later awarded the REP-Hirsch Gold Medal.) As Malina wrote to his parents,

> Tsien really should have had his name on it as he helped with many of the ideas. He is truly a brilliant fellow. I wish I could work as persistently as he does. The

last ten days he worked night and day on a tough problem and then found that the solution of the mathematics was not satisfactory. He left the office; afterwards he said [he would] work up his courage to start over again. He has what it takes.

—◦—

There was much in common that Tsien shared with this young man from Texas. Both had been pampered from birth by their parents. Both shared a passion for art and classical music. And, finally, according to those who knew them at Caltech, both had an arrogant streak in their personalities: a stubborn conviction of their intellectual and ideological superiority to others. Frank Malina, who idolized Leonardo da Vinci, considered himself a Renaissance man ("brilliant at art, brilliant at literature, brilliant in science and to be able to do all of these things and juggle them back and forth," Malina's first wife remembered), while Tsien, aloof, quiet, and elitist, was quickly acquiring on campus the nickname, "the Son of Heaven."

(Arrogance, of course, was nothing new at Caltech. As William Sears wrote in his autobiography, *Stories from a 20th-Century Life*: "The Caltech fellows were a very cocky bunch, clearly convinced that Caltech was the greatest and most demanding college in the world and that they, its graduates, must be the smartest students.")

While Tsien was clearly proud of his mental powers and proud of his association with Kármán, he seemed especially sensitive about any insult or derogatory comment directed toward his homeland of China. He became sensitive with good cause. One time, a patron in a movie theater demanded that the usher eject Tsien from his seat because he did not want to sit next to an Asian. Tsien was so enraged he would remember that episode for the rest of his life. On another occasion, some American students derided China as being a poor and ignorant country. "I can't help the fact that China is poor," Tsien reportedly told them. "But let's have a competition. I'll be China, you be the U.S. At the end of the term, let's see who has the better grades!" Tsien won the contest, to the great relief of the other Chinese students. Many admitted years later that they would not have dared to make that challenge themselves. He was also, it appears, hard on his compatriots, making him not at all popular among the Chinese student community. "He was very stubborn, very individualistic fellow who criticized others quite often," remembered his roommate, Shao-wen Yuan. "He always thought he was right, and usually, he was. But he made a lot of enemies."

Tsien, it appears, had little time or inclination to deal with what others thought of him. He preferred to spend his spare hours with a few close friends he trusted and admired. And in 1938 probably no one his age was closer to him than Frank Malina, whose wide range of intellectual pursuits matched his own.

He began spending more time socially with Malina. The rocket group, probably under Malina's urging, threw Tsien a surprise party when he passed his oral exam for his doctorate. Malina also owned an old, dirty-gray Chevy with a rumble seat, and on weekends he and Tsien would drive to Los Angeles. Sometimes Andrew Fejer, an aeronautics graduate student from Czechoslovakia, would join them, and he, too, became Tsien's close friend. This was before the era of superhighways, and the car would travel the little roads along the foothills of Pasadena and Glendale until it reached Hollywood.

Music appears to have been the dominant theme of their excursions. They would frequent the Philharmonic Auditorium to listen to performances of the Los Angeles Symphony conducted by Otto Klemperer. They spent long afternoons poring through music stores for good classical records, buying dozens of albums of Russian composers like Igor Stravinsky, Dmitri Shostakovich, and Serge Prokofiev. Tsien especially loved the chamber music of Bach and the symphonies of Beethoven. Afterwards Tsien sometimes went to Malina's or Fejer's home, spending the entire evening listening to records without saying a single word. His love for classical music was so profound it was almost religious, Fejer remembered. "I wouldn't be surprised if he had at home the printed sheet music," he said.

Decades later, when Andrew Fejer and his wife, Edith, were asked to describe Tsien, they recall the image of a carefully dressed young man with perfectly groomed hair, delicate, long fingers, and manicured nails. ("He was *very* elegant," Fejer remembered.) Sensitive and cultured, Tsien would frequent museums to look at impressionist and modern art, noting the connections between art, music, and science. ("He found the physical image of a solution of a mathematical problem, by definition, beautiful," he recalled.)

Tsien's sensitivity, however, did not extend to those he deemed mentally beneath him. "He intimidated people by talking about things they didn't know anything about," Fejer said. "It was basically intellect that attracted Tsien to others. He was not interested in contact of any kind with people who were intellectually inferior."

In 1938 Malina introduced Tsien to an intellectual discussion group at Caltech. The members met for evenings of music and political discussion at the homes of some of the older, married scientists. Sometimes the meetings were held at the two-story, white stucco home of Jacob Dubnoff, a Caltech biologist who owned an excellent high-fidelity music system. But most of the time they gathered at the home of Sidney Weinbaum, who worked as a research assistant in the chemistry department.

Sidney Weinbaum had sharply aquiline features and humped shoulders, as if permanently bent from reading too many books. He looked unathletic and in poor health, appearing dark and gloomy in his blue serge suits. But in Tsien's eyes, Weinbaum was something of a Renaissance man. A Ukrainian-born Jewish scientist in his forties, Weinbaum had fled from his homeland after the Russian Revolution, emigrating to Los Angeles in 1922. An accomplished concert pianist and chess player, he taught music to pay for his undergraduate tuition at Caltech and won the Los Angeles chess championship twice in the 1920s. He worked as a technical assistant to chemist Linus Pauling while earning his doctorate in physics at Caltech in 1933. His astonishing range of talents fascinated Tsien.

Once every few weeks, after supper at about 8:00, guests began to arrive at Weinbaum's gray bungalow on Steuben Street. Inside some twenty or thirty Caltech students sprawled out on the furniture and chairs of the living room. Tsien would come neatly dressed in vest, tie, and polished shoes; he was often accompanied by Malina, a tall, lean figure with a razor-fine moustache and immaculate dress. They were a dignified contrast to the more sloppily dressed Bohemian crowd around them. Frank Oppenheimer, the brother of the famous physicist Robert Oppenheimer, was dressed in a manner more characteristic of the group. He was a tall, awkward figure with tattered sleeves, cigar ashes sprinkling the front of his shirt, and fingers still dirty from the laboratory.

This was a group of Caltech intellectuals drawn together by concern for a number of international crises: the Great Depression and the rise of Nazism and fascism in Germany, Italy, and Spain. After witnessing widespread unemployment and hunger in the United States, many students wondered if the predictions of Karl Marx would indeed soon come to pass: that capitalism as a system was doomed to break down into global chaos. Consequently, they saw the rise of socialism in Russia as a fascinating experiment. As pacifists, they were also alarmed when Germany occupied Austria, broke the Munich agreement by seizing all of Czechoslovakia, and assisted Franco in establishing a fascist state in Spain.

Tsien found the group sympathetic to the plight of China. In July 1937, Japanese and Chinese forces had clashed at the Marco Polo Bridge near Beijing, a conflict that quickly escalated into full-scale war. (For Tsien, the impact of the war was not only emotional but financial: the scholarship stipends to all Boxer scholars in the United States were cut by half, from one hundred to fifty dollars a month.) Then in December 1937, Japanese troops moved into Nanjing, where they perpetrated one of the worst orgies of rape and massacre in world history. During the weeks that followed, the Japanese killed between two and three hundred thousand Chinese in the city. The atrocities received worldwide coverage, and Tsien followed the events closely in the newspapers. Here were students who shared his feelings of outrage and despair, and it must have been soothing to be around them.

The discussions at the Weinbaum home followed a certain procedure. The gatherings were somewhat more formal than a party and somewhat less formal than a university discussion class. On a coffee table in the living room were a spread of publications, many of them Russian. Once every two to three weeks, a few attendees would prepare book reports that would be read before the group. The books included the works of Karl Marx, John Strachey, Joseph Stalin, and Vladimir Lenin. A lively question and answer session followed, as the audience debated the theories and viewpoints presented in the literature.

The meetings would end with music and refreshments. There was a large community of musicians in Hollywood in the 1930s, many of whom had worked in theater orchestras during the era of silent film. Some spent the day recording the tunes of composers like David Raskin for the major studios and on occasion relaxed at night by playing classical music at the Weinbaum home. When professional musicians were not present, the Caltech scientists would form their own amateur string group. Malina, son of a high school band leader, would join Weinbaum at the piano. Frank Oppenheimer would play the piccolo with such consummate skill that he could have passed as a professional. Tsien was learning how to play the alto recorder, and sometimes at these parties he would sit in a corner and pipe away.

Ever the loner, Tsien remained in the background. Sometimes, he sat completely silent, deep in concentration as he matched wits with Weinbaum over a game of chess or over the more complex war game of *krigspiel*. He preferred to listen rather than talk, sitting on the sidelines as political arguments broke out and continued into the night. And, as Malina recalled, the evenings almost always ended with a stimulating argument.

We used to meet to listen to music—we'd get into arguments—we would get together at night and sometimes we'd be arguing about rockets, and the next thing we'd be arguing about was what is happening in Spain. . . . We were a peculiar, closely knit group with many cross-interests which threw us together; the discussions would swing from music to rocketry to political questions to social organizations.

Unlike Tsien, Malina was vociferous during these debates. There was nothing he loved more than a good argument. A staunch liberal and idealist who believed in the possibility of world peace, Malina was not afraid to be controversial and often said things that offended his more conservative friends. One such person was A. M. O. Smith, who was appalled at some of the discussions that took place at the Weinbaum home. As Smith recalled: "Malina would actually blame all of the ills of the country on its political system. I didn't see that at all. He was radical."

As 1939 approached, the rocket research project slid into the doldrums. Tsien devoted his time to finishing his Ph.D. thesis, which would include papers he had written with Malina and Kármán. Now he was tackling some important problems in compressible fluids. Malina spent more time in the wind tunnel conducting tests on the shelter belts of trees, designed to reduce soil erosion. Engaged to marry Liljan Darcourt, an eighteen-year-old aspiring artist, Malina now prepared for his future responsibilities as a family man. The group lost two members: Weld Arnold left for New York City and Smith took an engineering job with an aircraft company. Impatient to see results, Parsons and Forman launched a few black-powder rockets on their own, and then, short on money, went to work for the Halifax Powder Company. The thousand dollars that had fueled the rocket experiments was fast dwindling away.

Membership in the Suicide Squad seemed ill-defined and constantly in flux. Individuals seemed to float in and out of the group, leaving permanently to take other jobs or working elsewhere briefly and then returning. There were young men in National Youth Administration jobs who assisted Malina in some of the rocket experiments, yet they are not credited by Malina to be full-fledged members of the group. Strangely enough, Martin Summerfield, Malina's former roommate who observed the activities of the Suicide Squad and worked on many theoretical problems with Malina and Tsien, was also not officially considered a member of the squad. It is difficult to get reliable information about the day-to-day dynamics within the group, since Malina, Parsons, Forman are dead, Tsien, in China, is uncommunicative about this subject, and Smith, still alive and well, left the group in 1938. There are hints of hostility and politics within the Suicide Squad about which little is known. According to Forman's widow, Jack Parsons and Ed Forman thought Malina was unbearably cold and arrogant toward them, while Malina recorded in his memoirs that he felt they were unreliable as researchers. "Jack and Forman would disappear for weeks or months sometimes with me trying to keep the project alive," Malina claimed later. "Tsien and I for one period anyway must have been the only two left." The rift in the group seemed to draw Malina and Tsien closer.

Just as the rocket program was saved two years earlier by Caltech student Weld Arnold, it would be rescued by still another Arnold—General Henry Arnold of the U.S. Army Air Corps. During the spring of 1938, General Arnold paid a surprise visit to Caltech. Intrigued by the rocket research he saw, he looked into the possibility of using rockets for national defense. Arnold's interest came at a time of escalating tension in Europe: in Munich, signatories ceded German-speaking portions of Czechoslovakia to the Third Reich. That fall, he invited Theodore von Kármán and Robert Millikan to Washington, D.C., to attend sessions of the National Academy of Sciences Committee on Army Air Corps Research. There, Caltech was given a choice of one of five Air Corps projects, one being research on rocket-assisted takeoff for heavy bombers. Jerome Hunsaker, head of the aeronautics department at MIT, had passed over the rocket project in favor of building de-icing mechanisms for the windshield of aircraft. "Kármán," he said, "can take the Buck Rogers job."

Caltech accepted the rocket propulsion project eagerly. When Kármán returned to Caltech, he asked Malina to draw up a presentation for the National Academy of Sciences (NAS). Traveling to Washington in December, Malina testified before an NAS committee that, based on the current level of technology, rockets lacked the propulsive power to significantly assist aircraft. He proposed a future study to develop solid and liquid rocket fuels and motors that could withstand the intense heat of combustion. As a result of his talk, in January 1939 NAS awarded Caltech one thousand dollars to conduct a preliminary investigation and to prepare a more detailed grant proposal.

Malina had mixed feelings about the grant. He was torn between his dream of building rockets for peaceful scientific exploration and his distaste for building them as instruments of war. "My enthusiasm vanishes when I am forced to develop better munitions," he wrote. But at the same time, Malina was worried about the spread of fascism and the plight of democratic countries in Europe. Before long, the rocket team—now consisting only of Tsien, Malina, Parsons, and Forman—was working overtime to produce results. The U.S. military, alerted by the German invasion of Czechoslovakia and rumors that the Nazis were building their own rocket program, wanted to see the first report by June.

In March, however, during the rush to get things done, there was a near-fatal accident. Somehow, propane leaked into the oxygen tank, causing the entire rocket apparatus to explode. A piece of metal knocked a hole in the wall inches above a chair where Malina would have been sitting if he had not been out running an errand; if he had been there, he surely would have been killed. After the accident, Caltech banished all rocket experiments from campus. Robert Millikan, the president, almost shut down the project entirely, but Kármán convinced him to let the young men remain in the basement of Guggenheim to work on theoretical problems. They repaired the apparatus and moved it to the secluded Arroyo Seco, where they build crude test stands along the west bank of the canyon.

Finally, in June, they got the response they were waiting for. Malina's report to NAS resulted in a ten-thousand-dollar grant for the fiscal year 1939–40 to build an experimental station and purchase materials to pursue the rocket fuel problem. The money would also pay the salary of additional research assistants, such as Homer Joe Stewart, one of Kármán's graduate students in aeronautics, and Martin Summerfield, a physics major who was Malina's friend and former roommate. The goal was to develop solid and liquid rocket fuels that could aid the "superperformance" of aircraft.

It was an exciting moment for the group—and especially so for Tsien, who graduated on June 9, 1939. On a bright, warm day, he stood in his cap and gown, waiting to receive one of the thirty-two Ph.D.'s that would be awarded that year. The ceremony was held outdoors, near the administration building. The campus had expanded over the past year with the construction of the Crellin Laboratory, the East Kerckhhoff Biology Building, and the Arms and North Mudd buildings, signifying for Caltech the end of the Great Depression. Together with Frank Oppenheimer, who had completed his doctorate in physics, Tsien marched in a procession as the campus orchestra played Gounod's *Marche and Cortège for the Queen of Sheba*.

As the invocation slid into address, and one speech blended into the next, it would have been hard for Tsien not to reminisce about the changes he had witnessed in just three brief years at Caltech. Four years ago, he had just finished a tour of China's aviation facilities for national defense. Now his homeland was at war with Japan. Then, his only encounter with Kármán was through textbooks and scientific articles. Now he was a Kármán protégé—perhaps his best one. Earlier, he and Malina were struggling to build rockets with junkyard equipment in a project that almost perished for lack of $120. Now, they were flooded with funds and given the blessing of one of the most powerful generals in the country.

August 1939 saw the publication of Tsien's paper "Two-Dimensional Subsonic Flow of Compressible Fluids" in the *Journal of the Aeronautical Sciences*. Destined to become a classic, the paper represented some of Tsien's most important scientific work at Caltech. From the paper emerged the famous Kármán-Tsien pressure correction formula—an equation that would aid engineers for decades in the design of high-speed aircraft. According to A. M. O. Smith, now a distinguished and retired engineer: "The Kármán-Tsien formula was almost universally used until modern computers made inroads."

By 1939, airplanes were encountering a dangerous new phenomenon known as compressibility effects. As planes grew more sophisticated and achieved faster speeds, pilots also began to experience wild pitching motions in the air, known as buffeting and flutter. Sometimes there would be a sudden loss in lift, forcing the plane to plunge from the sky. In 1941 a Lockheed test pilot died when his plane dove out of control. Meanwhile, it was reported that a Curtis SB2C Helldiver fluttered so badly its tail was breaking off. This was because, as a compressible gas, air could be squeezed closer together so that its molecules occupied less volume. At low flight speeds, this behavior was negligible. But, as the plane approached the speed of sound, the air could bunch up, or burble, on the upper surface of the wing. Pressure could build up, causing the plane to stall. It became clear to aircraft engineers that more sophisticated formulas would have to be developed to account for this phenomenon at high speeds.

Before the age of high-speed computers, engineers would design airplanes by making a number of physical assumptions and creating a mathematical model of actual phenomena. The Kármán-Tsien pressure correction formula would help engineers correct some of these errors caused by simplification when estimating the pressure against an airfoil at high subsonic, or "transonic," speeds (velocities approaching that of sound)—a significant jump in progress from the previous theories of aerodynamicists who had calculated approximations for flows with velocities smaller than half of that of sound. (For instance, an older, less accurate formula was the Prandtl-Glauert formula—a significant contribution of Theodore von Kármán's former professor in Göttingen.)

William Sears recalls being present at Kármán's home when Tsien and Kármán were still working on the theory:

There seemed to be some confusion between them about the mathematical details; they had each written it up and there was a difference, seemingly nontrivial, between their analyses. Naturally I pricked up my ears, wondering whether the professor or his brilliant student had erred. They went back to the beginning and compared line by line; they were both right! As can happen when an approximate theory is constructed, they had taken different paths at a certain point and had arrived at results that were equivalent within the scope of the approximation.

—⸱⸱—

In fall of 1939, Tsien, now called a research fellow, became a member of the aeronautics staff, continuing his aerodynamical work with Kármán, and his work with Malina to find a solution to the solid fuel problem.

One of their objectives was to develop a rocket motor strong enough to propel airplanes during takeoff; that is, capable of delivering a thousand pounds of thrust for ten to thirty seconds. They had a choice of developing one of two kinds of fuel: liquid or solid propellent. A liquid-propellent fuel would be pumped into the combustion chamber with gaseous or liquid oxygen, while solid-propellent fuel would contain its own oxidizer in its chemical mixture. Both kinds had considerable problems. Liquid-fuel rockets were easier to design to provide thrust for a period of time but messier to handle. Solid fuels would be simpler to handle, meeting the field needs of the military, but no known solid-propellent engine at the time had ever burned longer than three seconds. Most experts said a longer-duration solid-fuel engine simply was not possible: pressure would build up inside the combustion chamber and shatter the entire rocket. In experiment after experiment, Parsons and Forman seemed to prove the experts right. They designed motors with black powder in numerous formulations, but all of them exploded immediately or did not burn.

All of this must have added to the existing skepticism about the value of their research in rocketry. The word *rocketry* itself carried such stong connotations of amateurism and science fiction pulp magazines that Caltech decided to use the euphemism *jet propulsion* instead. (Only a few months earlier, Caltech physics professor Fritz Zwicky had called Malina a "bloody fool." Didn't he know, thundered Zwicky, that rockets had to push on air to get any thrust?) There were also officials in the military who wondered if the NAS grant was money well spent. When Major Benjamin Chidlaw, an aide to General Arnold, visited Caltech in 1939, he asked Theodore von Kármán: "Do you honestly believe that the Air Corps should spend as much as ten thousand dollars for such a thing as rockets?"

Undaunted, the team continued to work on the solid-fuel engine problem. On some nights, Tsien worked at Malina's home. Malina now lived with his wife near campus in a small house at 1288 Cordova Street: a one-story white shingled cottage covered with large vines. A Cecil Brunner rose bush grew over the garage, diffusing a prickly, peppery scent. After dinner, Tsien and Martin Summerfield would arrive with notebooks, pens, and slide rules. They would join Malina in his office: a screened-in porch equipped with a single black phone, a table, wooden horses, and rattan chairs. Before long, the table would be covered with reams of paper scrawled with equations. As the night wore on and temperatures dropped, the boys plugged in the electric heater and kept on working. "You would hear them talking and carrying on," remembered Frank Malina's first wife, Liljan Malina Wunderman. "They would meet and compare notes and argue and 'this must be wrong; it can't be right' and Martin screaming 'It's not! It's not!' And then they would all start laughing. It sounded as if they were having a wonderful time."

On other nights, Tsien worked late in his office on the third floor of the Guggenheim building. As one graduate student from that period remembered:

> Early one day—it was a holiday, either Thanksgiving or Christmas, I went in to catch up on some studies and, thinking that I was alone in the building, turned on a record player rather loudly, playing, I distinctly remember, "Dance of the Hours," which has a rather tremendous crescendo. Part way through that crescendo there was an enormous banging on the wall. I had disturbed Tsien and learned that Chinese students studied harder and longer than Jewish students. . . . [Later,] he gave me several copies of his then most recent work on compressibility corrections at high subsonic Mach numbers—a sort of left-handed apology for having screamed at me during the "music" episode mentioned above.

By 1940 the group had reached some concrete conclusions about the rocket fuel problem. That summer, Theodore von Kármán developed four differential equations describing the operations of an ideal restricted-burning solid-propellent motor, which Malina solved, proving that it was theoretically possible to build such a motor as long as the ratio of the area of the exhaust nozzle throat to the burning area of the propellent charge remained constant.

For most Americans, 1940 would be remembered as the year the United States, though not itself belligerent, took sides in the war in Europe. With Hitler's air attack on England, the U.S. government, acting through the U.S. War Department, released stocks of arms, planes, and munitions to the British while implementing an embargo of aviation fuel, steel, and scrap metal to the Japanese. Nothing could dam the flood tide of defense spending that surged forth. President Franklin D. Roosevelt called for the production of fifty thousand combat planes a year. Congress voted for $17 billion in defense funds, up from $1.9 billion the previous year. Not surprisingly, the NAS increased its grant to Caltech, awarding the rocket group $22,000 for the fiscal year 1941.

To Tsien, 1940 had additional significance. It appears that in that year he received a letter from Lieutenant Colonel Tsoo Wong requesting that he return to China to fulfill his obligations as a former Boxer Rebellion scholar. The maximum length of his stay in the United States was supposed to be only three years, and Tsien had already been in the country for five years. He had graduated from Caltech, and his student visa would soon expire. The services of an American-trained aerodynamicist from Caltech could be put to use immediately in a country crumbling under the repeated air raids and bombings of the Japanese.

There is no reason to believe that when Tsien left China three years earlier he had gone with any intention other than to return to China once his studies were over. Clearly, however, something had changed, particularly after his move to Caltech.

It was not just a question of the opportunity for a first-class mind to address cutting-edge problems in a supportive environment, although Tsien knew better than anyone else that once he returned to China, his days of breaking new scientific ground were over. China needed all its scientists to use their talents to defend the country, not to sit for hours considering the mathematical solutions to space travel. Nor could it have been solely a question of community, although Tsien surely knew that his friendships with Malina and other Caltech scientists were very satisfying parts of his Caltech years. But anyone who has ever been part of a winning team knows there is something very compelling about those few moments in life when their team's effort seems to be paying off. The decision to return would be wrenching.

And yet his homeland needed him, his family missed him, his sense of honor compelled him.

Malina captured Tsien's feelings with a pen-and-ink cartoon. In the drawing, Tsien is holding an egg inscribed with "US" on one end and "China" on the other. He flips the egg back and forth, and back and forth, looking utterly perplexed.

Tsien decided to try to remain in the United States for at least another year. As a peace offering, it seems that he promised to consult for China's Bureau of

Aeronautical Research. He asked Kármán to plead his case to the Chinese government. In a letter to Tsoo Wong dated April 20, 1940, Kármán wrote:

> I should like to emphasize that it is not my intention to deter Dr. Tsien from doing his duty to his country, however, as you say, I believe one can serve his country efficiently and faithfully in different ways. I have the feeling that it is not only in Dr. Tsien's interest but also in the interest of the Chinese cause that before returning to China he does research work in different fields of aeronautical engineering and aeronautical sciences. He already did excellent work in high speed aerodynamics and in structures. We are now engaged in an investigation of hydrodynamics of floats and boats. I have the impression that this is an important subject, and it will be in the interest of your organization to have a man thoroughly acquainted with the problems of planning surfaces.
>
> In accordance with these views, I should like to suggest that Dr. Tsien remain for one more year at the California Institute of Technology. To be sure, I appreciate both his abilities and pleasant personal qualities as a collaborator, but believe me Dear Colonel Wong, that my advise [sic] is not directed by selfish motives.

In December, Caltech officials prepared the appropriate paperwork for Tsien to stay in the United States. They asked the Department of Immigration and Naturalization to extend Tsien's visa for another two years, after which Tsien intended to return to China to join the Chinese National Research Council in Chungking. Meanwhile, Tsien sent the Bureau of Aeronautical Research in the city of Chengtu a Chinese-language paper, entitled "A Method for Predicting the Compressibility Burble," that contained his famous pressure-correction formula.

The war seems to have caused Tsien to drift away from his Suicide Squad friends for a few years. The NAS rocket project was classified, and Tsien's lack of U.S. citizenship must have prevented him from getting the necessary security clearances. Separated from Malina in research in 1940, Tsien found a new circle of peers at Caltech, most of them Chinese. He attended meetings of the Chinese student organizations on campus and wrote papers for the West Coast chapter of the Chinese Natural Science Association. "I listened to them talk to each other in Chinese and every tenth word would be 'differential equation' or 'substitution,'" Fejer remembered. "When they got into a scientific argument, they did it the easiest possible way, which was in Chinese."

Tsien became especially close to Chieh-Chien Chang, an aeronautical engineering graduate student at Caltech whom Tsien had first met at Qinghua University in 1934 during his tour of China's aviation facilities. That year, under the auspices and financial support of General Chiang Kai-shek, C. C. Chang had helped design one of the largest wind tunnels in China; in 1935, he became assistant to Frank Wattendorf, one of Kármán's protégés, who accepted the post of aeronautics professor at Qinghua University. When Kármán visited China in 1937, he offered Chang a scholarship to study at Caltech. Chang arrived in Pasadena in September 1940 and shared an office with Tsien on the third floor of the Guggenheim building. For more than a year, Chang would see Tsien practically every day. Chang remembered:

> Tsien and I were very close friends. He was a very quiet man, very conservative. He never showed emotions about politics. Only one, two or three times, maybe. We were both very serious about the Japanese invasion of China. We just despaired. We felt it was hopeless in China at the time. We were concerned but we felt we couldn't do anything about it.
>
> We were together every day, for lunch and dinner. Sometimes we worked so hard that after dinner we just went back to the office. And sometimes I left a little early—I left before midnight. He wouldn't stop, however. Even after midnight his room was still bright.

As a foreigner, Tsien fell under scrutiny by the U.S. government. In 1940 the Alien Registration Act was passed so that the U.S. government could better monitor immigrants from enemy nations. Although Tsien was from an Allied country, he was nevertheless required by law that summer to get his fingerprints taken and to receive an A-number issued by the Immigration and Naturalization Service, which he did on December 2, 1940.

Tsien continued his aerodynamical research in nonclassified areas. Between 1940 and 1942 he was preoccupied with the subject of structural buckling. The subject assumed greater importance as metals replaced wood and fabric in the design of aircraft, and as aeronautical engineers sought materials that were both lightweight and sturdy. It was important for engineers to understand the precise limits of different materials and to predict the locations in a structure where buckling was most likely to occur, so that they could strengthen those areas with reinforcements.

Over the span of two years, Tsien wrote several papers on the buckling of spherical shells, thin cylindrical shells, and columns. Some were coauthored with Theodore von Kármán and Louis Dunn, a recent Caltech graduate and member of its aeronautics staff. Their theoretical papers served as models for engineers working with different shapes and bodies of aircraft. Colleagues remember vividly one experiment conducted on campus that confirmed some of Tsien's theoretical predictions. Mathematically, Tsien had found that cylindrical shells tended to buckle in a diamond-shaped pattern. To test his finding, a metal shell about ten feet long and three feet in diameter was erected outside of the aeronautics building. Slowly, it was filled with water to pressurize the cylinder from within. Gradually, the shell warped and bent into a grid of diamonds. "Sure enough, the theory predicted exactly what happened," recalled one Caltech alumnus, Bernard Rasof. "Tsien was fascinated by the fact that his theory had been proven experimentally."

Tsien was also involved in the design of a small wind tunnel at Caltech capable of generating wind currents at supersonic speeds. Kármán had originally approached the Army Air Corps with the idea of building such a tunnel, but the Army turned his proposal down. "The answer he got," recalled Allen Puckett, then Kármán's assistant and later the chief executive officer of Hughes Aircraft, "was that the Army Air Corps was not interested in funding supersonic wind tunnels because airplanes would never fly supersonically." Next, Kármán contacted the Army Ordnance Corps and convinced it that it needed the wind tunnel for its supersonic artillery projectiles. ("How he sold them this bill of goods I'll never know," Puckett wrote years later. "They needed a wind tunnel like a hole in the head.") The net result was that the Army Ordnance Corps granted Caltech ten thousand dollars to build a wind tunnel with a tiny two-and-a-half-inch throat. Three people were involved in its creation. Tsien worked out mathematically a general outline of the design concept; Mark Serrurier, formerly the chief engineer for the Palomar telescope, focused on the mechanical design; and Allen Puckett was responsible for the theoretical part, such as the design of the nozzle shapes and performance calculations, and was then in charge of using it as an experimental tool.

By 1942 they had produced the first continuously operating supersonic wind tunnel in the country to reach Mach numbers above four. It served as the model for a larger wind tunnel that Puckett designed and built for the Ballistic Research Laboratory of the Aberdeen Proving Grounds: a fifteen-by-twenty-inch tunnel demanding thirteen thousand horsepower that became the first large supersonic wind tunnel of its kind in the United States.

Possibly because of this work, Tsien wrote a paper for the *Journal of the Aeronautical Sciences* entitled "On the Design of a Contraction Cone for a Wind Tunnel." A contraction cone is shaped like a funnel, so that the wind that rushes through the diminishing hollow of the cone is squeezed through a tiny opening at the end to increase its velocity. In 1942 Tsien observed that if the curvature of the cone is too large at certain points, the boundary layer of the air may separate from the wall. If the velocity of the wind is too high at the narrow end of the cone, the tunnel risks the dangers of compressibility shock. In his paper he worked out formulas that would keep the velocity of air flow below that of sound in a subsonic wind tunnel. How far Tsien had come from his first failed experiments at MIT building wind tunnels!

Tsien's isolation from high-level military projects would end abruptly on December 7, 1941—the day news from Hawaii stunned a nation. In their surprise attack early that morning, the Japanese bombed Pearl Harbor on the island of Oahu, destroying much of the U.S. Pacific fleet. No longer could Americans feel secure in their isolation between two oceans: technology had brought the enemy to their very doorstep. The reaction was immediate. Los Angeles braced itself for a possible invasion as local aircraft companies moved valuables inland and Hollywood set designers camouflaged airplane assembly plants as green fields and city streets. Generations of Japanese-Americans and Japanese immigrants, most of them completely loyal Americans, were herded into concentration camps. In this atmosphere of national emergency, the United States could not afford to let the brain power of Chinese foreign national scientists go to waste. Even before the Pearl Harbor attack, in August 1941, the U.S. Justice Department had changed Tsien's immigration status from student to visiting scientist so that he could continue his research. By 1942, Kármán so believed in Tsien's value to the U.S. government that he personally arranged to get Tsien clearance to secret military projects at Caltech. "I haven't the slightest doubt as to Tsien's loyalty to the United States," Kármán wrote.

Apparently, neither did the United States. After the standard checks, Tsien's clearance was approved on the first of December by Colonel M. S. Battle, chief of the personnel security branch of the internal security division of the Provost Marshal General's Office. Now Tsien could work on secret contracts—for the Army, Navy, Army Air Corps, War Department, and Office of Scientific Research and Development—and on higher levels of access than he had ever been permitted previously. In 1942 he again joined Malina on rocket research, rekindling a partnership that would produce some of America's first military missiles.

The Jet Propulsion Laboratory
(1943–1945)

The mid-1940s were undoubtedly some of the happiest and most fruitful years of Tsien's life. He was in his early thirties and at the very height of his creative powers. After four years of working as a research assistant, Tsien was asked to stay on as an assistant professor of aeronautics at Caltech, a post he accepted in the fall of 1943. Even though Tsien had made an agreement with the Chinese government to stay just one more year in the United States, China was dependent on U.S. efforts to defeat Japan and did not press the issue. At Caltech, Tsien would divide his time between teaching and research in aerodynamics and jet propulsion. And, just as Theodore von Kármán had done as a young professor in Aachen, Tsien would learn to move smoothly during the war years among the triple spheres of academia, industry, and government.

During the 1940s, Pasadena was undergoing growth so phenomenal that merchants referred to that ten-year period as "the Miracle Decade." These were the boom years of the aircraft industry for Los Angeles, when more than 100,000 airplanes would be manufactured for the U.S. government. The statistics were simply astounding. In 1939 there were 13,300 aircraft workers in the city. By 1941 that number grew to 113,000. In time, the industry absorbed more than 40 percent of all factory workers in the city. The need for workers was so great that Boy Scouts were hired to distribute employment applications from door to door, while radio broadcasts and neighborhood skits proclaimed: "It's fun to work in an aircraft factory!"

Everywhere on campus Tsien could see the changes brought on by the war. The undergraduate student body doubled as young men enrolled at Caltech under the Navy V-12 program, creating a severe housing shortage. Each morning, the campus awoke to the sounds of reveille as these students began their daily ritual of calisthenics, forced classroom attendance, mess hall meals, and inspections. Frequent military parades and marches were commonplace, as were students in blue Navy uniforms or olive-drab Army attire. Remembered one alumnus, "It was a no-nonsense undergraduate life, allowing little room for undergraduate pranks."

The Arroyo Seco had changed, too. Five years ago, only the Devil's Gate Dam stood sentinel over their experiments, its white concave face cold and desolate in the predawn air. Now, after the GALCIT group negotiated with the city to obtain a lease on seven acres of land in the Arroyo, buildings of corrugated metal rose on the western banks of the river bed, patrolled by guards twenty-four hours a day. Tsien needed a security pass to get in. The dry river bed where the Suicide Squad boys had first cheered and clapped and blown up scrap rockets had become the exclusive reserve of classified military research.

With clearance no longer a problem, Tsien was once again permitted access to the Arroyo. From the bumpy dirt road that led there he could see below him, in the gully of the Arroyo, a few scattered buildings, small and humble structures of redwood and corrugated metal. One of them was the tiny shack Malina had used years ago. Two were newly erected laboratories. A fourth building was an office with corridors so narrow that if someone opened the door while someone else was walking down the hall "you were liable to lose your teeth," an architect observed. (During the early stages of construction, housing was so tight Martin Summerfield was forced to use his car as an office.) Rammed into the side of a hill were a string of liquid- and solid-propellent test cells, some triangular if viewed from above and covered with steel siding and railroad ties.

The rocketry experiments had nearly knocked down parts of Caltech. The cells were constructed so the experiments on rocket motors could be conducted in a much safer environment. Engineers stood outside these cells and watched experiments involving liquid- or solid-propellent motors through telescopes or peepholes shielded with shatterproof glass. Upon ignition, the rocket motor would vibrate, shaking the test shed and giving rise to a low, medium-pitched roar that drowned out all conversation and reminded the engineers of Niagara Falls. Sometimes the engineers jumped back involuntarily when an explosion sent nozzles and shrapnel ricocheting against the walls and floors of concrete. "Half the time they [the rocket motors] blew up and blew pieces of steel casing all over the place," one observer remembered. Nearby homeowners complained of the strange noises and explosions rebounding from the canyon, although they were at a loss to describe their cause. It was not until well after the war was over that they learned what was taking place in the Arroyo.

Much had happened in the Arroyo during the years when Tsien was excluded from the action. The rocket team had worked on America's first jet-assisted takeoff (JATO) engines. In the beginning they were primitive little rockets, only a foot long, with two-pound units of black-powder propellent packed into small cylindrical shells. Initial tests were unsuccessful because the JATOs had a tendency to explode upon ignition. Later, Jack Parsons figured out that the JATOs had to be fired immediately after creation, before cracks could form in the powder. One day in August 1941, twenty-four JATOs were rushed to March Field near Riverside, California, and fastened to a small Ercoupe, a low-wing, lightweight monoplane piloted by Lieutenant Homer Boushey. When the rockets were ignited, "the plane shot off the ground as if released by a slingshot," Kármán remembered years later. "None of us had ever seen a plane climb at such a steep angle." That year, the NAS not only renewed the grant to Caltech but increased it to $125,000 for the fiscal period beginning 1942.

Then the Navy gave Caltech a contract to develop an even better JATO engine. This time, the rocketeers were expected to make one that delivered 200 pounds of thrust for a duration of eight seconds. A better fuel was needed, and Parsons invented one in 1942 by heating paving asphalt and oil to 350°F, mixing them with potassium perchlorate, pouring it into casings, and cooling the combination. It looked like hardened paving tar. It was superior to the black-powder propellent in both thrust and burning time, delivering 2,000 pounds of pressure per square inch and an exhaust velocity of 5,900 feet per second (with the right nozzle). It survived storage in both hot and freezing temperatures.

Engineers who worked in the Arroyo remember how during the 1940s solid-propellent fuel evolved from sticks of fuel burning from one end to another, like a cigarette, to hollowed-out cylinders of fuel burning along a lengthwise

surface. The goal was to maximize in a short time the surface area of the flam-
ing propellent, thereby increasing the force of rocket thrust during the first crucial moments of takeoff. Researchers initially punched out a thin cylinder of
propellent within a tube of propellent and later, outside of GALCIT, developed
a more sophisticated method of removing a long, star-shaped section of propellent, which vastly increased the burning surface area within the propellent.
However, during the early 1940s, solid propellents were still too volatile for
practical military use. (One danger posed by solid propellents is that the oxidizer is mixed within the fuel itself, so once they are ignited nothing can extin-
guish them until the propellent burns out completely.)

The group turned its attention to liquid fuels. It tried several combinations
of gasoline fuels with red fuming nitric acid, but this caused the motor to throb
uncontrollably and explode. Finally, it tried aniline, which not only ended the
throbbing but ignited spontaneously with the nitric acid, allowing the group to
eliminate an auxiliary ignition system. On April 15, 1942, the team tested two
JATOs supplied with liquid fuel on a Douglas A-20A bomber at Muroc Field in
the Mojave Desert. When the pilot, Major Paul Dane, ignited the engines, the
20,000-pound plane with 2,000 pounds of JATO thrust shot up "as though
scooped upward by a sudden draft." Delighted, Malina wrote, "We now have
something that really works and we should be able to help to give the Fascists
hell!" It was the first time a plane had taken off in the United States with a permanently installed rocket power plant. It marked, according to Kármán, "the
beginning of practical rocketry in the United States."

Three years was all it took for Malina and his team to produce such stunning
results. Three years—and there seemed to be no limit to the numbers of people they could hire, or the amount of money they could get from the government. By 1943, more than eighty employees worked for Malina—a considerable leap from the original five founders of the Suicide Squad. Some labored
over a hydrobomb, an Army torpedo that could be dropped into the sea from
a high-speed airplane. Others worked on a variety of projects, which included
the development of different liquid and solid fuels. In 1943 the team was
informed that NAS funding would grow to $650,000 for the next fiscal cycle.

The impact of Malina's leadership extended far beyond the immediate
reaches of the Arroyo. In 1941 Malina suggested the Caltech rocket team start
a company to manufacture and sell JATOs to the military. Tsien was peripher-
ally involved when the company was founded: he and Malina had helped Kármán work out some mathematics for Andrew Haley, a lawyer friend of Kármán's, to free Haley from a complicated legal case involving a license to

construct a dam in Iowa. In 1942 Haley came to Pasadena to help the rocketeers launch the Aerojet Engineering Corporation. Kármán became the president, Malina the treasurer, and Parsons, Summerfield, and Forman vice presidents, each of them putting up $200 apiece to reserve their shares of company stock. (Haley put up $2,000.) Tsien would be called from time to time to do some consulting work. The company first opened in a room on East Colorado Street that was formerly used by a manufacturer of fruit juice extractors. Meanwhile, JATO test stands were constructed on an acre of riverbed land in the citrus town of Azuza, minutes away from Pasadena. By December 1942 the company had expanded to 120 employees. In 1943 Aerojet got its first big break when the Navy gave it large contracts to construct JATOs to be used on carrier-based aircraft. (After being bought by General Tire and Rubber Company, it would rapidly grow to become one of the largest manufacturers of rockets and propellents, launching the country's first rocket to probe space and building the earth's biggest rocket motor by the 1960s. In 1994 it was part of the corporate conglomerate GenCorp and generated more than $594 million in sales.)

In the midst of all this activity, Caltech began to hear strange and disturbing rumors about the nocturnal and weekend life of John Parsons. Intense and brooding, with a dark slash of moustache, Parsons was described as "an excellent chemist and a delightful screwball" by Kármán in his book *The Wind and Beyond.* "He loved to recite pagan poetry to the sky while stamping his feet." In June 1942, Parsons moved into a large Norwegian-style redwood mansion at 1003 South Orange Grove, in a wealthy neighborhood known as "Millionaire's Row" in Pasadena. The front porch held a tuxedoed mannequin bearing the sign "The Resident" next to a big sack to receive all the junk mail so addressed. The back porch exhibited an open canister of gunpowder that Parsons boasted was enough to blow up the entire city block. ("It was always open," remembered one guest. "The can was never closed, because that would ruin the thrill of it all.") Inside, guests remembered years later, the house gleamed with gold leaf wallpaper and stunning fixtures. One room was equipped with expensive sound equipment, walls of carved leather, and a giant portrait of Aleister Crowley, whom the London tabloids once described as "the most evil man who ever lived."

Crowley was the head and the "great beast" of the Ordo Templi Orientis (OTO), a cult devoted to the practice of black magic. After 1943 the local southern California chapter of the OTO came under the leadership of none other than Parsons, who conducted the rituals in a secret chamber in the mansion designated as the "OTO temple." Ed Forman, Parson's best friend and Suicide Squad colleague, became one of the members. At night the local OTO brotherhood would don their robes and enter the chamber, from which faint chants and the strains of Prokofiev's Violin Concerto would float past locked doors. The members obeyed the creed of the cult: "There is no law beyond do what thou wilt. Do what thou wilt shall be the whole of the law."

Some rituals were performed during large parties, when the mansion was packed wall to wall with people. (The FBI recorded Parsons's wild parties as early as 1940, before he moved to South Orange Grove.) Sometimes, in the middle of the attic, two women in diaphanous gowns would dance around a pot of fire, surrounded by coffins topped with candles. "All I could think at the time," recalled one guest, "was that if those robes caught on fire the whole house would go up like a tinderbox." On other occasions, naked pregnant women jumped through hoops of flame. At these parties, guests would urge Parsons to recite the poetry of Aleister Crowley. "Do the ode to Pan!" his friends would cry. "Yes! Yes! The ode to Pan!" And Parsons would take a deep breath and recite the poem, his oratorical voice rounding out each verse and punctuating each "Pan!" with the stamping of his feet:

> *Oh come! Oh come! I am numb*
> *With the wanton lust of devildom!. . .*
> *Oh Pan! Io Pan! Io Pan! Io Pan Pan Pan!*

The wealthy and weird alike flocked to the mansion on South Orange Grove, eager to witness acts of sexual magic. Many decided to move in. The Parsons home soon attracted an opera singer, several astrologers, and well-known writers such as Lou Goldstone, a science fiction author, and L. Ron Hubbard, a young naval officer who later penned the best-selling book *Dianetics* and started the religion of Scientology.

No one knew what to expect from Parsons. One time, probably on November 4, 1941, Malina received a phone call informing him that one of the men from the GALCIT rocket project was in jail. As Malina recalled:

He was a mechanic working with Jack, and it seems that he had gone to Parson's house. They'd had a seance—what they were doing I don't know—anyway [the mechanic] had a gun and he found a car on the street that was parked nearby. There was a couple necking in it. He forced them out at the point of the gun, took the car, drove to Hollywood, evidently not quite knowing what he was going to do. And then, after a certain amount of time, he drove back to Pasadena. When he arrived at the flagpole by the Colorado bridge, the police were waiting for him.

I went to the jail to talk to the fellow and asked him what exactly made him do a stupid thing like that. Well, he was very vague and I couldn't get anything out of Parsons or Forman as to why this had happened. . . . It then became quite evident that whatever it was that Parsons and Forman were playing with had certain worrisome aspects.

Parsons, who left the Jet Propulsion Laboratory by 1944 and Aerojet by 1945, died at the age of thirty-seven. In preparation for a move to Mexico in June 1953, Parsons was packing his lab equipment in a garage underneath an apartment at 101T South Orange Grove. He had been mixing fulminate mercury in a coffee can when it apparently slipped from his fingers and exploded. People today still speculate about the circumstances of his death, debating whether it was an accident, suicide, or murder.

After obtaining his security clearance, Tsien worked on War Department, Air Force, and Army Ordnance contracts labeled with ten- to twenty-digit code numbers. His schedule became increasingly hectic: in April 1943, he finished a study on the high-speed pressure distribution over an XSC2D cowl; in July a report on the possibility of using the ejector action of a jet as a source of power for driving liquid-propellent pumps; in October some research on the effects of metallic solids on performance when added to solid propellents; in May 1944 a report on the impact of twisted blades in a compressor or turbine; and in August a report for JPL on heat transfer from a flat plate to an airstream of high velocity.

That year, Tsien, along with many others, taught a special group of Air Force and Navy officers who had been sent to Caltech by the government to receive their master's degrees in aeronautics. They were men in their mid-twenties and thirties, mostly married with children and possessing bachelor's degrees in engineering and superb leadership skills. Handpicked for this several-months-long program at Caltech, they were being groomed to become the future leaders of research and development for the U.S. military-industrial establishment.

Officially, Tsien taught them two courses, one on mathematical principles in engineering and another on the theory of jet propulsion, although he seemed to be an omnipresent instructor in their lives. On Saturday mornings, when the officers drove to the Arroyo Seco for a day of practical laboratory work, Tsien was often there to give a lecture. One story current then was that a student who found Tsien lecturing first on one subject, later in the day on another subject, and still later on a third, exclaimed in exasperation: "Why! On Sunday, I expect, he'll be teaching Bible class!"

Like many professors who resent the need to shift gears from cutting-edge research to teaching, Tsien clearly did not see the education of his students as his primary objective. He was as diligent as ever in presenting the material, but his resentment at having to teach—or at having to teach these bright students whom he felt were not as bright as the typical Caltech student—clearly came through. "If someone asked him a stupid question, Tsien wouldn't answer," recalled Homer Joe Stuart, a colleague at Caltech. Nor would he permit them to take his classes. A. M. O. Smith remembered one unfortunate graduate student who wanted to enroll in Tsien's course, but Tsien did not think he was good enough and flatly told him so. The students in the special military program, however, didn't have a choice. For them, Tsien's course in mathematical methods in engineering was required. "Most students," remembered Chester Hasert, then an Air Force officer from Wright Field in Dayton, Ohio, "lived in dread of it."

The students remembered vividly a typical day in Tsien's math class, held in the third floor of the Guggenheim Building. "He'd always be a couple minutes late for class and we'd always be sitting there wondering if he was coming," Hasert recalled. "He'd dash in and without saying a word he'd write on the blackboard. He was a very tense person but his math required an intensity of thought." Silently, Tsien would fill up the blackboards in his small, precise writing. "There was no erasing of errors, or going back to change even a sign," former student Webster Roberts remembered. "He always came out with the right answer, just before the end of the class period."

In class, Tsien could be cutting and cruel in his remarks. One time a student raised his hand and said, "I don't understand the third equation on the second board." Tsien did not reply. Another student asked, "Well, are you going to answer his question?" Tsien merely said, "That was a statement of fact, not a question."

Another time, a student asked if the method Tsien had just outlined was foolproof. Tsien gave the student a cold stare and said in an acid tone: "Only fools need foolproof methods." (Some remember a less eloquent utterance: "Can't make so no fools can do!" Recalled Hasert: "Sometimes he came out with a sentence like that so that it took a few moments to understand what he was driving at.") Then Tsien put his chalk down and stalked out of the classroom. Soon, students felt too intimidated to ask Tsien any questions at all.

There was no way for the students to gauge their understanding of the course work. No quizzes or midterms or homework were given throughout the entire semester. In the evenings, students gathered in small groups to decipher the meaning of Tsien's notes. As one student put it: "His notes were bare bones mathematics with very little explanation. He would just write down the equations and he made up for his lack of fluency in the language by leaning more heavily on the mathematics than the written word. The better you knew the math, the better you got along with Tsien."

Then came the final exam. Some students remembered that one of Tsien's exams contained a single problem: analyzing the impact and structural vibration forces of an airplane mounted with a heavy machine gun. "I think half of the class got a flat zero," Hasert recalls. "It was a real bone crusher." On another occasion, students recalled, it took Tsien forty-five minutes to write out the questions on the board, wasting about one-third of the three hours allotted for the exam. "There were blackboards that went around all three sides of the room and he filled them all up with his handwriting," said Robert Bogart, class of 1944. "He wrote like a Chinese: in tiny letters. He could have just handed out the questions if he had written them down on paper."

People wondered if Tsien was making the class unnecessarily difficult to show off his own mental superiority. As one student recalled:

> One day I happened to walk into his office, when he wasn't expecting anybody. And there I saw him reading a book by Philips on vector analysis. I went to the library and got a copy of Philips. Tsien was lecturing right out of this book. But he was using as a textbook the worst book he could find. Every time I heard him speak, I used to think to myself, what marvelous lectures he has given, he knows the subject so well—much better than the author of the book. And he was doing this deliberately, to show how smart he was compared to the author of the book. I learned since that this is a favorite technique used by many professors.

The students retaliated. A group formed a committee and went to Kármán to complain about the difficulty they had in Tsien's class. When Kármán suggested that Tsien change his approach to teaching, Tsien was said to have responded: "I'm not teaching kindergarten! This is graduate school!" Tsien "would not budge an inch," as students recalled. Eventually Kármán asked Rasof, then a young teaching fellow, to replace Tsien as instructor for the class.

Rasof was more popular among the students because of his sympathetic and nurturing personality—so soothing in contrast to Tsien's. Years later, he said of the class: "Many of these students had been shot down [over the Pacific Ocean as fliers] many times, not just once. They had nervous stomachs and trouble with eyesight and all of them had the problems of people who had been in terrible danger. They had terrible nightmares of being shot down and floating on the water for hours and hours before being rescued. But Tsien was so tough on them! I'm an easy-going guy and I understood that many of these officers had been shot down. I would start at the beginning, as if they had forgotten everything."

Meanwhile, Tsien was well on his way to becoming one of Caltech's most detested professors. "Students were scared stiff of him," remembered Hans Liepmann. "He had the bearing of a Chinese emperor."

And yet, Tsien could be surprisingly generous with his time and energy with people he saw as intellectually capable. In the Arroyo, one project (not related to the military group) ran into technical problems in water hydrodynamics. "Out of the goodness of his heart, Tsien gave us a free unlisted seminar in hydrodynamics using a text by Sir Horace Lamb," remembered Leonard Edelman, one of Tsien's students and an engineer at JPL. "This required many hours of preparation and of the fifteen or so hours given as lecture by Tsien I am sure he prepared five or ten times that many hours to give the lectures. He did this because of his dedication to his students, without pay." Edelman also wrote:

> I recall going to see him one day. . . . I said, "I am getting depressed about spending all my time on work designed to blow someone's head off. How do you maintain such a cheerful and enthusiastic attitude?" Tsien answered, "I get up each morning and do the best work I can do that day no matter what the subject matter is, and when I put my head on my pillow and I feel that I have done a good day's work, that is an end in itself. I am happy." Although fifty years has passed, I will always remember Tsien for that. Perhaps no teacher in man's history has taught a more valuable lesson than that.

The source of Tsien's impatience seemed to be dealing with students whom he deemed incompetent. "The biggest crime you could commit in Tsien's eyes was stupidity," remembered his friend Andrew Fejer. Years later, Fejer would recall the time when Kármán said over lunch to Tsien: "A good lecture is when one-third of the people understand in detail what you are talking about, one-third have a pretty good idea of what you are talking about, and one-third doesn't know anything about what you are talking about." Tsien was said to have responded: "I'm interested only in lecturing to people who understand everything." He made it clear, in both speech and in action, that he preferred to associate with only the best and the brightest students at Caltech.

Kármán was probably the only man at Caltech toward whom Tsien was always deferential. Sometimes, Tsien and his students would watch Kármán give a rare lecture. While Kármán worked out some math on the blackboard, Tsien bent over a notebook, furiously writing it all down. "Von Kármán would start developing something on the blackboard that he had never done before," remembered Bill Davis, one of Tsien's students. "He'd get into some tremendously difficult equations. At the end of the period, he'd look at his watch and say, 'I think this probably comes out like this' and throw a formula on the board. A week or so later, Tsien would come in, having worked through the whole damn thing, and say, 'Kármán was right! It *does* come out like this!" On those occasions, Tsien was like a student again: awestruck by his professor's brilliance.

Though Tsien lacked Kármán's charisma, he made a concerted effort to imitate the elderly professor's style of mentorship. "Tsien attempted to do to his students like me the same kind of thing that he had from von Kármán," remembered Joseph Charyk, formerly Tsien's graduate student and later the undersecretary of the Air Force. "They would discuss a problem and Kármán would say, well, it's very complicated, but why don't we emphasize this or emphasize that and see if we can simplify the thing. That's exactly the way that Tsien would then work with his students. Take a very complicated thing and try to reduce it to the basic elements."

Tsien also hung on Kármán's every word, almost as if he were in the presence of a deity. As one of Tsien's former students, Leonard Edelman, wrote: "I guess that his goal was to be as good or better than Kármán, who at that time was probably the most highly regarded in the world along with perhaps Ludwig Prandtl at Göttingen, who was one of Kármán's teachers." Another student, R. B. Pearce, later testified: "Tsien thought Kármán was such a compassionate and great man. He almost worshipped the guy."

During the summer of 1943, the U.S. Army Air Forces sent Kármán some top secret aerial photographs taken of the north coast of France. The pictures depicted some strange architecture that looked vaguely like concrete ski jumps. Did Kármán know what they might be? asked military officials. Kármán speculated that they were rocket launch pads, but he had never before seen any so large. Then, the Air Force sent him three British intelligence reports on German rocket activity. Although much of it was inaccurate, the reports tipped off the

U.S. military that the Germans were building rockets and missiles on a grand scale.

Alarmed, the military establishment took action. In September 1943, Army Ordnance created a rocket development liaison department for its research division and dispatched to Caltech two liaison officers: Colonel W. H. Joiner and Captain Robert Staver. Joiner asked Frank Malina to write a report assessing the ability of U.S. rocket engines to propel long-range missiles.

The response came in a memorandum that Kármán, with technical support from Malina and Tsien, submitted to Joiner in November. They concluded that, given the present level of technology, the U.S. military would be unable to build rockets achieving ranges of 100 miles or more. However, they proposed a new research laboratory, the Jet Propulsion Laboratory (JPL), engage in government-sponsored research with the expressed purpose of building rockets with greater ranges and larger explosive loads. They also suggested in a separate study that JPL investigate the possibility of the ramjet (called at the time an "athodyd"): an air-breathing propulsive system that they felt might power a long-range rocket.

Expanding on the Malina-Tsien analysis, Kármán suggested a four-stage research program. First, they would conduct tests of a 350-pound, restricted-burning solid-propellent rocket missile that could carry a 50-pound explosive load for 10 miles. Second, they would design a 2,000-pound, liquid-propellent rocket missile that could travel for 12 miles with a load of 200 pounds. Third, they would conduct theoretical studies of ramjet engines. And fourth, they would construct and test a 10,000-pound missile with a 75-mile range. This new report, authored by Kármán with Malina and Tsien's help and dated November 20, 1943, bore the very first mention of the name *Jet Propulsion Laboratory*. "I believe," wrote Kármán years later, "our proposal was the first official memo in the U.S. missile program."

Oddly enough, the Army Air Force turned the proposal down. The word from Air Force material command was that the project would not elicit immediate results. But, fortunately, Colonel Joiner had anticipated this rejection. Determined to see Caltech become a new national center of missile research, he had urged his colleague, Captain Staver, to send a copy of the report to Army Ordnance, along with a powerful letter of recommendation.

Once again, Army Ordnance proved to be more supportive. Years before, it had approached Vannevar Bush, director of the Office of Scientific Research and Development, with the possibility of entering the field of guided missile research. Bush rejected the idea, arguing that missiles were too inherently inac-curate to hit their targets consistently and that the development of more accurate missiles would absorb too many researchers who were needed in other areas of war work. Ordnance officials now saw the November 1943 proposal as an opportunity to achieve their own goals in rocket research. Colonel Sam B. Ritchie flew to Pasadena to meet with the GALCIT group personally, and was so impressed that he urged them to submit a more comprehensive proposal. Ritchie encouraged them to expand the scope of the new laboratory so that it would encompass not only the research of missiles but their actual development as well: building prototypes of working rockets and designing the intricate guidance and control technology.

A few months later, Kármán received a letter from Colonel G. W. Trichel of Army Ordnance dated January 15, 1944, a portion of which read:

> The Ordnance Department is very anxious to initiate a development of long-range rocket missiles as expeditiously as possible. . . . We are prepared to furnish the large sums of money needed to cover such a project . . . not more than three million dollars ($3,000,000) on the one year program.

Three million dollars. "The scale of operations threw us into a proper dither!" wrote Frank Malina years later in his memoirs. The following month, in February, the Caltech Board of Trustees approved the project, which was named ORDCIT for Ordnance Contract to the California Institute of Technology.

JPL began to draw up rough blueprints of how manpower would be organized for the long-range rocket research program. There would be an executive board consisting of Kármán, Malina, Clark Millikan, and possibly a few others. Under a section entitled "Caltech research and basic design," there would be four divisions: ballistics, materials, propulsion, and structures. In the preliminary organizational chart, Tsien would be the head of the propulsion section and run the ballistics division with Homer Joseph Stewart.

In March 1944, Tsien conducted for JPL an important comparative study of various jet propulsion systems. Together with Kármán, Malina, and Summerfield, he worked out a detailed analysis of the benefits and drawbacks of different kinds of rockets: air-breathing rockets, pure rockets, high- and low-speed models, rockets launched from the ground and from planes. They concluded that a combination of turbojet and ramjet engines might be best used in the early stage of rocket flight, possibly after rocket-boosted takeoff. They saw three basic reasons for this choice: first, a combined turbojet and ramjet vehicle would be recoverable; second, during the early stages it would take its oxygen from the atmosphere, permitting the rocket to carry less oxygen and thus be lighter; and third, it would be difficult for an enemy to shoot down the rocket because it could be launched from a moving target.

In June Caltech received a contract from the Army to build the ORDCIT missile, obtaining $1.6 million with a $3.6 million continuation contract. On July 1, 1944, guided missile work officially began under JPL, which now had four major research areas. There was JPL-1, the original AAF engine research program; JPL-2, a project directed at testing solid-propellent underwater missiles; JPL-3, the AAF ramjet program; and finally, JPL-4, the ORDCIT contract, which was the largest and most heavily funded program of all. The ultimate goal of JPL-4 was to build a guided missile with an explosive load of 1,000 pounds, range of 150 miles, and accuracy within three miles of the target. The founders of JPL envisioned building first a small solid-propellent rocket called the Private, and then a heavier liquid-propellent missile named the Corporal. They planned to improve each missile and give it a higher rank until it reached the Colonel.

Mass construction overtook the face of the Arroyo as new buildings were hastily erected from World War I Army surplus metal shipped from Albuquerque, New Mexico. Within a year, manpower doubled to more than two hundred employees. And, during the summer of 1944, Tsien found himself chief of the first research analysis section of JPL—his first position as the head of a major scientific team.

As a JPL section leader, Tsien was fast becoming recognized as the world's foremost expert on jet propulsion. And his timing couldn't have been better. His work and expertise in the field was especially significant to the U.S. military after June 1944, when the Germans sent their V-1 and V-2 missiles raining down on England. From summer to winter that year, Tsien directed research at JPL on the Private A, a small solid-propellent missile based on his and Malina's calculations in their report JPL-1. About a dozen scientists worked under Tsien's direction on the Private A, approximately half of them theoreticians skilled in applied mathematics, the other half specialists in electronics. Many of them were already affiliated with Caltech, such as Homer Joe Stewart of the aeronautics faculty; Chia-chiao Lin, who received his doctorate in aeronautics from Caltech that year; and Wei-zhang Chien, a Caltech postdoctorate researcher.

On Wednesday afternoons in either the aeronautics or astronomy building, Tsien, clad in suit and tie, participated in meetings with other section leaders. "The key problem that we finally recognized was that the lead people had to be sufficiently familiar with each other's specialities," Homer Joe Stewart said. The meetings were crucial to their success as well as the presence of every group leader during rocket launches. "That's why every group leader was required to go to field test operations," Stewart said. "You slept in the same lousy tents and ate the same lousy food until you understood their part as well as you did your own." There were nine sections in all: research analysis, underwater propulsion, liquid propellent, solid propellent, materiel, propellents, engineering design, research design, and remote control.

In December 1944 the Private A was ready for testing in the Mojave Desert at Leach Spring, Camp Irwin, near Barstow, California, and Tsien was on site to watch. Trucks with equipment arrived in the desert, tents erected. The missile stood eight feet tall between four guide rails of a thirty-six-foot steel launcher, equipped with a main engine that provided 1,000 clustered pounds of thrust for thirty-four seconds and four standard armament motors that gave the rocket 21,700 pounds of initial thrust in less than one-fifth of a second during take-off. Four foot-long tail fins provided the rocket with directional guidance. If the test worked, the Private A would be the first missile in the United States with a solid-propellent engine to perform successfully.

The news that day was all thumbs up. The Private A did everything expected of it. After twenty-four flights, the maximum range of the Private A was 11 miles, with a peak height of 14,500 feet. Concurrently, Tsien helped work out the preliminary designs of several other missiles that would be tested years after he left Caltech. One of those was the Private F, which proved to be a failure. It had been built on Tsien's prediction in 1943 that a missile's range could be increased by 50 percent if wings were added to the body. On April 1, 1945, the Private F was launched in seventeen tests in Fort Bliss, Texas, but spiraled out of control each time because of the lack of an effective rocket guidance system.

Another rocket for which Tsien had provided some of the basic ideas was the WAC Corporal. It was a sixteen-foot tall, 655-pound liquid-propellent sounding rocket that reached a height of 230,000 feet, or more than 40 miles, when it was launched in the White Sands Proving Grounds in New Mexico on October 1945. Not only was it the first manmade object to escape the earth's atmosphere, but it also served as a model for the Corporal E, a long-range guided missile that was successfully launched in June 1949.*

In late May 1944, Kármán went to New York City for stomach surgery and did not return to Pasadena until about mid-September. Recovery was slow because

---

*The Corporal E was launched for the first time on May 22, 1947, but was able to be launched reliably only after June 1949.

of two surgery-caused hernias, and as Kármán recuperated in a sanitarium in Lake George, New York, he continued to conduct business for JPL. In August 1944, Malina and Tsien, who relied on Kármán's guidance, visited him in New York to discuss missile remote control.

Tsien and others were concerned by Kármán's long absences from Pasadena. Most notably, hostility had reached a dangerous boiling point between Frank Malina and Clark Millikan. Malina had long disliked Millikan because the latter had almost killed the entire rocket project at its inception in 1936. (On one occasion on October 25, 1945, Malina was so angry at Millikan that he slammed down his papers at a JPL meeting and threatened to resign.) There is some evidence that Tsien shared Malina's feelings. "Tsien did not like Clark Millikan," his friend Hans Liepmann remembered. "He once said to me, literally, 'Why don't we both leave and blow up GALCIT?' He was very unhappy when Kármán wasn't at Caltech anymore."

In early September 1944, General Henry Arnold arranged a secret meeting with Kármán alone in a car at LaGuardia Airport in New York City. There, Arnold told Kármán that he wanted a study that could describe the future possibilities of aerial warfare, air power, and guided missiles for the military. "The war was not over," Kármán recalled years later. "Yet Arnold was already casting his sights far beyond the war, and realizing, as he always had, that the technical genius which could help find answers for him was not cooped up in military or civilian bureaucracy but was to be found in universities and in the people at large." General Arnold said he needed Kármán to gather a group of scientists and send them to the Pentagon to draw up a blueprint for air research for the next twenty to fifty years. Immensely flattered, Kármán agreed—but on one condition: "that nobody gives me orders and that I do not have to give orders to anybody [else]," the professor said. Arnold assured Kármán that he would be his only boss. Kármán arranged for a leave of absence from Caltech and officially became an Army Air Force consultant in October 1944.

When it was rumored that Kármán would not be back for another eight months, a flurry of panic set in among Chinese graduate students and their protégés. In a letter signed by Tsien, C. C. Lin, W. Z. Chien, and Y. H. Kuo on November 7, 1944, they spelled out what can only be described as an ultimatum and a list of demands. If Kármán planned to stay away indefinitely, then they wanted him to help introduce them to faculty jobs at other universities: Tsien a position, possibly, at the University of California; Kuo, a spot in the Princeton University shock wave group; Chien, a research position elsewhere; and Lin, a position, perhaps, at Brown University. Without Kármán's guidance at Caltech, they wrote, it was difficult to continue their research. His absence, they said, also hindered them from working in an atmosphere of "inspired leadership and warm personal relations."

Within a few weeks, Kármán made an interesting request of Tsien. He asked him to join him in Washington, D.C., and to work with him as part of a three-man staff and also as a member of the Scientific Advisory Group, which would aid the chief of staff of the Army Air Force in examining all possible options of air conflict in any future war. Tsien would work with two of Kármán's close associates, Hugh Dryden and Frank Wattendorf, as his staff, and would also belong to an elite team of some three dozen leading scientists and engineers. Near the end of 1944 or early 1945, Tsien resigned his position as head of research analysis, JPL-1, and transferred the responsibilities to his colleague Homer Joe Stewart. The young Chinese student who arrived in the United States only ten years earlier was now on his way to Washington.

Washington and Germany
(1945)

Tsien's time in Washington would be exciting but brief. He found the city still reeling from four years of wartime chaos—during which time the number of federal employees nearly doubled and the number of records created in those four years surpassed the total from its entire previous history. Issued a gold badge and a top-secret clearance, Tsien worked as a scientific consultant in the Pentagon, the gigantic structure built only two years earlier to bring the headquarters of the Army, Navy, and Army Air Force under one roof.

In December and January, General Henry Arnold held several Pentagon meetings to outline his mission for the Scientific Advisory Group (SAG). Listening to his speeches were some three dozen young men in their thirties and forties, soberly dressed in suits and ties. "For twenty years, the Air Force was built around pilots, pilots and more pilots," Arnold said. "The next twenty years is going to be built around scientists."

Arnold urged them to search nationwide and abroad for developments that would make American air power the very best in the world. Don't worry about the expense, he told them. Forget the past and look twenty years into the future. Look into the possibilities of supersonic flight, pilotless aircraft, bombs with increased explosive power, and aerial reconnaissance—even atomic energy as a source of propulsion. "Regard the equipment now available," Arnold said, "only as the basis for [your] boldest predictions."

To fulfill Arnold's directive, Tsien traveled to other laboratories across the country. Between February and April 1945, he toured RCA Laboratories, the National Advisory Committee for Aeronautics (the precursor to NASA), the Jet Propulsion Laboratories (JPL), and other research facilities to assess the direction of U.S. aircraft development. "It was not unusual to say, 'Let's go out to California and have a meeting with the West Coast bunch,' grab an Air Force airplane at an hour's notice and our toothbrushes and fly out there the same day," recalled Chester Hasert, Tsien's former student and colleague at SAG.

When he was at the Pentagon, Tsien usually worked on reports. In the mornings, he and Hasert would write a little, discuss their ideas, and exchange drafts of their writing with other experts. ("All that we were writing was controversial because we were trying to predict the future," Hasert said.) It was during this time, Hasert recalled, that Tsien wrote the outline of *Future Trends of Development of Military Aircraft*, a long-range report that would describe different methods of propulsion, control, and high-speed aerodynamics. Then they would go to lunch together at the Pentagon cafeteria. "I really got to know Tsien best at some of these lunches," Hasert remembered. "He was much more personable there than as a professor. . . . He was really a fine gentleman, very polite and very friendly."

In March 1945, near the end of the German defeat, General Arnold suggested to Kármán, "Why not go to Germany and find out first hand how far the Germans actually have gotten in research and development?" Arnold wanted a team of top scientists to interrogate the Germans and to inspect their research and development facilities. He also wanted such a team to gather indirect information on the Germans by interviewing aerodynamicists in neighboring neutral countries such as Switzerland and Sweden.

Obviously, Kármán wanted Tsien to come along. As the only alien in the group, he still possessed a student 4E visa from 1936 and a passport from the Republic of China. Getting out of the United States was no problem but Tsien worried that he might encounter difficulties with the Immigration and Naturalization Service when the European mission was over and not be able to get back in. "All I need," wrote Tsien in a telegram to the Pentagon, "is assurance from the U.S. immigration service that I can get into this country again." On

April 17, Fredrick Glantzberg wrote to the INS and tried to obtain for Tsien a U.S. exit and reentry permit so that he could resume his student status upon his return to the United States. The military was concerned about the legality of such a permit, but under wartime circumstances it was confident it would be granted. A few days later, on April 23, the INS granted Tsien a special waiver so that he could return with the same status he possessed before departure. The Air Force also promoted Tsien to the assimilated rank of colonel and gave him the title of expert consultant.

Tsien undoubtedly spent April as others in his group did: getting immunizations and military passes, packing dark Army-green wool shirts, caps, and battle jackets along with his civilian clothes and neckties. A memo from a top Pentagon official suggested that the SAG group bring, in addition to personal toiletries, a small medical kit, flashlight, sunglasses, pocket knife, and, believe it or not, "appropriate gifts for blondes, such as silk stockings, rouge, nail polish and extra U.S. insignias." They would travel under the code name Operation Lusty, which Kármán found "unlikely but pleasant." Then, at the end of April, Tsien boarded a C-54 transport at Gravelly Point, Virginia, and flew to Europe.

If Tsien had any doubts about the strenuous work ahead, they vanished within the first few days. His life became so hectic and his schedule so unpredictable that it seemed no sooner had he risen from bed than he found himself in a jeep headed for some strange destination. Rather than a smooth line of travel from one city to the next, his path on a map was a tangled web. He was constantly on the move, touring Europe in his Army uniform, staying in laboratories, schools, and private homes.

One of the very first German rocket scientists Tsien interrogated was Wernher von Braun himself. Formerly the technical director of the Peenemünde Army Research Establishment (the secret facility in northwestern Germany on the Baltic Sea which had been the birthplace of the V-1 and V-2 rockets), von Braun had surrendered to the Americans after Peenemünde fell to the Soviets. On May 5, in the village of Kochel, Tsien met with von Braun in person, although no record exists of the meeting.

One important paper did, however, emerge from their encounter. In Kochel, Tsien asked von Braun to prepare a report that would describe his past work in rocketry and his predictions for the future. The resulting report, entitled *Survey of Development of Liquid Rockets in Germany and Their Future Prospects*, spelled out von Braun's vision of rockets traveling between Europe and the United States in less than forty minutes; satellites orbiting the earth in one and a half hours; laboratories in space constructed by "men who would float . . . wearing diving suits." (His most fantastic suggestion was to suspend an enormous mirror in space with a huge net of steel wire and use the reflected sunlight to burn enemy cities and to dry their lakes and seas.) The report would eventually capture the attention of the Navy Bureau of Aeronautics, ultimately stimulating the American development of earth satellites.

During one of his visits to Kochel, Tsien also interviewed Rudolph Hermann, a famous German aerodynamicist who did much of the theoretical work on the V-2 rocket and who had headed a group in the process of designing a hypersonic wind tunnel. Years later, Tsien was the only scientist in the SAG group to be mentioned by name in a draft of Hermann's memoirs:

> I remember one of them, Dr. Tsien, von Kármán's closest associate, because he had written the paper about the "Pressure Distribution on a Cone in Supersonic Flow." He was the only scientist who had ever written a complete theory [on the subject]. We knew about his theory, because it was published about two years prior to the end of the war. We had used his theory and tested it in our tunnel exactly. I found out that nobody so far had tested Dr. Tsien's theory in his country. We did it, because we had the equipment, we had the supersonic tunnel, the scientists and the engineers.

—☞

Perhaps the most exciting discovery during the mission was of a secret aerodynamical laboratory in a pine forest near Volkenrode, a village outside the city of Braunschweig. It was the Luftfahrtforschungsanstalt Hermann Goering: in English, the Hermann Goering Aerodynamical Institute. "The whole thing was incredible," Kármán marveled in his memoirs. "Over a thousand people worked there, yet not a whisper of this institute reached the ears of the Allies."

Some fifty to sixty red-brick two-story buildings had been concealed by branches and trees planted on top of the roofs. A layer of ash hid an airfield from aerial view. The facilities included an eight-meter wind tunnel, a high-speed wind tunnel, two supersonic wind tunnels, an armaments laboratory, and a workshop. Most of the buildings had been looted but the majority of the equipment was intact. In addition, Americans used metal detectors to retrieve thousands of top-secret documents that had been stored in steel boxes and buried in the forest. The group worked and slept in the rooms of the Hermann Goering Institute. Tsien, his colleagues remember, was interested in papers on the subject of rocketry and explosives.

A model of an airplane with triangular, arrow-shaped wings found in the institute aroused a great deal of discussion. Only a few months earlier, a National Advisory Committee of Aeronautics (NACA) aerodynamicist by the name of Robert Jones had worked out a controversial theory that claimed that such an airplane could bypass the great effects of drag and fly right through the sound barrier. Tsien and his friends had spent many hours debating whether such a plane was feasible. Now the arrow-winged model indicated that the Germans had not only worked out a similiar theory but probably had experimental test results as well.

In 1945, scientists were still grappling with a phenomenon popularly known as the sound barrier or "sonic wall." A plane flying smoothly up to speeds eight-tenths the speed of sound would suddenly encounter a rapid increase in drag, which could cause the plane to stall. The situation was analogous to certain difficulties that face a high-speed motorboat. Just as a motorboat will slow down in the wake of pressure created by its own waves, an airplane will hit a cone of disturbance caused by the pressure of air resistance when approaching the speed of sound.

Jones discovered that wings swept back in a V-shape could smooth out the stream of supersonic airflow that might stall a fast-moving airplane with perpendicular wings; a craft with such wings might even break through the sound barrier. Tsien himself had unwittingly played a role in the early development of Jones's theory. In 1945, Robert Jones had carefully studied Tsien's 1938 paper *Supersonic Speed over an Inclined Body of Revolution,* in which Tsien demonstrated that certain slender projectiles exhibited little influence of drag when revolving at high speeds. Jones adapted some of Tsien's formulas for high-speed airplanes and found, to his amazement, almost no effect of drag for very slender wings.

In late April 1945, shortly before Tsien flew to Europe, Jones submitted for publication a report of his swept-back wing concept to NACA Langley's in-house editorial committee. The reviewers, however, were skeptical. One even asked him to replace the "hocus-pocus" with some "real mathematics," calling the entire theory a "snare and a delusion." During a meeting, Jones complained of the situation to George Schairer, one of the SAG members, who in turn discussed it with Tsien while flying over the Atlantic to Germany. After much discussion, Tsien and Schairer agreed that Jones was theoretically correct but should back up his findings with experimental results.

The Americans found these results near the Hermann Goering Institute, in a dry well where the Germans had hidden hundreds of secret documents on rocketry and aerodynamics. The papers, slightly damp, were retrieved and taken to the city of Göttingen, where American scientists (such as Tsien's "Suicide Squad" partner A. M. O. Smith) worked around the clock to scan, index, and microfilm the papers so that they could be shipped to the United States. The papers revealed that a well-known German aerodynamicist by the name of Adolf Busemann had worked out the "arrow-wing" (*Pfeilflugel*) theory ten years before Jones came up with his swept-back wing theory but was not taken seriously until 1942, when German scientists sought methods to achieve supersonic flight.

Someone had to go back to the United States right away to relay the arrow-wing discovery to the proper military authorities. Because the group didn't trust mail or telegraph, which could be intercepted, they designated George Schairer as the courier. In July he returned to the United States, bearing a couple thousand feet of microfilm. Soon after World War II, Schairer designed for Boeing Aircraft the B-47, the first swept-back-wing bomber in the United States.

In Germany Tsien saw the human capacity for both creative genius and mindless annihilation. Sometime during the first two weeks of May, Tsien may have visited the Dora concentration camp near Nordhausen. Those members of the SAB who were there recall that on one side of a hill above the camp stood a small building with a brick smokestack, next to which lay some two dozen stretchers stained with blood. Inside the building were big piles of shoes and clothing, and chisels to remove gold fillings from teeth. The concentration camp survivors, emaciated or dying, belonged to all nationalities: Polish, Russian, Czech, and other groups from East Europe. Through interpreters, the skeletal figures communicated that they wanted to show the Americans "something fantastic . . . underneath the mountain . . . important. . . ."

What they were referring to was the elaborate V-2 rocket factory hidden under the base of the hill. A railroad track ran through the entire underground tunnel system, laid with miles of production lines. In one tunnel lay an entire V-2 rocket and boxcar after boxcar of rocket parts. There were enough parts to build seventy-five V-2s, which American troops carted away in three hundred large railway wagons. There, slave laborers had manufactured the deadly V-2 rockets that had rained terror on London in 1944. A Nazi war criminal once called Dora the "Hell of all concentration camps." A Jew sent there from Auschwitz was heard to comment, "Compared to Dora, Auschwitz was easy."

Wrote one eyewitness: "They were told when they went into the tunnels that the only way they would come out was in smoke." Two large furnaces served as crematoriums where more than twenty-five thousand missile slaves had died within a year and a half. Of the sixty thousand people sent to Dora during the war, thirty thousand did not survive.

While it is likely but not certain that Tsien visited Dora, it is known that he did accompany Kármán to the city of Göttingen. It was the first city they saw in Germany that was not in ruins. There, the research leaders from the university were lined up and interrogated. Kármán's old professor, Ludwig Prandtl, was one of them, his once chestnut-brown beard completely white. "This was not a handclasping, welcoming get-together, as I recall," remarked Colonel Paul Dane, Kármán's former student who was present at the meeting. "In some ways Kármán was a little bit cold." Years later, Kármán would write: "Nordhausen was fresh on my mind. I do not believe I smiled once."

Kármán was furious that the professor and mentor of his youth did not seem at all concerned about German war atrocities committed in the name of science. Prandtl seemed to be angered only by the fact that the Americans had accidently blown the roof off his house. Other people in Prandtl's group appeared similarly devoid of remorse. One commented, "If we had to be conquered, I'm glad it was by Eisenhower. With a good German name like that, things will be all right." Prandtl even asked Kármán where his research funds would come from in the United States. Wrote Kármán of the meeting, "I couldn't tell whether Prandtl and his colleagues were horribly naive, stupid or malicious. I prefer to think that it was naivete."

Tsien seemed to share Kármán's revulsion for the Germans. A few years later, when he refused to attend a luncheon at which German scientists would be present, Tsien was heard to comment: "I'll learn from the Germans, but I won't eat with them."

All during May and June, Tsien inspected wind tunnel facilities in Germany and throughout Europe and wrote several reports on the state of jet propulsion in Germany. A group of SAG members prepared to go on to Japan to inspect the aerodynamical facilities there. Tsien was to be one of them, but for reasons unknown to surviving scientists of the team he did not journey to Asia. Instead, Tsien made preparations to return to Washington on June 20.

Upon his return from Germany, Tsien resumed his research and teaching at Caltech. After years of tireless work and investment of time, he was beginning to see dividends. In November 1945, Tsien moved up in the department from assistant professor to associate professor of aeronautics. Meanwhile, during the academic year 1945–46 his wartime technical contributions were published in three major publications: the book *Jet Propulsion*, the *Toward New Horizons* series, and the paper "Superaerodynamics." All three publications had lasting effects for the military and academic community. They also firmly established Tsien as a leader in the field of aeronautics, perhaps second only to Theodore von Kármán.

Jet Propulsion was a book of some eight hundred pages written by the staff of JPL and GALCIT for the Air Technical Service Command. Tsien served as editor, compiling chapters on topics such as liquid and solid propellent rockets, thermal jets, motors, jet-assisted takeoff, thermodynamics, combustion, and aerodynamics. The book included the weekly mimeographed notes that Tsien and other Caltech faculty handed out to their military students during the academic year 1943–44. Two years later, in 1946, the volume was distributed on a classified basis among military engineers. Just as the early collection of papers written by the Suicide Squad was dubbed the "Bible," so was the book *Jet Propulsion*. For many years, it would be, Tsien's colleague Allen Puckett testified, "the most authoritative anthology on jet propulsion in this country."

While Tsien was putting together the volume on jet propulsion, Theodore von Kármán was coordinating a similar effort for the Army Air Force. In 1945, Kármán wrote two introductory volumes—*Where We Stand* and *Science, the Key to Air Supremacy*—for a nine-volume series entitled *Toward New Horizons*.

In these two volumes, Kármán's predictions for the future of military technology were exciting, if not frightening. He foresaw the emergence of supersonic aircraft. He also described pilotless rockets that could travel in high altitudes or beyond the atmosphere at speeds of 17,000 miles per hour and for distances exceeding 10,000 miles—one of the earliest military projections of an intercontinental ballistic missile. He even stressed the possibility of using rockets propelled by nuclear power. Kármán urged the military to devote more funding to open new centers to pursue such projects.

The series *Toward New Horizons* expanded on all of Kármán's points in detail. It included topics in aerodynamics, aircraft design, aircraft power plants, aircraft fuels, rocket propellents, guided missiles, pilotless aircraft, explosives, terminal ballistics, radar communication, and aviation medicine and psychology. The reports, written by SAG, provided a blueprint for the development of a modern air force. Remembered Joseph Charyk, one of Tsien's students and later undersecretary for the Air Force, "It was a comprehensive scope of all of the developments that the Department of Defense would be thinking about in the years to come."

Tsien provided his own insight and ideas for *Toward New Horizons*. As part of his contribution for the series, Tsien wrote at least six reports on high-speed aerodynamics, aeropulse engines, ramjet engines, solid and liquid rockets, and jet-propelled supersonic winged missiles. He summarized his findings in Germany and Switzerland, devoting considerable space to describing different wind tunnel facilities, the swept-back wing concept, and propellent fuels. He also elaborated on the theoretical analyses conducted at Caltech or JPL during the war years.

One of Tsien's most intriguing reports, entitled *Possibility of Atomic Fuels for Aircraft Propulsion of Power Plants*, was written shortly after the atomic bomb was dropped in Japan. Tsien calculated that the energy released from nuclear reactions was about a million times that of conventional fuels, and speculated on the feasibility of using atomic fuels for thermal jet power plants and rockets. "Such an enormous increase in the heat value," Tsien wrote, "would mean that in all engineering practice the 'fuel' consumption would be reduced to a negligibly small quantity and the range of an atomically powered aircraft would be almost infinite." The superiority of such a system, Tsien concluded, warranted further investigation by the military.

Kármán was so delighted with Tsien's work that he vouched for his protégé's membership in the newly formed Scientific Advisory Board of the Air Force commanding general. "At the age of 36," wrote Kármán in his memoirs, "he was an undisputed genius whose work was providing an enormous impetus to advances in high-speed aerodynamics and jet propulsion. For these reasons I nominated him for membership on the Scientific Advisory Board."

The U.S. military establishment was equally impressed with Tsien. In December 1945 General Henry Arnold, chief of the Army Air Corps, gave Tsien an official commendation for his "excellent and complete" survey of ramjet and rocket performance and "invaluable" contribution to the field of propulsion and nuclear energy. Tsien was similarly praised by James Conant and Vannevar Bush of the Office of Scientific Research and Development. The following year, Army Ordnance lauded Tsien for his "outstanding performance" during the period between September 1939 and September 1945, when Tsien contributed his expertise to the development of JPL.

On May 20, 1946, Tsien submitted to the *Journal of the Aeronautical Sciences* a paper entitled "Superaerodynamics, Mechanics of Rarefied Gases." Published in December that year, it was perhaps Tsien's most famous paper in the United States. The paper forced aerodynamicists to reexamine the behavior of air flow over a wing in the highest reaches of the atmosphere. Air can be seen not as a perfect gas but as an aggregate of rapidly moving particles that are constantly colliding with each other. At low altitudes, the air molecules are so plentiful the gas can be perceived as smooth and continuous. But at high altitudes, when the air thins out, the molecules become more sparsely distributed, so that the collisions between them are much less frequent. What Tsien did was to craft an entirely new set of formulas that took into account the very molecular structure of air and the average distance between air particles. In doing so, he revolutionized the way aerodynamicists thought of high-speed flight at high altitudes.

"Tsien's article was one of the early ones that said to everyone, 'Wake up! Wake up! We can't continue to think of fluid as a continuum,'" said Albert de Graffenried, an aerospace engineer who remembered the influence of the paper. "It called people back to fundamentals, that the air is not a continuum, it's made up of little ping-pong balls that are bouncing around. It was the precursor to the type of work that would be coming shortly thereafter, as we went up into the stratosphere and then out into interstellar space."

The paper attracted so much attention and was cited so frequently that it firmly established Tsien as one of the most prominent theoretical aerodynamicists in the United States. But even before its publication in 1946, Tsien's alma mater MIT decided to lure him back. Its department of aeronautics offered Tsien an associate professorship with the promise of tenure. It seems that Tsien vacillated at first when he received the offer. "I believe C.I.T. put considerable pressure on him to reconsider," wrote Jerome Hunsaker to James Killian, president of MIT, on June 14, 1946. But eventually Tsien accepted.

Tsien's decision to go to MIT was an astute one. It was encouraged, and in some cases required, for academics who aspired to full professorship at one institution to gain some experience in another. "It was a crucial transition, and Tsien took it," remembered Homer Joe Stewart, an aeronautics professor at Caltech. "Tenure is an extremely important step if you want to make a career in your field."

There were other compelling reasons for Tsien to accept MIT's offer. His friend, C. C. Lin, suspects that Tsien returned to complete his growth as a rocket scientist. "Tsien had this vision—that in order for him to develop rockets he had to know something beyond what Caltech was good at," Lin said.

"There was quite a bit of rivalry between Caltech and MIT at the time. Caltech was good in structure and aerodynamics and other aspects. But at MIT, Tsien could learn more about instrumentation and control systems."

The summer of 1946 found Tsien preparing for his move and future obligations. On June 17, 1946, he, Kármán, and two dozen other consultants attended the first meeting of the Scientific Advisory Board in the Pentagon. A few months later, in August 1946, Tsien officially resigned from his position at Caltech. He also terminated his employment at JPL, transferring leadership of research analysis, section one, to his colleague, Homer Joe Stewart. Then he headed east, to the university that had welcomed him and then rejected him more than ten years earlier.

Return to MIT (1946–1947)

It was in the cool, blustering month of September that Tsien arrived in the Boston area, looking for a place to live. He settled on a large, red-brick Georgian Colonial home on 5 Hobart Road in the prosperous suburb of Newton, Massachusetts. The neighborhood was quiet, the streets lined with gold and crimson maples, oaks, and gingkos. The brilliant colors of autumn, however, were lost on Tsien, who began to miss California almost immediately upon his arrival on the East Coast. After ten years of living in a virtual paradise, Tsien had to adjust, once again, to the shock of changing weather.

"It rained here yesterday all day," Tsien complained in a letter on October 1, 1946. "Today it is almost cold! I imagine in [Pasadena] it is hot." Nor was the chill in Boston restricted to the weather. "I have yet," Tsien wrote, "to break down the icy attitude of my landlady."

Tsien lived about thirty minutes away from MIT by car. As he drove from Newton to campus, the landscape changed from suburban houses and well-groomed lawns to the massive brick and concrete apartment complexes of Brighton, and then Boston University, marked by the cold, clean lines of modernist buildings. Here, during his commute, Tsien would have driven past the brick rowhouses of the Back Bay, a familiar sight during his student days as a patron of a nearby opera house. Eventually, he would have seen before him Harvard Bridge spanning the length of the Charles River against the skyline of Boston. Within minutes, he would reach the familiar Guggenheim building, building number 33 at MIT.

As a new professor, Tsien now had his own office on the third floor of Guggenheim, furnished with three desks, a conference table, a few drafting tables, and many bookshelves. From the window in his office Tsien could see the wind tunnel below him and, further east, the new wooden Sloan automotive engine laboratory and gas turbine laboratory. Further beyond, offices and laboratories with yellow brick spandrels and long bands of glass stood connected to the chemical engineering building. His view was bounded by the stretch of parking lot and main MIT complex on his right and the hydraulic laboratory and high-voltage generator to the left.

Everywhere Tsien could see the concrete, rectangular modern buildings that had been hastily thrown up during the war. There were new buildings along Vassar Street that contained a cyclotron, nuclear research facilities, and a generator. Further east, there were some three-story wooden barracks for the newly established radiation lab. Where the track and field used to be now stood the athletic center, a yellow brick and glass structure with a swimming pool, small garden, and squash courts. There was housing for war veterans. Wrote one architectural historian, "Research contract activities created an unprecedented demand for floor space with little time for reflection upon planning and architectural subtleties. Buildings appeared on the main campus like mushrooms after a spring rain."

During the war MIT had seemed more like a military base than a university. Enrollment plummeted far below 1930s levels, leaving vast lecture halls and classrooms empty. Students who remained on campus were required to take ROTC. Freshmen and sophomores went to drill and arrived in class in khaki uniforms, while Navy V-12 students were dressed in sailor habit. Meanwhile, a new population of scientists arrived on campus. Some two thousand Army and Navy officers came to MIT for training in ultra-high-frequency radio techniques. The school engaged in some four hundred defense contracts that at times brought the total population of MIT to nearly ten thousand.

The end of war filled the corridors and classrooms of MIT with students different from most of their predecessors. Many of them were veterans of the war: older, serious men in their middle and late twenties who had seen much action and life and death during the war and were now anxious to start careers and families. In 1946, the MIT population swelled to an all-time high, with more than two thousand undergraduates and eight hundred graduate students. In addition, there was a profusion of government research money that made it possible for professors to assemble not only teams of graduate students and research assistants but postdoctorate engineers as well.

Perhaps nowhere was the postwar expansion felt more at MIT than in the department of aeronautical engineering. Many of those who had served as instructors or professors at MIT when Tsien was a student were still there—Jerome Hunsaker, Shatswell Ober, Joseph Bicknell, Charles Stark Draper, Otto Carl Koppen, Joseph Newell, John Markham, Manfred Rauscher, Edward Taylor. But there were also a number of newcomers on the faculty: Tsien, Rodney Smith, and Horton Guyford Stever in aerodynamics; Walter Gale and Raymond Bisplinghoff in structures; Rene Miller and Frank Bentley in design; Augustus Rogowski and Pei-Moo Ku in engines; Walter McKay, Roger Seamans, James Forbes, William Weems, and Robert Mueller in instrumentation and control. The usual prewar enrollment was about 150 students, but during the 1946–47 academic year the number had jumped to 425. Aviation was the field of choice in the 1940s, as glamorous as the space program was later to become in the 1960s. Moreover, Tsien's appointment at MIT made the major all the more appealing for incoming students. As it turned out, his reputation had preceded him. "There was considerable excitement when we learned that Tsien was coming, because he was a rising star," remembered Bob Summers. "He was treated as an important new addition and phenomenon at MIT."

In October 1946, Tsien applied for security clearances to work on two classified Navy projects: one for Project Meteor, a Navy Bureau of Ordnance contract with MIT to build an airbreathing missile with a solid rocket motor, and another for a Navy Bureau of Weapons contract to build a supersonic wind tunnel at MIT. He filled out numerous forms, giving information about his parents, his previous addresses, his memberships in professional societies and other topics so that intelligence officers could conduct thorough checks on his background. Tsien also applied for top secret clearance for projects originating from the Army Air Force and from the Manhattan District—the latter being related to activities concerning the atomic bomb.

Exactly what research Tsien did for the Manhattan District project remains unknown. It was evident, however, that during this time Tsien became fascinated with nuclear physics and envisioned a society in which atomic energy would be used for beneficial and practical purposes. In 1946, the *Journal of the Aeronautical Sciences* published Tsien's explanatory paper, "Atomic Energy," which gave a lucid and detailed description of Einstein's theory of mass and energy, atomic structure, nuclear fission, and the curve of binding energy. "That paper didn't have a practical influence in American development of peaceful application of nuclear energy, but I think it sketched essentially the course of nuclear development that actually took place," said Yuan-cheng Fung, who knew of Tsien at Caltech. "Its foresight was fantastic." Tsien also prepared a series of lectures on nuclear-powered rockets that he later delivered at MIT and the Johns Hopkins Laboratory in Silver Springs, Maryland—lectures that so captivated the imaginations of his listeners that they were remembered for decades. In these lectures, Tsien would discuss the problems with constructing a nuclear-powered rocket. ("The difficulty of constructing a simple nuclear fuel rocket," Tsien asserted, "is . . . the enormous temperature developed in the combustion chamber which would then disintegrate in an instant.") He would also outline possible solutions to these problems.

There was every indication that Tsien's rise at MIT would be just as rapid and spectacular as it had been at Caltech. Within months of his arrival at MIT, the faculty considered promoting Tsien to full professor immediately rather than waiting several years. When department head Jerome Hunsaker asked Kármán in February 1947 to provide a recommendation for Tsien's promotion to tenured full professor, the latter wrote:

> Dr. Tsien is certainly one of the leading men in the field of application of mathe-
> matics and mathematical physics to problems in aerodynamics and structural elas-
> ticity. . . . I believe he has the maturity required for a full professorship. I believe
> he is a good teacher and that he also has a talent for organization. His intellectual
> honesty and sincere devotion to both science and the institution which gives him
> the opportunity for working scientific research represent great assets, which I am
> sure you will appreciate.

During the spring semester of 1947, Tsien taught his first course at MIT. It was a fundamental course on compressible fluids for some thirty graduate students in aeronautics. The topics in the class included two- and three-dimensional flows and viscous flows of compressible fluids. It also gave a thorough descrip-tion of some of his own work, such as the Kármán-Tsien pressure correction formula, along with the work of other giants in the field: Prandtl, Glauert, Rayleigh, Janson.

His audience was surprised when Tsien first strode into the classroom, for the dimunitive Chinese man seemed no older than the students themselves. "He seemed very young," remembered Jim O'Neill. "He looked a lot younger than he was at the time, because he was small and slightly built." At five feet seven inches and 125 pounds, Tsien was also shorter than most of his students. They remembered him as a little man who wore nothing less formal than a suit and tie. He seemed almost vulnerable with his small, delicate features, high-pitched voice, and a trace of hesitation in his heavily accented speech. More than one student recalled the "excitement" at the prospect of taking one of Professor Tsien's classes, nationally renowned as Tsien was as Theodore von Kármán's protégé and finest student.

The excitement, however, soon gave away to dread. Although students came to MIT prepared for stiff competition (they readily bought felt banners pronouncing "Tech is Hell" in silver-gray letters against a cardinal-red background) nothing had prepared them for Tsien. So much fear and awe did the small Chinese professor arouse in his classroom that his students were able to recount almost half a century later, in painstaking detail, the intensity of his teaching style. They remembered the problems on his impossibly difficult exams, the marks they received, and most of all his scathing comments—which left the students limp in their wake.

"He had the reputation for being an egotistical loner," wrote James Marstiller. "He appeared ill at ease socially, and to most of the students, aloof and arrogant," recalled Daniel Frank. "He appeared to be a loner—no personality," remembered Rob Chilton. "He was not liked by the students." "He was a very cold and unemotional person," wrote Frederick Smith. "He stood out as the only aloof, remote, impersonal, boring professor I ever had, who seemed to go out of his way to make his subject unattractive and his students disinterested," wrote Leonard Sullivan. "He was an enigma about whom I knew or cared little." "As a teacher Professor Tsien was a tyrant," testified Claude Brenner. Students "generally disliked and even feared him," remembered James van Meter. Added Robert Wattson, "At least one good man of my acquaintance . . . left school literally in tears over his experience with Tsien."

Before each class, Tsien would shut himself in his office and write out his notes furiously on the blackboard. Those who walked by could hear the frantic squeak of chalk for hours. Then Tsien would stride purposefully down to the classroom, where his students were waiting.

The most colorful depiction of Tsien's teaching style was published in an article for the *Saturday Review* by one of his former students, Edgar Keats, two decades later. "There were no texts or notes available and there was no laboratory," Keats remembered. "All we had was Dr. Tsien and the blackboards on the four sides of the room. Dr. Tsien used the blackboards copiously, fully, rapidly, and we copied as fast as he wrote. There were no words—nothing but mathematical symbols."

He would walk quietly into the classroom almost precisely two minutes after the bell, approach the left end of the front board, mumble something like, "Let us start with . . . " and write an equation with a clear, firm hand. Then, consulting his own notes, he would add a line below, and another, and another, until he reached the bottom. Of course, we could not see what he was writing because he was in the way. When he moved to the next board, we copied the one he had uncovered, and so on around the room. On the second round, he erased one panel at a time.

Once in a while he would give us a hint of what was going on by saying "integrating" or "differentiating," and I would try to tack that note on the appropriate line when I reached it, but I never was sure whether he was referring to the equation he had just written or to one he was about to write.

After maybe twenty minutes, he would stand back and stare at the board and then say, "Here is an important relationship." But he never would tell us why, and before we could catch up he was off again, stopping only when the bell rang. Then he would walk out of the room without a word, leaving us to complete our copying.

There were no homework assignments, just the equations from our classroom notes to try to decipher. That was not easy, and Dr. Tsien was no help. He did not suffer fools. One of my classmates interrupted him at the beginning of the third class to announce: "Dr. Tsien, I was not able to follow your derivation of the pressure-volume relationship."

"Did you study your lesson?" he asked.

"Yes."

"Then you should have," he replied, and turned back to the board.

From time to time he would be absent. He offered no excuses. "The class will not meet on Wednesday," he would say, and that would be all. No one ever substituted for him. One theory held that he would not give up his notes; others maintained that no other professor would dare take over. Rumor had it that when he was absent he was presenting secret technical papers at high level meetings, but he never mentioned the papers or the meetings.

As the end of the term approached we knew there would be an exam. He avoided all questions relating to it. "If you understand, you will have no trouble," he advised us.

Results bore him out. My mark was 12 on a scale of 100. The high mark in the class, but a brilliant man who later became a professor at MIT [received] 22. Dr. Tsien gave me a passing grade, but I have never dared to try to use anything from that course.

Perhaps in Chinese it all makes sense.

Sometimes Tsien would try to lecture without any notes at all. What often ensued was a marathon session to see how much he could cram on the board before he forgot his material. During the break between the two hours of the lecture session, recalled his former student, Holt Ashley, "Tsien would rush up to his office on the third floor and slam the door, so he could look at his notes for the second hour and come back down and lecture again for an hour! It was very impressive. I never had personally another instructor who could do that, and as I say, we were all quite scared of him."

As devoted as he was to teaching, Tsien would have probably accomplished more had he simply been more relaxed. The intensity of his teaching style and the demanding nature of his personality terrified his students. A driven person, he kept his emotions wound up inside like a coil, taut and ready to spring. "He had an intense look when you looked at him," remembered one of his students, Bob Summers. "Actually, he didn't like to look at you, unless you were in class and asked a question, in which case you got a piercing look and were cut down to your knees. Boy, he could really cut you down to your toenails." When a number of students refused to sign up for his class the following semester, he seemed to take it personally. "Shortly after the start of the second semester, I met Dr. Tsien on the staircase in the Aero building," wrote Mrs. Robert Postle, then one of MIT's few female graduate students. "Our conversation consisted of two sentences: his declaring me to be one of the incapable students who'd lost out on dropping his course and mine saying that I enjoyed my new course."

He was unduly hard on his students during exams. They complained that his problems were so difficult that they would floor even a top rocket scientist or doctor of mathematics in the field. "You had to be clever enough to see the twist in his problems," one student recalled. "Otherwise, they would be impossible." It was said that during Tsien's tenure at MIT, only one doctoral candidate of fourteen candidates in aeronautical engineering passed his orals. One of Tsien's students gave an example of how difficult the tests could be. Jim Marstiller pointed out that the semester he took Tsien's course, only one student, a doctoral candidate, received the passing grade of 73 on the final exam. The second highest score, a 58, went to Holt Ashley, "a genius in his own right" who later became professor at both MIT and Stanford and also gained a reputation for being one of the world's leading structural dynamicists. Ted Pian, who later became another professor at MIT, netted a score in the 30s or 40s. The average score was in the 20s, and the mean was only 14.

Another factor that made the course daunting was Tsien's pronunciation of English. Students struggled to understand his lectures in "masamatics" and his occasional high-pitched outbursts, such as "Where did you get this handbook formwula?" He had a strange accent—almost a mixture of German and Chinese, students thought. Wrote one student, "I had often wondered if he had learned spoken English from von Kármán, who used to say, 'I vill now speak to you in die unifersal Lankwitch: bad English.' "

Left to their own devices, the students often stayed after class to "copy down every smidgen" of mathematics left on the chalkboard and then held group study sessions at night to decipher the notes. No textbook was available during the first year because the technology was too new. Worse, there were practically no other textbooks in compressible fluids at the time—only a few outdated copies in German. So, remarked Edwin Krug, "I ended up with a book full of notes I didn't understand."

Tsien, however, was preparing a set of lecture notes that would be permanently available for future students of compressible fluids. Assisting him in the preparation of these notes was Leslie Mack, a tall, slender doctoral candidate at MIT with slightly stooped shoulders. Although Tsien was his thesis advisor, Mack was initially so intimidated by Tsien that he worked out his entire master's thesis without consulting him once for advice.

Mack remembered Tsien as a dedicated professor who devoted himself singlemindedly to his work. Tsien expected his students to share this dedication, and when they appeared not to, he flew into a rage. One time, Tsien asked Mack to work out some calculations on a fan jet turbine. "I was writing away, doing his calculations, and lunchtime came along," Mack recalled. "So I went off to lunch. And when I came back, Tsien was furious. He said, 'What kind of scientist do you intend to be, when you go off to lunch in the middle of a calculation!' "

"He was an *extremely* hard worker," Mack remembered. He cited the time Tsien was asked to write part of a textbook entitled *Fundamentals of Gas Dynamics*, edited by H. W. Emmons of Princeton University. "Each week he would do it all at home and each week he would come in with a chapter that was handwritten in his very nice handwriting. To write technical material at that rate, on a new subject, was quite an achievement. . . . The book was supposed to come out in 1950, but it came out in 1958, which is usual for these multi-author things. Only people like Tsien actually had the articles prepared on time."

On campus Tsien was something of a mystery. Outside of class, faculty and students caught only fleeting glimpses of him in the Guggenheim Building, and only one professor, Rene Miller, remembered going to his home at all. When students sought him out in his office with questions, he tended to brush them off with a comment like "That seems okay" and shut the door. There were times when he refused to see anyone at all. Remembered Larry Manoni, "He would sit in his office with the door closed and locked, and when anyone knocked at the door he would shout, 'Go away.' This happened several times to me when I had an appointment with him to review my thesis work."

"Inevitably," wrote Claude Brenner, a former student of Tsien's and now president of Commonwealth Energy Group, "the students found small ways to express their feelings."

> Because we westerners had difficulty getting our tongues around the aitches and esses in his name, someone one day simplified it to "Choo Choo Train." This was not derisory. It was simply a way of cutting him down to size, in fact, of humanizing him, to give life to this remote, uncommunicative, unapproachable academic. To some it may also have been a metaphor for the speed at which he conducted his lectures. Because above all, he enjoyed mighty respect. We knew that we had to learn what he was trying to teach us, as poor a teacher as he was.

It was a tradition for the students at the MIT Graduate House to invite a professor each month to dine and discuss career opportunities available in their field. The students would reserve a private dining room for the professor, who would sit at the head of the table. After the meal, during coffee, the students would turn to the professor with questions. The aeronautics graduate students had already invited Jerome Hunsaker, Joseph Bicknell, and other faculty before they considered Tsien.

"Dare we invite Tsien? Would he come?" wondered Brenner and his friends. "We invited him (with trepidation)—he came. He talked (to our amazement) freely and comfortably during dinner. One could almost describe him as affable. He was open and helpful in the career discussions that followed—not a hint nor whisper of the sarcasm we dreaded." So Brenner and the other students discovered at the end that Tsien "was quite human after all."

Among all the criticism, there remains a smattering of positive comments from those who recognized the importance of Tsien's lectures. Judson Baron, a former student of Tsien who is now a professor of aeronautics at MIT, said that he brought an applied mathematical approach to problem-solving based on the German school of thought—something that must have come from his years of working under Theodore von Kármán. Alumni also claim that outside of Caltech no university offered a course in compressible fluids that had the depth and theoretical rigor of Tsien's at MIT. "Tsien's courses," recalled Leo Celniker, "were invaluable for the first ten years of my career."

In May 1947, the *MIT Technology Review* announced that Tsien had been promoted from associate professor to full professor at MIT. Nationwide, most professors had to put in more than twenty years of teaching, counseling, and administrative drudgework before they could be appointed to the permanent faculty. Tsien was only thirty-five years old. "For Tsien to get tenure that early was nothing less than phenomenal," said one commentator on the academic scene. He was one of the youngest professors to receive tenure in the history of MIT.

But at the same time, Tsien was considering the possibility of leaving. In 1947, a year after Tsien had accepted the MIT professorship, he received a rare opportunity to teach in his homeland. Information about this offer remains scanty, but it appears that Chinese Nationalist officials had approached Tsien about the possibility of becoming president of his alma mater, Jiaotong University. Ever ambitious, Tsien decided to consider their offer. He made plans to visit China that summer, his first visit in more than ten years.

This was a move Tsien must have considered carefully. Quite possibly, it would mean sacrificing everything he had worked for in the United States. Still, the idea of running an entire university must have been tempting. Tsien could influence generations of top Chinese engineering students the way Cheng Shiying, his MIT-educated professor, had influenced him. He could cultivate new schools of thought in China, perhaps even bring about a revolution in the aircraft industry there.

Now that the Japanese had been defeated and the war was over, China would surely begin a period of rapid reconstruction. The country would soon be in desperate need of top engineers and scientists, and few American-trained Chinese professors had Tsien's depth and breadth of training in aeronautics. Tsien might be more than just another university president in China; he could become a legend.

There was another incentive for Tsien to consider the position: the rift forming between him and the rest of the MIT faculty and Tsien's growing disenchantment with the department. Unfortunately, Tsien was apparently no more popular with his peers in the department than he was with his students. These conflicts came out during MIT seminars.

During these seminars, which MIT intended as a friendly exchange of ideas among students, faculty, and visiting scholars, Tsien would often sit in the back of the room, reading magazines. Whenever the speaker made a mistake, Tsien would shout out the error across the room. "One of the things Tsien brought to MIT was the Caltech seminar tradition, which is to be rather severe towards the speaker," Tsien's assistant Leslie Mack remembered. "It wasn't to humiliate the speaker, but Tsien had very high standards. He felt free to express himself and unfortunately, whether he intended to or not, sometimes rather sarcastically."

He was especially critical of those whose work he felt lacked theoretical rigor. This population included many of the senior members of the aeronautical engineering department, some of whom had a bachelor's degree or no college degree at all. They belonged to a generation in which aviation engineers tended to be explorers and pilots rather than mathematicians, to an era in which the field was so new the university curriculums on the subject did not exist. "The old-timers like Shatswell Ober and Otto Koppen weren't at all scientifically inclined," remembered Mack. "They thought the whole von Kármán school of thought had nothing to do with building airplanes. There was that intellectual gulf, plus the personality problems. I doubt that Tsien was particularly happy at MIT." Tsien's faith in using mathematical analysis to predict the physical phenomena put him in "a different class of guy altogether," Jim O'Neill, another MIT alumnus said. "He wasn't really an engineer, he was a scientist."

Although neither Mack nor O'Neill would have been able to answer the question, one can only wonder if Tsien's rudeness, particularly toward the non-theoretical members of the department, came out of his memories of his year as a student in the aeronautics department, when all the mathematical skill in the world could not protect him against the requirement that he be able to build something useful.

But leaving MIT was one thing; returning to China quite another. Even as he made plans to go see for himself what China had in store for him, he considered the possibility of staying permanently in the United States. In 1947, he applied for permanent residency status, which is the first step in obtaining citizenship. In order to obtain the visa, he had to exit the country and reenter under a new immigration status. During the spring of 1947, Tsien flew to Montreal and entered the United States through Rouse's Point in New York on April 2. Then he flew to California, spent a few weeks with friends in Pasadena, and left for China in July.

Summons from China (1947)

One can only imagine the emotions Tsien must have felt upon setting foot in his homeland. More than a decade had elapsed since the day he boarded the *President Jackson* for the United States. For Tsien, there were the surprises that come with meeting old friends and relatives after a twelve-year absence; there was also the shock of the overwhelming changes in China.

The person Tsien most wanted to see in China must have been his father, who was living in an apartment in the international settlement in Shanghai. (Tsien's mother had died of typhoid, which she had caught while traveling between cities.) There was every indication that the elder Tsien was lonely and yearned to live out his remaining years with his son. If Tsien took the presidency of Jiaotong, he would be able to assist his father in his waning years.

The first few weeks Tsien spent in China were promising enough, for he was treated like a minor celebrity. He visited three major cities, speaking in packed auditoriums at Jiaotong University in Shanghai, Zhejiang University in Hangzhou, and Qinghua University in Beijing. So eloquent was he that several enthralled listeners fortified their ambitions to come to the United States and study engineering.

In his speeches, Tsien urged young engineering students to think of themselves as scientists rather than high-paid technicians. He advised them to take courses not only in engineering but also in mathematics, chemistry, and physics. The training of a competent engineering scientist, Tsien said, is a long process that takes seven or eight years of education after high school: at least three years in a good engineering school, another three to master the math and science, and one or two years focused on a specific problem under the guidance of an experienced master in the field. He outlined the merits of a university doctoral program: "The unhurried academic atmosphere in an educational institution is certainly conducive to thinking, which is, after all, the only way to gain wisdom."

Tsien also stressed the potential the engineer had in bringing about technological innovation that would benefit society. He foresaw engineering applications in the fields of medicine and agriculture. He pointed out that even the creation of deadly weapons such as the atomic bomb "contributed much to the victorious conclusion of World War II on the side of Democracy." Tsien concluded his speeches with a quote from Professor Harold C. Urey: "We wish to abolish drudgery, discomfort and want from the lives of men and bring them pleasure, leisure and beauty."

He had come to America a young man and in the intervening years, largely free of strife or obligation, Tsien had been given the opportunity to do something special—in his case, to think about complicated problems in mathematics and aerodynamics until he could truly say he understood and could predict the forces at work.

Though something in him could not give back to other students—either at Caltech or at MIT—the nurturing and the attention he got from Kármán and others, his words to young Chinese scholars seemed to be those of a mentor looking back on all that had been given to him. Clearly Tsien was ready to give something back, but it was only in his return to China that he thought he had found worthy recipients.

But alas, these were optimistic words, spoken during a time of plummeting Chinese morale. The war had devastated the economy and population of China, causing an estimated three to fifteen million deaths. There was widespread hunger. Wrote historian Lloyd Eastman in his book *Seeds of Destruction*: "The famine in China during late 1945 and early 1946—affecting as it did some

33 million people—was probably the most severe and extensive crisis of its kind anywhere in the world during the immediate postwar period."

Tsien must have heard of the food shortages that plagued the cities during the war. The salaries of students and professors had failed to keep up with runaway inflation, and they rushed to the stores to buy rice once they received their paychecks or stipends. Many fell ill and died during this period, suffering malnutrition on meager diets of vegetables and almost no meat. The situation was even worse in rural China. During the war the GMT imposed strict taxes in grain from peasants to feed the soldiers, and often there was not enough food to go around. In some areas, men sold wives or children for two pounds of rice and subsisted on leaves, bark—even human flesh.

The soldiers who received the grain had fared no better during the war. Most of them came from poor families, because more privileged young men eluded the draft. The Nationalists used forced conscription to draft peasants into the army, often tying them together like convicts to prevent them from running away. They were poorly fed, for it was a routine practice for grain tax collectors to steal portions of the supply and dilute the rest with water, gravel, and weeds to make up the difference in weight. By the time it reached the soldier's plate, it was inedible. Indeed, one in ten soldiers drafted died before reaching his assigned post.

The country had no sooner stumbled out of one war when it plunged into another. As soon as the Japanese were defeated, fighting broke out between the Chinese Communists and the Nationalists. Despite efforts from the United States to reconcile the two parties, the conflict soon escalated into civil war. At first the Nationalists occupied one town after another in Communist-occupied territory. But in 1947, the year Tsien arrived in China, the tide was turning.

The Chinese Communist Party drew its power from the peasants. During the war, the party had penetrated regions under Japanese control and recruited young peasants to fight against the Japanese with guerrilla tactics. By April 1945, they had built up an impressive military force, claiming to have an army of nine hundred thousand and a militia of more than two million. By the time Tsien arrived in China in 1947, the Nationalists had lost half of Manchuria to the Communists. Deserting troops and casualties further cut the GMT forces in half. Moreover, the Communists were maneuvering their troops toward the Yangtze River.

If Tsien had any doubts about staying in China, the decision was made for him. Under mysterious circumstances, the Ministry of Education denied Tsien the presidency of Jiaotong University that had been promised to him earlier. The official reason given later was that Chen Lifu, the Chinese Minister of Education, rejected Tsien as too young for the position. But it was whispered later that education officials who distrusted Tsien's loyalty to the GMT had conspired to cancel his appointment. The truth has never been fully revealed.

The trip to China, however, was not altogether a disappointment. That summer he began to court an opera singer whom he had known since childhood: a young woman named Jiang Ying.

Jiang Ying

It was only natural that Tsien would be attracted to Jiang Ying. She had his passion for music. She was elegant and intellectual. And she belonged to the ruling elite of China—with a background even more impressive than Tsien's.

Ying probably would never have met Tsien had her father and Tsien's father not been good friends. She was the daughter of Jiang Fangzhen, a military strategist for the Nationalist government of China who had grown up in Haining, a city not far from where Tsien's father had lived. It is not known exactly how and when they met, but it is likely their friendship began when they were boys. They both went to school in the city of Hangzhou and later abroad in Japan during roughly the same period.

Early in his life, Jiang distinguished himself as a brilliant and passionate scholar. He had a penchant for stirring controversy everywhere he went. Like Tsien's father, he attended the Qiushi Academy in the city of Hangzhou, but was expelled for expressing views hostile to the imperial government of China. Forced to finish his education elsewhere, he went to Japan and enrolled in the renowned military academy, Shikan Gakko. He proceeded to embarrass the administrators there by graduating at the top of his class, and thenceforth Chinese and Japanese students were placed in separate sections to prevent a recurrence of such a situation.

Years later, when serving as director of the Chinese military academy at the city of Baoding, Jiang grew so frustrated with the politics of the corrupt Yuan Shikai regime that he shot himself in the stomach—right before a stunned audience of assembled cadets. It was because of this attempted suicide that his daughter Ying was later born. While recuperating from the wound in a Baoding hospital, he fell in love with a Japanese nurse, whom he later married. She bore him five daughters, who were reportedly so lovely neighbors referred to them as "the five flowers." Ying was the third daughter, born in Beijing in 1920.

By the age of three, Ying was a vivacious child who could sing, dance, and act, to the delight of her doting family. She so entranced Tsien's father that he begged Jiang for permission to adopt her. This was probably more or less an informal act of betrothing Ying to the young Tsien Hsue-shen, and to this Jiang gave his blessing. Ying moved into Tsien's home with her nurse and took on the new name Tsien Hsue-yin (or, in pinyin, Qian Xueying). The twelve-year-old Tsien became Ying's older brother and teacher. He told her stories and gave her lessons in science and before long, Ying felt her "brother" knew everything. How long Ying stayed at Tsien's home is not known, but within a few years both of them went their separate ways.

In the 1930s, while Tsien was a student at Jiaotong University, Ying entered the Zhongxi Women's School, an exclusive private high school in Shanghai. As a teenager, Ying enjoyed all the privileges of wealth and power. She mastered the sports normally reserved for the well-born, and became a skilled equestrienne and swimmer. She learned to sing and play the piano. Her father encouraged her in her musical ambitions, for he loved classical music himself and sang Wagnerian operas at home.

There was probably no better place in Shanghai to gain a Renaissance education than in her own home. It was a haven of culture, a place where she read poetry from the Tang dynasty while her father wrote books about Western history, constitutional practices, and tactics of war. He edited a literary journal, which put him in constant communication with the leading writers and artists of the day. Periodically, Jiang invited world-class authors such as John Dewey, Bertrand Russell, and the Bengali poet Rabindranath Tagore to China to give lectures, and Ying was no doubt exposed to their ideas. Ying developed the sophistication that years of contact with high society can give, mingling with locally and internationally prominent people.

In 1936, Ying, then sixteen, joined her family for a three-month grand tour of Europe. Her father, now a senior advisor to Chiang Kai-shek, was dispatched on a mission to study the structure of national mobilization in Italy, Germany, and other countries. He was deeply impressed by the importance that European leaders attached to air power and urged the Nationalist government to establish a department of defense unifying three separate military services: the army, navy, and air force. While Jiang scrutinized the defenses of Europe, Ying was dazzled by the architecture and music. In Italy she visited the Vatican, studied Renaissance sculpture and painting, wandered through the ruins of ancient Rome.

Upon their arrival in Berlin, Jiang enrolled Ying at an aristocratic German high school famous for its discipline. Ying decided to remain in Germany to attend college and major in vocal music, and she matriculated in the music department at the University of Berlin to study under the distinguished baritone Hermann Weissenborn.

Her early years in Berlin were perhaps the most idyllic of her life. During the day she practiced foreign languages, sang, and played the piano; at night she went to concert halls to listen to operas, symphonies, and recitals. "After I entered college I felt as if I had fallen in an ocean of knowledge," she remembered decades later. Here in Berlin was a stimulating, poetic environment of music and language. And here Ying herself had became a luminous presence on campus. A photograph from her youth reveals lustrous hair gleaming like lacquer, delicate cheekbones, and unblemished velvet skin. At the university, she sang at parties, carefully made up and dressed in *qipao*—a long silk sheath with a side slit.

International tensions, however, soon put an end to this charmed existence. In 1939 the Germans invaded Poland, thrusting Europe into world war. And in the midst of the confusion and fear that seized Berlin, Ying read her father's obituary in a German newspaper. (He had died of a heart attack while traveling in China.) Though she longed to return to China to see his remains, the war kept her in Germany. The following year she fled to Switzerland, where she resumed her studies under a Hungarian vocalist at a small conservatory in the town of Lausanne. While fighting raged through Europe, Ying persisted in her musical studies, and later transferred to a conservatory in Munich to practice Wagnerian opera.

When the war ended and the Mediterranean was open to navigation in 1946, Ying returned to China and reunited with her friends and family after an eleven-year absence. She decided to start her musical career afresh in Shanghai, and debuted with a recital in the Lanxin Theatre. Her voice created a sensation in the local press, and critics lauded her as one of the best sopranos in the country. Ying received invitations from presidents of Nanjing and Shanghai conservatories to teach music in their programs. With a single performance, Ying had burst upon the Chinese music scene as a rising star.

It was shortly after her triumphant performance in the city that she also met Tsien. Her "older brother" and childhood mentor was now a distinguished scientist who wanted very much to marry her.

Unfortunately, the art of courtship did not come easily for Tsien. In his usual blunt manner, he said to Ying: "How about it? Will you go with me?" Ying, hoping for something more romantic and also reluctant to be again separated from her family, rejected him. A few days later, Tsien made a second attempt, as direct as the first: "How about it? Have you thought it over? Let's get married. Will you go?"

Tsien had already made up her mind that it would be Ying or nobody—if she spurned him, he would remain single for the rest of his life. Ying recognized his earnestness in this matter and finally accepted his proposal. The wedding was held on September 17, 1947, in Shanghai.

Tsien had to return to the United States a week later, and Ying made plans to join him in a few months. Going to America must have been a bittersweet choice for her. A large opera house had invited her to perform in Italy (it is not clear whether this was for a position or merely a one-time engagement), but she turned the offer down because of her marriage to Tsien. There was no guarantee that she would be given a similar opportunity in the United States.

Tsien, meanwhile, seemed moved by China's condition. During his three-month visit, Tsien had written a long, graphic letter to Theodore von Kármán describing the suffering under Chiang Kai-shek's regime. It appears that Tsien also urged his friends not to remain in China or go to China because the political situation was too unstable. There was incompetence, infighting, waste, and mismanagement of funds. And now civil war, with no way of telling how it would all turn out.

14

Ascent (1947–1948)

In December 1947 Ying arrived in the United States. Tsien was back teaching and doing his own research at MIT. They lived in an apartment on 9 Chauncy Street, which stood across the Charles River and within walking distance of Harvard Square. With Ying, Tsien now had a companion with whom he could explore the musical culture of Boston. He joined the American Academy of Arts and Sciences in Boston and bought season tickets for the Boston Symphony Orchestra. Both he and Ying avidly collected records and attended the symphonies, concerts, and operas given in the city. They also went to art galleries together. Occasionally, they spent an evening with Tsien's friend Rene Miller and his wife.

The early part of 1948 was an industrious time for Tsien. He had a huge teaching load at MIT and gave two courses in aerodynamics, one course in rockets, and seminar lectures on structures and stability. In January he presented a paper on wind tunnel testing problems at a conference in New York, and between March and September he coauthored some journal articles with graduate students and colleagues. He also served on the aerospace vehicles panel of the Scientific Advisory Board (SAB), the purpose of which was to inform the chief of staff of the U.S. Air Force of the latest scientific developments. Within the SAB, Tsien also belonged to a committee that agreed to draw up a program for the future use of the Bell XS-1—the first manned aircraft to break the sound barrier. Together with Theodore von Kármán, one of the founders of the SAB, he traveled frequently to military installations across the country.

Upon his return to Boston, Tsien told his friends at MIT that he believed Mao Tse-tung would soon win the civil war and drive out Chiang Kai-shek. Now there was every indication that his prediction would come true. The decisive year of the civil war was 1948: the Nationalists lost four hundred thousand men and the Communists captured Shantung and beseiged Manchuria.

It was also a year of escalating inflation in China. A bag of rice that sold for 12 yuan in 1937 sold for 6.7 million in 1948. Prices jumped even as customers walked from one store to another with sacks and wheelbarrows of worthless paper money. Desperate to end the inflation, the GMT introduced a new currency system and passed laws to force the Chinese population to sell their gold, silver, and foreign currency for the new yuan at a fixed rate. But despite Draconian measures, the GMT failed to stop the black-market trading and the printing presses. By October 1948, businesses shut down, people turned to barter, and confidence in the GMT sank to a new low.

With all the trouble in his homeland making headlines in American newspapers, Tsien must have felt relieved that he and his wife were now in the United States. Their lives seemed charmed. On October 13, 1948, the Tsiens were blessed with a son, whom they named Yucon. During the same period, Tsien received an offer from the Daniel and Florence Guggenheim Foundation that would forever change his life.

The foundation was an granting institution with a long tradition of funding research in aeronautics. In the 1920s it created the Guggenheim schools of aeronautical engineering. In the 1930s it supported the work of Robert Goddard, who labored in solitude in Roswell, New Mexico. Then in 1948 it decided to establish two new centers of jet propulsion research at Caltech and Princeton. Tsien was offered the directorship of both.

Both universities avidly recruited Tsien. At Princeton, Tsien's former student Joseph Charyk urged him to take the position. Concurrently, Lee DuBridge, the new president of Caltech, offered Tsien a tempting package: a stipend of ten thousand dollars a year and additional funds to support junior scientists and assistants. DuBridge told Tsien it was likely that the grant would be renewed every seven years. But even if the grant were not, Tsien would still be welcome to stay at Caltech with tenure as the Robert Goddard Professor of Jet Propulsion. Wrote DuBridge to Tsien on September 29, 1948: "All of your many friends here at the Institute hope most sincerely that you will be willing to accept this opportunity of returning to Pasadena."

After carefully weighing his options, Tsien decided to go to Caltech. His decision was no surprise, for he had friends in Pasadena and it was Theodore von Kármán's home. After Tsien formally accepted the offer, the Board of Trustees at Caltech authorized the appointment in October. Tsien began to lay out plans to return to Pasadena in the summer of 1949.

It was about this time when Tsien received a card from his old friend Frank Malina, now living in Paris. Malina had resigned from his position as acting director at JPL in 1946 and started a new life: in 1947 he took a position with the United Nations Educational, Scientific and Cultural Organization (UNESCO). The personal problems arising from a divorce and political troubles with Clark Millikan may have been factors in his decision to leave Pasadena, although Malina later claimed that he left JPL mainly because he was mentally and physically exhausted from the war and had become disillusioned with military research. "I had long been convinced that war between or by advanced technology [sic] nations was a form of national insanity," Malina wrote years later. "It seemed to me that ideas and effort were really needed now to find ways for sovereign nations to function in peace together rather than to develop better means of destroying themselves."

Malina had the financial means to leave his career in rocketry behind. The enormous wealth he acquired from his shares of Aerojet company stock gave him the time to explore other interests. As a young man at Caltech he had dreamed of becoming an artist, and in Paris he set to work on this goal. (Years later, he was to invent kinetic art, a form intended to combine art with science, and crafted some intricate gear-driven shapes that moved and threw flecked patterns of light.)

Two days after Christmas in 1948, Tsien excitedly wrote to Malina about the Caltech offer. He was glad to leave MIT, he wrote. "The atmosphere here is too business-like and too much old-style engineering to fit in with my Caltech training or Kármán-training," Tsien wrote to Malina. "Frankly, I am not happy here. I do not think Hunsaker is going to feel badly about my leaving either. I just don't fit in this old Department of Aeronautical Engineering."

Now, Tsien wrote, he would have not only the Robert Goddard position but a younger man to assist him, three postgraduate or postdoctoral fellowships each year, and funding for research. In a strange twist of fate, it was Tsien, not Malina, who was now enjoying the fruits of the Suicide Squad research that

Malina had so enthusiastically promoted years before. Tsien pointed it out himself:

> So it is really the sort of thing you wanted years ago. I think the fact that it now arrived is certainly to a great measure due to your effort during the war years with the JPL at Caltech. In accepting Caltech's invitation, I could not help but think back of the days we worked together on the rocket project with the personal contribution of that Arnold. I could not help to feel somewhat strange to do this job without your participation. Of course, you may not at all [be] interested now in such matters.

In a paragraph more prophetic than he realized, Tsien wrote: "The whole situation in the Orient is now fast-moving. I am really not too sure about my own future. But perhaps, nobody is sure of his future."

Through the end of 1948 and well into 1949, Tsien followed the events of the Chinese civil war in the newspapers. The Nationalists were losing. The People's Liberation Army had entered Beijing and by April 1949 took Nanjing without a fight. In May they seized Shanghai, China's largest city. Fleeing before the victorious troops led by Mao Tse-tung, the forces of Chiang Kai-shek retreated to Taiwan, ninety miles off the south coast of China.

It was then that Tsien came to grips with perhaps the single most important decision in his life. Sooner or later, he would have to decide where to put down roots for himself and his family. He was proud to be Chinese; he would always be ethnically Chinese; but his future, he knew, lay in the United States. In 1949, he took the final step: he applied for U.S. citizenship.

Caltech (1949)

When Tsien arrived in Pasadena in the summer of 1949, he displayed every indication of settling there permanently. It was known at Caltech that he wanted to buy a house, but that racism and even legal restrictions would make it difficult for him to do so. The "covenant codes" of house deeds in certain affluent areas of Los Angeles prohibited residents from selling their homes to nonwhites. Consequently, in June 1949, Tsien decided to rent for his family the home where he used to live in the 1940s: a one-story redwood clapboard and brick house surrounded by a large lawn and eucalyptus trees.

The house rested at the very end of East Buena Loma Court, a quiet cul-de-sac in the unincorporated residential area of Altadena. With little traffic, it was the ideal neighborhood for raising children. Inside, a long hall bisected the house, separating the dining room, living room, kitchen, and laundry from three bedrooms. The living room was comfortably furnished and had a phonograph; Tsien and his wife spent long hours listening to classical music there.

Tsien's home became the center of a tiny social circle, a small core of people at Caltech or JPL with whom Tsien was becoming close. Among his friends were Frank Marble, a thirty-one-year-old fluid dynamics expert who joined the

Jet Propulsion Center as assistant professor in 1949; W. Duncan Rannie, an associate professor of mechanical engineering who had known Tsien since their graduate school days in the 1930s; and Frank Goddard, a 1949 MIT graduate whom Tsien had recommended for the position of chief of the high-speed wind tunnel section at JPL. They remembered the elaborate dinners that Tsien hosted at his home: Tsien, with gusto, cooking vegetables before his guests while Ying, who often spent the entire day in the preparations, sat modestly by his side.

Perhaps his closest friend was Luo Peilin, whom Tsien had known since his college years. Luo was now a Caltech graduate student in electrical engineering, and he spent almost every weekend at Tsien's home. Decades later, Luo remembered those evenings fondly. Ying would pour him a drink in a glass cup set with quartz, and the three of them would listen to classical music together: Béla Bartók, perhaps, or the string quartets of Beethoven.

Ying appeared to have a softening effect on Tsien. She was more outgoing and took a drink now and then, while Tsien looked on with disapproval. Occasionally, she would tease him about being stuffy. With her, Tsien gradually eased into the domestic role of family man. Friends described the marriage as serene and happy. By the end of 1949, she was pregnant with their second child.

The future was rich with promise for Tsien. He looked so young and healthy he could have easily been mistaken for one of his own graduate students. He was thin, with a full head of shining black hair. Though he was over forty, he looked twenty-five. And his wife, with her sophistication, beauty, and perfect voice, drew even more admiration from the predominantly male Caltech community. (Gushed one friend of Tsien's in a letter to Theodore von Kármán: "We are all in love with Mrs. Tsien!") He had everything, it seemed, that a man could want: a lovely wife, a son, a fabulous career, and lifetime security. The little Chinese student who arrived fifteen years earlier was now the living embodiment of the American Dream.

At Caltech, Tsien now had a reputation almost equal to that of Kármán in the 1930s. And like Kármán, Tsien now wore the different hats of scientist, teacher, and administrator.

His scientific work would be balanced between his own theoretical projects and consulting. Outside of Caltech, he planned to consult for JPL, which was working on the Sergeant test vehicle (a large solid-propellent rocket) and had just successfully tested the Corporal E, a full-sized surface-to-surface liquid-propellent missile with an axial-cooled motor. Tsien also made arrangements to consult three Mondays a month for the Aerojet Corporation, which had grown from a staff of a half-dozen when it was founded to that of a couple thousand. In preparation for this, he applied for security clearance to work on secret and aeronautical research contracts for the Air Force, the Navy, and Army Ordnance.

Tsien carried a heavy teaching load at Caltech. During his first semester there, Tsien taught a class in liquid- and solid-propellent rockets and another in high-temperature design problems. He also taught, with JPL staff members, two courses on jet propulsion power plants and jet propulsion research. The curriculum was virtually unchanged from the courses he had taught at JPL during the war. The composition of the student body remained as heavily military as during the war years. As late as 1953, nine out of the ten students who graduated with an engineer's degree from the jet propulsion center were service officers.

Most of them left Caltech at the end of a five-year master's degree program or a six-year program to get the professional degree of aeronautical engineer. A few proceeded with independent research and several more years of study to get a doctorate. Throughout the program, they were taught according to Kármán's principle of the unification of theory and practice. Undergraduates received a solid grounding in mathematics, physics, chemistry, and mechanical engineering. First-year graduate students took basic courses in rockets, thermal jets, and chemistry problems in jet propulsion, while second-year students advanced to specialized topics in stability and control, high-temperature design problems, and physical mechanics. On tests students were sometimes asked to design, within the space of an hour or two, a conventional sounding rocket or missile—underscoring the ultimate goal of equipping the students with skills to launch a device into outer space.

To his students, Tsien emphasized the importance of mathematics, for which he had an almost religious respect. Reported a visiting scientist who attended one of Tsien's lectures: "His advice was that—whenever possible—one should calculate the phenomenon associated with those extreme conditions rather than attempt to measure them. Perhaps he had more faith in the validity of the applicable equations of physics than in most people's ability to get good data from a challenging experimental setup."

Once again, Tsien proved to be a stern taskmaster, unwilling to tolerate mental laziness such as rote memorization of formulas. One time, when Tsien was writing out an equation, a student pointed out that he was using a different symbol from the standard one for the density of gas. This irritated Tsien greatly. He said that the class should understand the equations regardless of which symbol stood for what property, and from that day forth he deliberately used a completely different set of symbols for every physical property. Recalled one student, "It made us realize that we didn't understand the equations as well as we thought we did! We looked at equations from then on in an entirely different light."

Tsien also was tough during seminars, speaking in an abrupt, crisp manner. He had high standards for the students and faculty and even higher ones for Chinese scholars, for reasons of national pride. It had never been Tsien's intention to be popular. He seemed to adhere to the Machiavellian principle of preferring to be feared rather than loved, especially when popularity clashed with his standards of scientific truth. Stories of Tsien's outbursts during these seminars filtered into Caltech legend.

Once, when a student was critiquing a flawed scientific paper, Tsien continuously interrupted him with impassioned exclamations. "But it's wrong!" Tsien cried. "It's *wrong!*" It had been the student's intention to point out the errors in the paper himself, but he sensed that Tsien would have had more respect for him if he had simply torn up the paper and walked out of the room in disgust. Another anecdote depicts Tsien impatiently flipping the venetian blinds in one direction and then another before yelling at the speaker: "*Stop!* This is a bunch of mathematical chicanery based on unsound mathematical principles!"

If Tsien had been shy in the 1930s, he had broken out of his shell of reserve by 1949. He quarreled frequently with his colleagues—one of them remembering an argument so loud that the people in the Guggenheim building could hear it down to the bottom floor. Liked or not, Tsien was now clearly a character at Caltech to be reckoned with.

His personality seemed to clash with his administrative responsibilities as director of the Jet Propulsion Center: to promote the peacetime and commercial uses of rockets and let the world know of the existence of the center. It was a role better suited to a public relations officer than a scientist, but surprisingly, Tsien excelled in it. It soon became apparent that Tsien, like his friend Frank Malina, possessed the gifts of promotion after all. By 1950, he seemed well on his way to becoming a celebrity.

In December 1949, Tsien received a major boost to his reputation after giving a presentation at the American Rocket Society conference at the Statler Hilton Hotel in New York City. There, Tsien presented his blueprint of a transcontinental rocketliner: a futuristic mode of transportation that would carry commuters from New York to Los Angeles in less than an hour. It would be shaped like a fat sharpened pencil, eighty feet long, nine feet in diameter.

The ship would take off vertically and move along an elliptical path across the atmosphere for the first 1,200 miles, traveling 10,000 miles an hour. After reaching a height of 300 miles it would arc downwards toward the earth, break through the atmosphere, and glide 1,800 miles to its destination. Tsien showed the audience a sketch of the rocket and announced that the Army and Navy were experimenting with designs for such a vehicle. It was, Tsien claimed, "not at all beyond the grasp of present-day technology."

The presentation was the highlight of the conference. Newspapers described in detail what life would be like inside the rocketliner: how anything not tied down would float around, how people might need special pressure suits, how being shot up in the air would be like going up in a fast elevator but a hundred times more intense. *Popular Science* and *Flight* magazines ran picturesque spreads of the rocketliner. The *New York Times* carried a long description of Tsien's idea as did *Time* magazine, which published a photograph of Tsien. "The rocketliner," the *New York Times* reported on December 2, 1949, "could give virtually immediate reality to the oft-expressed wish: 'I wish I were in California now.'"

During his first year at Caltech, Tsien made an even more daring prediction. He announced that a trip to the moon would be possible within thirty years and that the journey could be accomplished in a week. In May 1950, *Popular Mechanics* ran a drawing of astronauts on its cover and cited Tsien's belief that engineers could start construction of a moon rocket immediately. This idea generated far less publicity than the rocketliner idea because of the widespread skepticism that it would ever be possible. When Tsien once mentioned his idea to some women at a party, they thought he was drunk or mentally unbalanced.

Still, at this point in his career, Tsien had established himself as a major spokesman in his field, with the potential to be as famous as Wernher von Braun. Nothing, it seemed, would deter Tsien from taking his place in the burgeoning rocket program in the United States—from becoming one of the scientists who would one day launch a person into space.

Then came an event that would irrevocably change his life. In the summer of 1950, Tsien received a visit from the FBI.

Suspicion (1950)

June 6 was a cloudy, drizzling day. Two agents from the Federal Bureau of Investigation came to Tsien's office. Tsien now occupied Kármán's old room on the second floor of the Guggenheim Building. It was a large room with Chinese journals on the tables. Behind his desk were blackboards covered with mathematical equations. A big window opened to a view of the courtyard. What the FBI wanted to know was simple: Was Tsien, or had Tsien ever been, a member of the Communist Party?

The FBI claimed that several people Tsien had befriended at Caltech in the 1930s were Communists. The social gatherings held at Sidney Weinbaum's home were, they said, in reality meetings of Professional Unit 122 of the Pasadena Communist Party. Tsien's name had somehow appeared on 1938 membership lists linked with the alias "John Decker." The agent interrogated Tsien to learn more about his relationships with the party and with Weinbaum.

Tsien repudiated all the charges. He denied that he had ever been a Communist. In fact, he insisted he was philosophically opposed to the idea of Communism. Russian Communism, in his opinion, was nothing more than a totalitarian form of government, and relative to democratic or free government it was "evil." As for his name appearing on the party roster, Tsien said he had no idea how that happened. He had never even heard of the name John Decker.

Moreover, Tsien told the FBI that he believed his friend Sidney Weinbaum was loyal to the government. Weinbaum, like some of his other friends, had expressed left-wing tendencies at times, but Tsien couldn't say for sure whether they were Communists. According to the FBI report: "Tsien said as a scientist, the only yardstick he has in measuring a person's worth or loyalty is fact, and that since facts cannot be applied to such intangibles as a person's loyalty or political beliefs, he couldn't ever speculate upon them."

But the U.S. government had already taken precautions against Tsien's possible breach of loyalty by revoking his security clearance. The very day the FBI talked with Tsien, Caltech administrators received a hand-delivered letter from the headquarters of the Sixth Army, at the Presidio in San Francisco, informing them that Tsien would no longer be permitted to work on classified military projects. This was no trivial matter for Tsien. He had planned to consult for JPL and the Aerojet Corporation, as well as take on secret defense work contracted to Caltech. Now that would be impossible. With ninety percent of all research at JPL under classification, a security clearance for a consultant was a practically a necessity. However, Tsien knew that the government's action would not hamper his theoretical studies, which did not require a security clearance.

The Caltech faculty and administration were incredulous when they heard the news. Tsien, a Communist? Some believed Tsien to be the most aristocratic person they knew. And wasn't he married to the daughter of a military strategist who had worked under Chiang Kai-shek? No, Tsien was the last person— the very last person—the Caltech establishment would have suspected as being Communist.

No record remains of Tsien's immediate reaction when the FBI left his office, but one can surmise that he was shocked. Friends remembered later that during this period Tsien expressed bewilderment that he was the target of such an accusation, and that he was deeply hurt. Just how much so would become quickly apparent, for two weeks later Tsien made a stunning announcement: that he was resigning from Caltech and returning to China.

When the FBI learned about this through an informant at Caltech, they contacted Tsien again, this time by telephone. Tsien said he felt it was "the only thing left for me to do" but agreed to meet with the FBI again.

On June 19, Tsien gave the FBI a prepared statement. He said he had been a welcome guest in the United States for more than ten years, and he had prospered. He believed that his visit was of mutual benefit because he had given the United States much in scientific advancement during the war years. "Now, he says, this original status which he has enjoyed as a welcome guest no longer exists and a cloud of suspicion hangs over him; therefore, the only gentlemanly thing left to do is to depart." Tsien told the FBI he had made the same statement to Fred Lindvall, chief of the division of engineering, and Ernest Watson, dean of faculty at Caltech.

Exactly why Tsien decided to leave at this point is not known. Speculation about his motives floated about Caltech for years, and the differing accounts of his behavior given later make the story a confused and murky one. To further complicate matters, Tsien himself changed or fabricated the narrative of the situation once he returned to China. In retrospect, this sudden decision to leave only confirmed in the minds of many what the FBI initially suspected about him: that he was a Communist and possibly a spy. But the real reason, judging from what his friends remember at the time, appears to have been a combination of pride, anger, confusion, and fear, all emotions consistent with the person Tsien had become.

One must consider that Tsien was an extremely proud person to begin with and his recent string of intellectual successes had made his arrogance at times nearly intolerable. He took the denial of clearance as a personal insult and felt he had "lost face." Theodore von Kármán, writing of the situation, observed: "[Tsien] didn't believe that he should have to prove to the authorities that he wasn't a Communist. I believe my own reaction would have been about the same as Tsien's if someone had falsely accused me of Communism as a result of my brief association with the short-lived Béla Kun government of Hungary."

There was another, more compelling reason behind Tsien's abrupt decision to leave. Shortly after the fall of China to the Communists in October 1949, Tsien had begun to receive letters from his father, each increasingly urgent. His father was urging him to return, as he was about to undergo a serious stomach operation. Apparently, he also wanted to spend time with Tsien's children, whom he had never seen.

Before his security clearance was lifted, Tsien had discussed these letters with friends and had seemed guilt-ridden over them. "He was confused as to what to do," recalled his friend Martin Summerfield, who had been teaching at UCLA before leaving to accept a professorship at Princeton. "He was torn between loyalties, but I think he was determined to stay in the United States. He wanted to stay in the United States, he wanted to be a citizen, but he had to figure out some way to pacify his father."

Tsien half-suspected that his father was being pressured by the Communists to write these letters in an attempt to lure him back to mainland China. At some point he discussed these fears with Lee DuBridge and a few other friends, one of whom secretly shared this information with the FBI. Whether his father was indeed the subject of such pressure will probably never be known. But there were cases of open harassment by the Chinese Communist government of Chinese scientists living in the United States. Like Tsien, many such Chinese nationals received letters from relatives urging them to return to mainland China or even found themselves the target of letters printed in American Chinese-language newspapers. One student was said to have suffered a nervous breakdown and ended up in an asylum as a consequence of the pressure put on him. Such episodes, however, were not well documented or well known.

But the denial of his clearance changed his perspective, prompting him to question his allegiance to the United States and to wonder if his work had caused him to neglect his filial duties. Perhaps now, it would be a good time to go back for an extended visit to China. Tsien also considered the possibility of trying to take his father out of Red China and into Hong Kong, but he acknowledged to his friends that the plan had little chance of success. Instead, he decided, he would take a temporary teaching position in China, possibly until his father passed away.

Perhaps what Tsien needed more than anything else at that moment was consolation and advice from his trusted mentor, Theodore von Kármán, but Kármán was in Paris, working as a scientific advisor for the U.S. government.

Within weeks, other events must have intensified Tsien's resolve to leave. One was the arrest of Sidney Weinbaum at his home on June 16. The government charged him with perjury for lying to Army officials in September 1949 in telling them that he had never been a member of the Communist Party. At the time, Weinbaum was undergoing a security investigation in connection with his application for the position of mathematician for the materials section of JPL. Tsien had originally recommended Weinbaum for the job. Kármán noted in his autobiography that Tsien was asked to testify against his old friend but he refused, which later "turned the FBI's attention to him."

Another possible inducement to his decision to return to China may have been the outbreak of the Korean War in late June, which heightened tension between the United States and the Communist countries of Asia. Tsien's friend Luo Peilin wanted to return to China immediately because his wife and children were still in mainland China. Given the worsening relations between the

United States and Communist China, Luo feared that if he waited any longer he might never be able to return. Tsien, Luo remembered, harbored the same fears.

Shortly after the birth of his second child, a daughter named Yung-jen, Tsien began to make his plans for his departure openly. He wrote to the State Department and even visited Washington to secure official permission to leave. He tried to make reservations to go to China by boat but was told that his reservations couldn't be confirmed. In early July, at the suggestion of some other Chinese scholars at Caltech, he wrote to the International Trade Services Association, which helped Tsien arrange a flight to China on a Canadian airline. He planned to leave from Vancouver and arrive in Hong Kong by plane.

Tsien's friends were shocked at his decision to leave. When Tsien casually remarked to aeronautics professor Hans Liepmann in the Guggenheim building staircase that he was going back to China, the latter thought he was "crazy." At some point, his old friend William Sears asked Tsien why he thought he needed access to classified material, to which Tsien replied "that he couldn't fulfill his responsibilities as Guggenheim Professor at Caltech without such access." Sears felt Tsien was exaggerating but knew that "the clearance matter had injured his pride."

Not surprisingly, Caltech officials intervened. They had no intention of losing one of their newest and youngest luminaries. They begged Tsien to reconsider. Lee DuBridge was especially active in working behind the scenes in an attempt to restore Tsien's clearance and so end the whole matter. In July, he wrote to Kármán and urged him to send word of Tsien's troubles to his friends in the Air Force. He also used his own contacts in Washington to pry into Tsien's file at Naval Intelligence, and got word back that the whole thing was nothing more than a witch-hunt. In all of his letters, DuBridge stressed that Tsien was a great scientist who was falsely accused of being a Communist and that the United States risked a double loss if Tsien went back to China. "This," he wrote, "is a ridiculous situation that one of the greatest rocket and jet propulsion experts in the country is not only denied the opportunity of working in his chosen field, but by such denial is forced to return to occupied China and his talents made available presumably to the Communist regime there." At least one Caltech professor expressed serious concern that Tsien might end up in "the hands of the Russians."

During this time, DuBridge also scheduled a hearing for Tsien before the Industrial Employment Review Board of the Army in Washington, D.C. This would force the government to explain exactly why Tsien was denied access to classified contract work. By now, Tsien had been denied clearance not only by the Air Force, Navy, and Army but also by Army Ordnance as well, which had cleared him a year before.

"The authorities at Caltech wished very much that I would remain," Tsien remembered later. He knew that Caltech officials believed that if they got back Tsien's clearance he might decide to stay in the United States after all. Seeing how important it was for Caltech not to have this issue hanging, Tsien reluctantly agreed to hold off his departure. He told them, however, that he wished to have the hearing scheduled as early as possible because he intended to leave the country by the end of August.

By now, Tsien was planning for the Washington hearing and his departure simultaneously. Toward the end of July, Tsien hired a packing company to put his belongings in crates and to ship them back to China. His possessions would be loaded on the *President Wilson*, which was scheduled to sail from Los Angeles the day after his flight to China. They would be delivered to an address in Hong Kong and eventually to Tsien's father's home in Shanghai. Tsien's wife and children would join him in China later.

By mid-August, time was running out. The hearing was scheduled in Washington on August 23, and Tsien had to leave on August 28. His house was in disarray as men from the packing company arrived to help Tsien pack up loose items. The company also received orders from Tsien to wrap up his books and papers at Caltech in boxes lined with waterproof paper. He seemed so pressed for time that when the packers arrived at Caltech, Tsien asked them to put the material in the boxes, carry them to the warehouse, and finish the wrapping in there.

There was every indication that Tsien's move to China would be a permanent one. Remembered C. Harold Sexsmith, owner of the Bekins Van and Storage Company of Pasadena: "The shipment appeared to be that of someone expecting to take up residence on a more or less permanent basis at destination. It consisted of a grand piano, articles of furniture, dishes, bedding, books, an office desk, radio combination, and other similar items—a washing machine—that a person would use in setting up a home." With Tsien packing up all of his worldly possessions, it didn't seem as if he was ever coming back.

On August 21, Tsien flew to Washington. His first stop was the office of Dan Kimball, a government official whom DuBridge had urged Tsien to see. Kimball was a balding, tall man with a hardened look about his eyes. Formerly the executive vice president and general manager of the Aerojet Corporation, Kimball had helped Truman get elected and was now the undersecretary of the Navy. During the war, he had followed Tsien's career and recognized him as one of the country's leading rocket experts.

In Kimball's office, Tsien started to tell him everything: the visits from the FBI, the humiliation of having his security clearance taken away. So distraught was Tsien that at one point in the story he lost his composure and burst into tears. Seeing Tsien's anxiety, Kimball reminded him that he had been offered a professorship in math at Caltech, which wouldn't require a security clearance. Nevertheless, Kimball was determined to see Tsien restore his clearance and referred him to a Washington lawyer, Paul Porter. "He was so overwhelmed," Tsien recalled later, "I half suspected he didn't listen very carefully. He just pushed me out of the office and said, 'You go see Mr. Porter and everything will be all right.' "

That afternoon, Tsien met with Porter. After listening to his story, the attorney decided that in order to make a good legal case for Tsien at the IERB hearing he would need time to do some research. Tsien then wondered if the entire hearing was necessary. He was planning to leave the country anyway. Porter agreed that the idea of the hearing seemed illogical and suggested that it be postponed. Tsien then went back to see Kimball again, who postponed the hearing indefinitely.

On August 23, Tsien met with Kimball again. This time, Tsien emphasized even more strongly that he planned to leave the country, feeling that Kimball had not seemed to take what Tsien said seriously the first time. Kimball reportedly said to Tsien, "You can't leave. You're too valuable." He added that if it were up to him Tsien would definitely not be allowed to leave. A strong supporter of Nationalist China, Kimball warned Tsien to think over the matter very seriously. Tsien said he was going back to California to do precisely that. That afternoon, he boarded a plane for Los Angeles.

But the moment Tsien left his office, Kimball called the Justice Department. He warned them that Tsien, with all the knowledge he possessed, must not be allowed to leave the United States. Kimball believed that the Chinese government wanted Tsien's technical expertise and that pressure was being put on Tsien's father to lure Tsien back to China. When Tsien arrived at the airport that evening, an immigration agent was waiting for him. As Tsien stepped off the plane, the agent handed him a paper from the government forbidding him to leave the country. One can only imagine his rage.

With no option of carrying out his plan, Tsien canceled his reservation on Canadian Pacific Air Lines to travel from Vancouver to Hong Kong and asked the packing company to withdraw the shipment. After talking it over with his wife, however, Tsien considered having her and his children go to China while he stayed in the United States, and he withdrew some savings from the bank for that purpose.

But unbeknownst to him, the Customs office had already inspected and impounded the luggage. On August 19, shortly before Tsien left for Washington, the books and papers were moved from his office to the packing room and unloaded in the morning. The packing foreman informed his boss, Harold Sexsmith, that certain papers of Tsien's were marked "Secret" and "Confidential." Sexsmith immediately contacted Customs official Roy Gorin, who told Sexsmith to proceed no further with the packing. Gorin also ordered Sexsmith not to discuss the matter of Tsien's luggage with anyone until Customs had had a chance to determine what action would be necessary.

From that point on, things moved quickly. On August 21 and 22, officials from Customs, the FBI, Naval Intelligence, Air Force Intelligence, Army Ordnance, and the State Department descended on the Bekins warehouse in Pasadena. The State Department officials recommended seizure of the documents and Tsien's detention while the U.S. attorney's office in Los Angeles wanted Tsien put under surveillance. Customs officials applied for a civil warrant of detention for his eight cases of papers on the grounds that Tsien might have violated the Export Control Act, the Neutrality Act, and the Espionage Act. A federal judge readily granted it. On August 25, immigration officials issued a warrant for Tsien's arrest.

All this was done, apparently, without Tsien's knowledge. The next morning, when the story appeared in the news, a panicked Mrs. Tsien called Sexsmith, the owner of the packing company, and demanded an explanation. Sexsmith said he had been ordered not to tell anyone what was being done to Tsien's luggage until the Customs people were ready to announce it. "I told Mrs. Tsien that I was rather surprised that she hadn't heard anything about the goods being held," Sexsmith recalled later. "She replied that the first word they had had of the matter was the story in the newspaper."

Apparently, Tsien had been unaware that special permission was required to export technical data in certain scientific categories. He spent the next few days answering questions from reporters about his luggage. "It was my personal property," he told them. "I was planning to go to China. Now I am not. I was told by the Immigration Service not to go. I don't know why they are inspecting. I don't know the complete story." He told them that he had planned to take care of some family problems in China and eventually return to the United

States. He claimed that he went through his personal papers very carefully, placed what was classified in a locked cabinet in his office and gave the key to another aeronautics professor, Clark Millikan.

"There are no code books, signal books and no blueprints," Tsien wrote in a prepared statement that was reprinted in the press. "There are some drawings and logarithm tables, etc., which someone might have mistaken for codes. I wished to take my personal notes, many of which were merely lecture notes, and other material with me for study while I was gone. I most certainly was not attempting to take anything of a secret nature with me, or trying to leave the country in any but the accepted manner."

It had been Kimball's original intention to stop Tsien quietly from leaving. But the unexpected news conference gave the story so much publicity that it spun out of Kimball's control. The story appeared in the Los Angeles papers: the *Times*, the *Mirror*, the *Examiner*, carrying headlines like "Secret Data Seized in China Shipment." The Associated Press and United Press International carried it across the nation, where it would be reprinted in the *New York Times* and countless other newspapers. Only a few months before, the media had hailed Tsien as a visionary. Now they branded him as a possible spy.

Arrest (1950)

The publicity concerning Tsien's luggage couldn't have come at a worse time for him. It coincided with the trial of Sidney Weinbaum, which began August 30 and lasted until mid-September.

The military had long suspected Tsien and his friends of Communist activities—as early as 1941. The accusation first reached Sidney Weinbaum sometime during that period, when he was working for the Bendix Corporation. During a big party at the Caltech aeronautics department, Professor Clark Millikan informed Frank Malina that he—Millikan—had heard that Malina and Weinbaum and two or three other people were members of the Communist Party. The FBI, apparently, had given Millikan the information. Recalled Weinbaum: "So I went to see a lawyer friend of mine and said, 'What can I do?' I was sure that with an accusation like that they were going to refuse me clearance. But no! I was cleared throughout all these years; from '41 to '49, when the trouble began, I was cleared for top-secret work." In the early 1940s, Army Intelligence heard accusations that Tsien was a Communist as well, but no effort was taken to suspend his clearance either. On the contrary, he was granted permission to work on projects classified "confidential," "restricted," and even "secret."

Apparently, the government decided that the scientific abilities of people such as Tsien, Weinbaum, and Malina far outweighed the potential security risks.

Initially, many people believed Weinbaum was innocent. The arrest of Sidney Weinbaum had shocked his friends and former colleagues at Caltech, most prominent among them Linus Pauling, the famous chemist and Weinbaum's former supervisor. Pauling told the newspapers that he had the "greatest confidence" in Weinbaum and saw "no reason to suspect him." Several scientists wrote letters attesting to Weinbaum's loyalty and honesty and a group of Caltech professors—James Bonner, Charles DePrima, Paul Epstein, Linus Pauling, and Verner Schomaker—even established a fund to collect money for his defense.

But when the trial opened, a number of witnesses testified that Weinbaum had not only been a Communist but had served as organizer and leader of a Communist cell in Pasadena. These witnesses consisted of former members of the cell: Frank Oppenheimer, a former physics graduate student at Caltech; Gustave Albrecht, a former chemistry graduate student and research assistant; Richard Rosanoff, a former undergraduate student; Jacob Dubnoff, a former biology graduate student. The picture that emerged was of Weinbaum as a passionate adherent of the Communist Party who actively handed out membership applications to students at Caltech and urged them to join. Equally damaging was a membership book issued to Weinbaum that the prosecution submitted as evidence. While Weinbaum was "biting his nails and furrowing his brow," jurors studied a five-foot-square photostatic enlargement of the book as an FBI document expert confirmed that the signature on the book was indeed made in Weinbaum's handwriting.

The most intriguing information to emerge from the trial was the use of code names among members of the Communist cell. For instance, Sidney Weinbaum's alias was Sydney Empton, Jacob Dubnoff's code name was John Kelly, and Frank Oppenheimer's was Frank Folsom. It raised eyebrows at Caltech and within the Pasadena community. Membership in the Communist Party was never against the law, so what was the need for code names? Party members later claimed that they had adopted the aliases to protect themselves from right-wing retaliation. The code names, however, only deepened the suspicion among the general public that the entire group might have been spying for Soviet Russia.

The trial ended badly for Weinbaum. Convicted on three counts of perjury and sentenced to four years in prison in September 1950, he was not even permitted bail because the judge pointed out that Weinbaum previously had access to secret technical information at JPL. "He might know things that would be of benefit to those who seek to destroy us," U.S. Judge Ben Harrison said. "We are too near an ocean port and the border for it to be safe to grant this defendent bail. He has no ties here except his wife and daughter. He has no job and no prospects of getting one until things cool down."

Meanwhile, government officials spent countless hours poring over Tsien's seized papers. The sheer bulk of the material was intimidating. There were more than one hundred unclassified books in the shipment, technical papers from scientific, government, and industrial sectors and material in Chinese, German, and Russian. All of it required review from top aeronautical experts summoned from all parts of the country. Simply organizing all of the material posed formidable problems. On September 5, 1950, three men arrived from Wright-Patterson Air Force Base and spent three days microfilming Tsien's papers. They returned with approximately twelve thousand frames worth of pictures. Investigators began drawing up detailed appendices and indexes of the papers. An outline inventory of Tsien's library filled twenty-six legal-sized sheets.

One of the most interesting things found in Tsien's luggage was a collection of nine extremely well organized scrapbooks, about four hundred pages' worth, representing "the expenditure of much time and effort." They consisted of newspaper, magazine, and scientific journal clippings on the U.S. atomic energy program and atomic espionage. Tsien had painstakingly created a chronological record of spy trials in the United States. He saved articles that profiled the people accused of either spying for the Soviet Union or of having sympathetic tendencies, such as Klaus Fuchs, or that discussed proposed legislation against espionage. This undoubtedly raised questions among the investigators. Why all of this interest in atomic energy, and espionage concerning that energy? His luggage also revealed a fascination with Russian, containing as it did notes on the language borrowed from Frank Malina. Could Tsien indeed have been a spy?

Then the investigators tackled the more difficult problem of evaluating his technical papers. It was hard to determine exactly what was classified in Tsien's libary. Much of it was outdated, or written by Tsien himself. The conclusions reached by different agencies revealed little agreement. Some agencies felt it was nothing more than an excellent library and that it posed no threat to national security. The Atomic Energy Commission, for example, concluded the papers were typical of what one of the world's foremost technical scholars in the fields of aircraft and missile design would accumulate after ten years in the field. Others held a darker view of Tsien's motives and found a few papers that they believed should be classified. Some officials believed Tsien's entire library should be classified if only for the purpose of keeping it out of the hands of a potential enemy.

A report from the Office of Navy Research perhaps hit closest to the truth. "The importance of the entire library to an enemy," wrote the agency, "would be negligible compared to the information which Dr. Tsien has probably accumulated in his mind during his contacts with the United States Military Establishment."

On September 7, 1950, the Immigration and Naturalization Service sent two agents to arrest Tsien at his home. The warrant had been issued weeks before; why they chose to wait so long before acting on it has never been explained. Government officials were worried that Tsien would try to sneak out of the country, especially after Sidney Weinbaum was convicted on charges of perjury the day before and was now facing a ten-year maximum prison sentence and twenty thousand dollars in possible fines. At one point, Tsien had apparently eluded FBI surveillance and disappeared. State Department authorities almost wired a telegram to the Mexican government in order to request their cooperation if Tsien fled across the border.

When the INS agents arrived at Tsien's home, it was clear that he hadn't tried to escape at all. Mrs. Tsien was holding her baby daughter Yung-jen in her arms when she opened the door, and Tsien's son Yucon was seen "cowering in a corner." Then Tsien appeared. One agent recalled years later that the expression on his face seemed to say, "Well, it's finally over."

They gave him a standard interview, gathering as much biographical data as they could. They also asked him about his relationship with the Communist Party; Tsien reiterated that he had never been a member. The agents then charged Tsien with concealing his membership in the Communist Party, thereby entering the country illegally when he returned from China in 1947. They served him with an arrest warrant by late afternoon.

News of Tsien's arrest spread swiftly, traveling even as it happened. That day, his friend Frank Goddard was driving up from San Diego and down around Long Beach when he and his wife heard over the car radio that Tsien had been arrested. "When we heard that, we drove directly to his house with the idea of 'Jesus, Tsien's in trouble, what can we do to help him?' " Parking in front of his house, the Goddards went and knocked on the door and Tsien let them in. "And suddenly," Goddard recalled decades later, "I realized that the place was surrounded by FBI people."

"There was a look on his face I had never seen before," Goddard said. "The way he looked at me and the way he spoke—he looked almost mortally wounded inside. He was deeply chagrined that I would see him in this state. But he was very gracious and in a few sentences made it clear: 'Thank you very much but it is best that you leave and go home.' So we left and he was arrested and taken over."

The INS agents escorted Tsien to a waiting car, searched him, and then drove south toward Los Angeles. They crossed a bridge to the suburb of San Pedro, east of which lay Terminal Island, in the harbor. Originally a narrow sand spit called Rattlesnake Island, in 1871 it grew as the Army Corps of Engineers dredged to build a jetty from the island and a main channel. Reservation Point, a small rectangle of land that jutted from Terminal Island into the Pacific, was government property, owned by the Federal Bureau of Prisons and the Coast Guard. In 1950, the island contained a federal correction facility, a lighthouse, government offices, and homes for government employees.

The car swung past canneries, crumbling shacks, and oil storage tanks on Seaside Drive of Reservation Point until it reached the wire gates of the Immigration and Naturalization Service detention center. It was a three-story beige stucco building with a red tile roof. The first floor contained administrative offices, the second and third floors the aliens. Most of the detained aliens were Mexican migrant workers caught sneaking over the border, and they were typically held in large rooms packed with rows of bunkbeds.

Tsien was not placed in these crowded quarters but in one of the private cells, which were furnished with individual bathrooms. Some cells had barred windows that provided the detainee a view of the Los Angeles channel and the residential areas of San Pedro.

One can only guess at Tsien's rage and fear as he was escorted to his room. Nothing had ever prepared him for an experience like this. And nothing he had accomplished thus far—his degrees, his awards, his role in starting JPL, or even his contacts with America's greatest generals—had prevented it. He was the great Robert Goddard Professor—a position for which he had worked hard and steadily for the last fifteen years. But he was also in prison. And that must have been the very last thing Tsien had dreamed would happen when he decided to move back to Pasadena.

Investigation (1950)

Years later, Tsien would give dramatic accounts of his incarceration. "For fifteen days," he told a reporter, "I was kept under detention. I was forbidden to speak to anybody. At night the prison guards would switch on the lights every fifteen minutes to prevent me from getting any rest. This ordeal caused me to lose thirty pounds during that short period."

It is difficult to say exactly how Tsien fared during his two-week stay at the INS detention facility. In all likelihood it was quiet and uneventful. Lee DuBridge remembered decades later that Tsien had a little cubicle that was "perfectly comfortable," with a desk and light. He was, however, cut off from most of his friends and colleagues. Theodore von Kármán tried to communicate with him by phone from Europe but was denied permission. Tsien's family, however, visited him almost daily. When they arrived, Tsien could be seen smiling and waving to them from his cell window.

Perhaps Tsien did suffer both physically and mentally during this period. The guards may well have deprived him of his sleep; it was standard routine for them to peer into cells with flashlights to check on detainees at night. One

Chinese student who spent a few hours in a detention facility remembered that the guards beamed a flashlight into his face to see if he were "still alive and hadn't committed suicide." And Tsien, already under stress, was less than enamored of the food the INS served and could have lost weight that way. Of all the discomfort he underwent, it was most probably his pride that hurt the most.

Caltech officials, however, worked behind the scenes to secure Tsien's release. On September 18, Tsien wrote a statement pledging that he would not leave the United States without getting written permission from DuBridge and Kimball first. Two days later, Tsien was interviewed at the INS bureau on Terminal Island. He was accompanied by Grant B. Cooper, a prominent Los Angeles attorney whom Caltech had found for Tsien. Among those present at the meeting were INS district director Albert Del Guercio, Customs agent Roy Gorin, and six other officials.

When asked about the impact of the loss of security clearance and the letters from his father, Tsien replied:

It had a rather drastic effect because I had been receiving letters from my father and because an elderly gentleman like him has always wished that I go back to China. His poor health was known to me for some time, but I just went along with my work, because I was so busy, and when you are always plugging along in your work you seldom give your personal situation much of a review or think of it. But it was quite a shock to me that I might not be cleared for classified work. And to my mind that brought me to think of my personal problems, and at that time I thought about whether to stay in this country or plan a trip back to China.

I was further disturbed when looking at this situation about the possibility that there would be an open hostility between the United States and whatever government is in the mainland of China. In other words, the mainland was practically controlled by the Communist regime.

If such a hostility took place, then I would have no means of sending money over to my father, who is completely dependent upon me and supported by me, and I was very much disturbed by this possibility. So my wish would be this: that some settlement could be made so that these difficulties of supporting my father would be removed or at least made fairly secure. . . . I actually informed Professor Watson that it is my wish that finally I would be able to come back to this country and continue my work here, but of course I made it very clear to him that Caltech would not expect of me a definite date when I would be back.

After the conference, Caltech officials held a conference about Tsien with authorities at the attorney general's office. Two days later, Tsien was released on bail, which was set at the unusually high figure of fifteen thousand dollars. His colleagues at Caltech had to tap a wealthy friend of Tsien's to secure the amount. Luckily, Tsien was freed just before the fall semester began.

He was able to joke about it later in life, recounting to a newspaper reporter that, "As compared to the ransoms of one thousand dollars and two thousand dollars in ordinary kidnapping cases, I really feel proud of myself." But at the time, he was deeply embarrassed and upset.

William Zisch, then the vice president of Aerojet, was one of the few people who saw how hurt Tsien was. Late one afternoon, he stopped by Tsien's home to talk to him briefly about the status of his security clearance. Zisch, who regarded Tsien as one of Aerojet's most valuable consultants, was now in the uncomfortable position of having to tell Tsien that he could no longer work there. He was astonished when Tsien, normally reserved and quiet, talked to him nonstop for four or five hours, his words coming out in a torrent of emotion.

Tsien, who emphasized that he was not a Communist, said he believed that his troubles stemmed from his friendship with Frank Malina. (Zisch later learned by examining Naval Intelligence files that the cause of the trouble for Tsien was not Malina but the Sidney Weinbaum affair.) But Tsien made it clear to Zisch that he was going to continue to assert to anyone who asked that he was a dear friend of Malina's. No investigation was ever going to change that.

Tsien, Zisch recalled, seemed divided by his love and duty to his real father, to his academic father, and to his homeland of China. He had hoped that he would be able to bring his father out of Communist China and into Hong Kong, where he would care for him until his father passed away. But he also wanted to fulfill his promise as the protégé of Theodore von Kármán, a man whom Tsien also loved and hoped to work with during *his* waning years. And finally, there was the lure of China itself, even under Communist rule. For Tsien said that the essence of China—the Confucian nature of its culture—would never be wiped out by Marxism or the Soviet Union. "China," Tsien said, "will always be Chinese."

With Tsien out of prison, everyone seemed optimistic that he would soon be vindicated. Other universities even expressed interest in hiring him. Within a week of Tsien's release, Robert Oppenheimer, the world-famous physicist and professor at Princeton University, wrote to DuBridge to inquire about Tsien's situation. He said that if Tsien continued to have problems with government officials in Los Angeles, he should think about coming to Princeton. Oppenheimer had discussed it with the brilliant mathematician John von Neumann, who "has a very high regard for and interest in Tsien's work, and sees a real appropriateness in his presence here at a time when the computer will be undertaking aeronautical problems." DuBridge later responded that Tsien was more interested in staying at Caltech, but thanked Oppenheimer for the gesture.

The best way to clear Tsien's name, DuBridge felt, was to schedule an appeals hearing from the Industrial Employment Review Board. He wondered, however, if it was wise to ask the board to restore Tsien's clearance. "The very fact that his father is still in China and subject to pressure from that end might in itself be cause for denying him security clearance," DuBridge wrote to a friend. Perhaps it would be better to prove that Tsien had never been a member of the Communist Party and that he never tried to sneak out secret information. After all, proof that Tsien was a member of the Communist Party was slight. DuBridge was also confident that Tsien had tried to remove all classified documents from his luggage before he prepared for shipment.

There was every indication that the matter would soon be cleared up and forgotten. But in October, the unthinkable happened. The INS decided that they wanted to deport Tsien under the Subversive Control Act of 1950—on grounds that Tsien had been, prior to his last entry in the United States, a member of the Communist Party.

19

Hearings (1950–1951)

It was a bizarre situation. The U.S. government, which months earlier had deemed Tsien too dangerous to send back to China, now ordered his deportation. And Tsien, the Chinese alien who had tried desperately to leave then, was now fighting for the privilege to stay.

An explanation given by the INS years later claims that there were two separate government policies working against Tsien. One was the 1918 Anarchist Act, which had been revised under the Internal Security Act of 1950. It had been created by Congress to expel those aliens who might subvert the United States political system. It was under this law that the INS hoped to deport Tsien. But at the same time, the Department of State was charged with preventing the departure of aliens whose technical training might be used by an enemy nation to undermine military defense. Tsien's experience in jet propulsion no doubt placed him in this category.

"Obviously, Dr. Tsien was caught between two contradictory policies," wrote Marian Smith, INS historian. "Other subversives of the time were ordered deported, but the majority of them were not Chinese scientists. Similarly, other Chinese scientists were prevented from departing, but most of them were not already under deportation orders."

—๑

The impending deportation hearings sent Caltech officials into a renewed flurry of activity. DuBridge wrote to Norman Chandler, publisher of the *Los Angeles Times*, in hopes that the paper would run an editorial in favor of Tsien. Clark Millikan wrote to Harry Guggenheim of the Daniel and Florence Guggenheim Foundation about what had befallen their Robert Goddard Professor. Meanwhile, Theodore von Kármán and other top aerodynamicists wrote letters to the government attesting to Tsien's loyalty and integrity. Caltech arranged for attorney Grant B. Cooper to provide defense for Tsien during the hearings.

Why did Tsien prepare to fight the deportation proceedings, when his original intention was to leave for China? Perhaps he planned to return to the United States one day, after his father passed away. Or maybe it was nothing more than an attempt to clear his good name, which by now was besmirched with accusations of being part of the Communist Party and lying about party membership during security investigations.

In early November, it seemed as if things were going to be all right after all. U.S. Attorney Ernest Tolin told the press that after a two-month examination of the papers it had been determined that Dr. Tsien was guilty of no crime. Tsien had violated the Export Control Act by trying to ship technical material overseas without informing Customs, but the government believed he had acted in "good faith." Therefore, he would be cleared of all charges related to the shipping of his luggage.

The hearings that would determine whether Tsien would be allowed to stay in the United States opened at 10:00 A.M. on November 15, 1950, in a small room in a building on 117 West Ninth Street in downtown Los Angeles. It was a typical government-issue room: pale green walls, venetian blinds, brown checkered linoleum floor. Tsien and his attorney arrived, immaculately dressed in suits and ties. Albert Del Guercio, the examining officer for the INS, presided over the hearings.

Unlike trials in a court of law, deportation hearings are heavily slanted against the alien from the start. The rights of a foreign national facing deportation are far less extensive than those of a defendant (alien or citizen) in a criminal case. Without U.S. citizenship, an alien in an administrative INS hearing can be presumed guilty until proven innocent; the burden of proof falls on the alien. Any kind of rumor, gossip, or innuendo can be introduced in the course of the hearings as evidence.

The hearings began with a detailed investigation into Tsien's family background, education, and career. Then two retired police officers were called to the stand. One was William Hynes, who had served as chief of the intelligence bureau of the Los Angeles Police Department in 1938. Another was William Ward Kimple, another member of the LAPD intelligence unit who had infiltrated the Communist Party and served as assistant to the director of the Los Angeles membership commission within the party. The police submitted as evidence a copy of what they claimed was Tsien's 1938–39 Communist Party card.

The retired policemen explained the procedure by which they had pilfered the card. Kimple's duties within the party were to keep track of people transferring in and out of different sections. Once a year the party would register members, issue new books, and check on payments due. Kimple said he would pick up the membership records from the home of the Resnicks, a couple in Boyle Heights, and take them to the home of William Hynes, where he and Hynes would make handwritten copies of the cards. Then he would deliver the cards to the local Communist Party headquarters.

One day in December 1938, the policemen claimed, they saw Tsien's registration card. Kimple recalled that they were curious about this card because it bore a Chinese name next to the alias "John Decker." Hynes was said to have commented at the time: "I wonder why he took the name Decker for a party name. I wonder if he knows Comrade Decker." At the time, they knew of a woman named Inez Decker who belonged to another Communist cell in Los Angeles. The policemen then made a copy of the card in their own handwriting, which they now submitted as evidence. The copy contained the notation "1938 Book #NM" and "NB," which the policemen said stood for "new member in 1938" and "no book."

The allegations from the police sounded serious, but in actuality they carried very little weight. They presented no direct evidence that Tsien was ever a member of the Communist Party. First of all, the card they possessed was not in Tsien's handwriting but a copy made in the infiltrator's handwriting. No one, in fact, had ever captured Tsien's handwriting on a card, a membership book, or even a piece of Communist stationery. Second, there was no proof that the cards the police had copied were membership lists at all, and not a list of potential recruits. Finally, and perhaps even more strangely, when the Resnicks were summoned to the hearings they did not recognize the two policemen, nor did the policemen recognize the Resnicks.

Tsien tried to explain to the INS the nature of his relationship with Weinbaum. In the late 1930s Tsien would drop in at the homes of his friends Malina and Weinbaum without calling them. Sometimes, Tsien recalled, there would be other people there whom Tsien did not know but assumed were other friends of Weinbaum's from Caltech. In retrospect, Tsien acknowledged that these could have been Communist meetings, but he had no way of knowing that. "They always argued," Tsien remembered. "Very often my opinions were solicited because I was Chinese." Yes, there were heated political discussions, but Tsien thought that was part and parcel of the college experience.

The hearings focused as much on Tsien's loyalties to the U.S. government as they did on his alleged involvement in the Communist Party. In one dramatic instance Del Guercio asked: "In the event of conflict between the United States and Communist China, would you fight for the United States?"

Tsien paused for a long time. Cooper argued that it could take him six months to answer the question, to which Del Guercio said sarcastically that he would wait six months. Finally, Tsien said: "My essential allegiance is to the people of China. If a war were to start between the United States and China, and if the United States war aim was for the good of the Chinese people, and I think it will be, then, of course, I will fight on the side of the United States."

Tsien repeatedly emphasized that he was not a Communist, in either belief or action. When asked whether he preferred the GMT or the Communist regime in China, Tsien responded that he owed his allegiance to neither government but to the people of China. As for his feelings towards Marxist-Leninist thought, Tsien said that one reason he went to Weinbaum's parties was to learn more about the philosophy. When asked, "Were you unfavorable to Marxism and Leninism at that time?" Tsien responded: "I am unfavorable to them, definitely, now. At that time I was still in the process of finding them out."

At one point the INS showed Tsien a photograph of a man named Richard Lewis, who was a former member of Communist Professional Unit 122—the one Tsien was supposed to have belonged to—and by the 1950s a professor of chemistry at the University of Delaware. Tsien claimed he did not recognize the name or the photograph but that Lewis looked vaguely familiar. In January 1951, INS officials began to question Richard Lewis in Philadelphia. At first Lewis refused to testify on grounds that he might incriminate himself. But he later changed his mind when the government indicted him for perjury for neglecting to mention his Communist Party membership on a security questionnaire four years earlier.

Undoubtedly confused and distraught, Lewis confided his fears about the Tsien case to a friend at Caltech. On January 8, he wrote: "If I am required to answer questions about Tsien's Communist Party membership I shall have to state that he did attend meetings regularly over the same period that I did, and that I considered him to be a member. Looking back now it seems possible that since he was an alien he might have been allowed to attend meetings without actually being a member. I shall never be able to state definitely that he was."

The following month, the deportation hearings reopened in Los Angeles. The INS paraded in several more former members of the Pasadena Communist Party, but not one of them gave any truly damaging information about Tsien. The treasurer of the cell, Caltech biologist Jacob Dubnoff, said he could not remember collecting dues from Tsien. Another witness revealed just how easily one could end up on a list of potential members when he commented that the Communist Party considered recruiting not only Tsien but also aeronautics professor Clark Millikan, who was widely known at Caltech as one of the most right-wing professors on campus. "It was so ludicrous," one former Caltech student recalled, "that I could not imagine anyone in his right mind listening to any more of that." The hearings then closed, presumably for good.

But to the surprise of all, hearings reopened in April. This time Richard Lewis was there. Among the testimony of all the witnesses, his was the most harmful. He said that he saw Tsien at a party meeting and that he believed that Tsien was a member. Still, he acknowledged that he had no proof of Tsien's membership and all of this was speculation. According to a letter from Frank Marble to Kármán, Lewis later told his friend Carl Niemann at Caltech that he had been under tremendous pressure from the INS to deliver evidence against Tsien. (Lewis has neither confirmed nor denied this story because the author was unable to locate him by the time this book went to press.)

On April 26, 1951, the INS reached a decision. They determined that Tsien was "an alien who was a member of the Communist Party of the United States" and was therefore subject to deportation.

20

Waiting (1951–1954)

Grant Cooper was not pleased. Convinced of Tsien's innocence, he was determined to fight the deportation order through appeal. An oral hearing was scheduled on September 17, 1951, in Washington, at which Cooper planned to argue his case. As Caltech officials mustered their forces, all Tsien could do was wait.

Tsien now had a sharply circumscribed life. He was cut off, of course, from all classified work. This created complications for some of the engineers at JPL who wished to discuss with Tsien certain concepts mentioned in one of his papers. Thomas Adamson, then one of Tsien's students and a coauthor with Tsien of the paper, "Automatic Navigation of a Long Range Rocket Vehicle," recalls that he had to field some general questions himself because the JPL engineers were not allowed to talk with Tsien. Likewise, Tsien refused to talk to Adamson about any aspect of JPL work on the Corporal or Sergeant missiles for fearing of getting Adamson in trouble. "He thought it was best just to divorce himself completely from it," Adamson remembered.

But that was not all. Tsien was also forbidden to travel outside the boundaries of Los Angeles, which barred him from attending numerous scientific conferences or even from going to the beach in Orange County. Once a month, he

Hangzhou in the 1910s. *(photo: The Sidney D. Gamble Foundation for China Studies)*

A recent photo of what used to be Tsien's family temple in Hangzhou, now a museum.

A recent photo of Tsien's childhood home in Hangzhou.

The Bao Su tower, constructed by one of Tsien's ancestors and a landmark of Hangzhou.

A group photograph of the teachers at one of the Beijing experimental primary schools in 1921. (*photo: Beijing Experimental No. 1 Primary School*)

Beijing No. 2 Experimental Primary School, where Tsien was a star pupil. The school, designed for gifted children, admitted only thjose boys and girls who could pass a rigorous screening process.

A recent photograph of what used to be Beijing No. 1 Experimental Primary School, then a school open to gifted boys for the remainder of their elementary school education if they could pass a competitive examination.

Tsien's dormitory at Jiaotong University in Shanghai.

Formerly the library at Jiaotong University, where Tsien studied railway engineering as an undergraduate student.

Tsien's graduation from Jiaotong University, 1934. (*photo: Jiaotong University Archives*)

Tsien and other Boxer Rebellion scholars aboard the steamship *Jackson* arrive in Seattle, September 1935. (*photo: Kai-loo Huang, Qinghua University Archives*)

Rocketeers lounging before a rocket engine test stand at the Arroyo Seco in Pasadena, California, in fall 1936. *left to right:* Rudolph Schott, a student assistant working for GALCIT and paid by National Youth Administration funds; Apollo Milton Olin Smith, also an NYA-funded student, wearing a pith helmet of his own design with a wind-propelled ventilation fan on top; Frank J. Malina; Edward S. Forman; and John Parsons. Smith, Malina, Forman, Parsons, and Tsien and Weld Arnold *(not pictured)* would eventually form the "Suicide Squad," the six-man team that began rocket research at Caltech. *(photo: NASA/JPL)*

Apollo Milton Olin Smith.
(*photo: Collection of Apollo Milton Olin Smith*)

Frank Malina.
(*photo: NASA/JPL*)

John Parsons. *(photo: NASA/JPL)*

Edward Forman. *(photo: NASA/JPL)*

The first buildings of GALCIT Project No. 1, built in Arroyo Seco in Pasadena, California, in 1941. *(photo: NASA/JPL)*

The Jet Propulsion Laboratory, 1992. *(photo: NASA/JPL)*

Tsien (*kneeling at left*) with Theodore von Kármán, Kármán's sister Pipö, and other Asian associates at Caltech. (*photo: Courtesy of the Archives, California Institute of Technology*)

A 1940 cartoon by Frank Malina depicting members of GALCIT. Tsien is shown as indecisive about whether to stay in the United States as he flips an egg marked "US" on one end and "China" on the other. (*photo: Courtesy of the Archives, California Institute of Technology*)

Theodore von Kármán, head of the Guggenheim Aeronautical Laboratory of the California Institute of Technology and mentor of Tsien Hsue-shen. (photo: Collection of Admiral William T. Rassieur)

Tsien at Yosemite, 1941. (photo: Liljan Malina Wunderman)

Tsien (center) with friends in Pasadena, 1943. Zhou Peiyuan, a physics professor at Caltech, is third from right. (photo: Family of Zhou Peiyuan)

Tsien *(second from right)* at Leech Springs, December 10, 1944. *(photo: NASA/JPL)*

A launching of the Private A, the first missile in the United States with a solid propellant to perform successfully. *(photo: NASA/JPL)*

The Private F, April 11, 1945. This missile failed because of the lack of a good rocket guidance system. *(photo: NASA/JPL)*

Malina with WAC Corporal, a liquid-propellant sounding rocket and the first object to escape the earth's atmosphere. *(photo: NASA/JPL)*

At the end of World War II, Tsien participated in a secret U.S. technical mission to interrogate top German scientists for aerodynamics information. Tsien *(right)* poses with Hugh Dryden *(left)*, Ludwig Prandt *(second from left)*, and Theodore von Kármán *(second from right)* on the steps of the Kaiser Wilhelm Institute in Göttingen, occupied Germany, May 14, 1945. *(photo: National Air and Space Museum [NASM Videodisc No. 2B-27530], Smithsonian Institution)*

Tsien with Prandt and Kármán in Göttingen, 1945. *(photo: From* The Wind and Beyond *by Lee Edson. Copyright © 1967 by Little, Brown and Company [Inc.]. By permission of Little, Brown and Comany.)*

Tsien (*right*) watching the interrogation of Adolf Busemann (*in dark suit*), the famous German designer of sweptback wings, Braunschweig, Germany, 1945. (*photo: From* The Wind and Beyond *by Lee Edson. Copyright © 1967 by Little, Brown and Company [Inc.]. By permission of Little, Brown and Comany.*)

Tsien, the newly arived professor, with colleagues at Caltech. *left to right:* Dr. W. Duncan Rannie, Dr. Howard Seifert, Dr. Tsien, and Dr. Frank Marble. This picture was probbaly taken at JPL in Pasadena, California, October 27, 1949. (*photo: NASA/JPL*)

Tsien demonstrating the flight of a theoretical jet in class at Caltech, 1949 or 1950. (*photo: Hearst Newspaper Collection, Special Collections, University of Southern California Library*)

was required to appear before the INS office in Los Angeles and report to the appropriate authorities. There was a cafe near the INS office, and sometimes Tsien consoled himself by buying some of his favorite coffee.

For the next few months, work, his oldest friend, once again became his steady companion. Frank Marble wrote to Kármán to tell him that Tsien was resigned to the fact that his case would not be resolved soon, and that while he was not really happy, he now had more peace of mind. Tsien tackled a wide variety of topics: linear systems with time lag, transfer functions of rocket nozzles, automatic guidance of long-range rocket vehicles, properties of pure liquids, takeoff from satellite orbit, a similarity law for stressing rapidly heated thin-walled cylinders. He turned out a scientific paper once a month for four consecutive months—an outstanding achievement in his field.

The future of space flight consumed his thoughts. On May 2, 1952, Tsien wrote a letter to Kármán in which he foresaw a day when people would travel in rocket passenger ships. The greatest problem, Tsien wrote, would be control of these rockets. A human operator could not be trusted to navigate at high speeds, but a computer could make instantaneous changes in the rocket's direction whenever atmospheric disturbances put it slightly off track. Indeed, Tsien was convinced that computers would soon lead to a revolution in engineering and industrial efficiency.

But while he focused on these problems, Tsien couldn't shake the feeling of being constantly watched. He later testified that government agents followed him in the street, opened his mail, and broke into his office and home. Different addresses on Tsien's correspondence during the 1950s indicate that he moved several times; Tsien himself later said that he moved his family to four different locations in order to avoid government surveillance. According to the book *The China Cloud*, a dark sedan could be seen parked outside Tsien's home for hours. Eventually, Tsien moved a couch inside the bathroom so he could work there without being observed.

The harassment continued over the phone lines. Friends who called him were later interrogated by the FBI. At least one of them received a strange call from a government agent warning him not to dial Tsien's number again. And yet Tsien's phone would ring repeatedly, sometimes ten or more times a day. The caller would hang up the moment someone answered. Tsien speculated that it was the FBI's way of checking to see if he was truly at home. One of his son's earliest memories was of his father angrily taking the phone off the hook.

Ying Tsien, his wife, was another victim of this surveillance. The woman who could have become a famous diva in Italy was now an American housewife, married to a man with a mark on his honor. And she, too, was followed about by government agents whenever she left the house. When friends drove her around Pasadena she sometimes flattened herself on the floor of the car so the FBI wouldn't see her. She was afraid to hire a housekeeper and thus risk employing an FBI plant. "We were," recalled Ying years later, "very nervous in those years."

Tsien must have known that this kind of life could go on for years: more investigations, more hearings, more monthly visits to the INS. He could not know how long it would last. He and his wife, however, were exceptionally good at concealing their emotions during this period. When visitors came the Tsiens often acted as if nothing was wrong. Few dared bring up the subject of his impending deportation, and Tsien rarely talked about it himself. In the evenings when guests were not present, Tsien and his wife found ways to keep themselves entertained. Ying continued to practice her music and sing regularly at home. Sometimes, they listened to classical music recordings—the symphonies of Beethoven and Mozart were their favorites—reflecting, perhaps, on the better times they had shared.

Tsien's initial efforts to fight the deportation order were unsuccessful. In February 1952, INS authorities rejected Grant B. Cooper's argument that the hearing officer had improperly accepted evidence to deport Tsien. They believed that the card the Los Angeles police had copied was authentic. They also believed that Tsien's behavior in the 1930s was consistent with that of a party member: he had, after all, tried to leave the country with documents that could endanger the safety of the United States, and acted evasively when questioned about what he intended to do with these scientific papers. Most significantly, the INS argued that although Tsien said he did not intend to remain permanently in China, he had failed to apply for a reentry permit that would have enabled him to return to the United States.

Lee DuBridge was miffed by the decision. On February 25, 1952, he wrote to Grant B. Cooper: "The conclusions are not based upon the assumption that a man will be declared guilty only if there is no reasonable doubt of his guilt. Rather the assumption seems to be that the man is guilty if there is any reason at all to doubt his innocence." He wanted Cooper to proceed with the appeal and emphasize that the documents of Tsien's that were thought to be classified turned out not to be classified at all. On March 6, Cooper sent word to DuBridge that he had appealed again to the INS and that if his appeal was rejected he would write to the attorney general. He prepared to argue Tsien's case in Washington in May 1952.

But these efforts proved to be futile. In November 1952, Tsien's last appeal was denied. The following year, Cooper warned Tsien that he could be picked up at any time and placed in custody. Meanwhile, DuBridge wrote to Assistant Attorney General Stanley Barnes and asked him to recommend people in the Department of Justice who could possibly help Tsien. DuBridge stated his willingness to go to Washington to visit these people personally. Wrote DuBridge: "I am personally convinced that he was never a member of the party; that he has loyally served his host country and that through all this injustice to him the United States is losing the invaluable scientific services of an unusually gifted scientist."

The response DuBridge received from Barnes was not encouraging. Barnes wrote back to say that he checked with Argyle Mackey, commissioner of the INS, who told him that all possible legal remedies had already been pursued on Tsien's behalf. DuBridge also learned through the grapevine that Tsien, growing embittered by the whole situation, was unlikely to follow through with any more action involving the U.S. government. At that point, others at Caltech seemed prepared to simply give up the fight. Suggested one professor: "Perhaps we should relax on the whole business."

With no other choice but to live one day at a time and to keep intellectually alive, Tsien continued to work and teach and await the next step, whatever it might be, whenever it might be. He turned to other fields of research, such as the study of games and economic behavior. In 1954, he published a textbook entitled *Engineering Cybernetics*, a book on systems of communication and control. It too would be well received.

Years later, Wallace Vander Velde, an MIT professor and renowned expert in cybernetics, would describe the book as "remarkable" and "an extraordinary achievement in its time." Wrote Vander Velde of the book:

In 1954, a decent theory of feedback control for linear, time invariant systems existed and servomechanism design was an established practice. But Tsien was looking ahead to more complex control and guidance problems—notably the guidance of rocket-propelled vehicles. This stimulated his interest in the systems with time-varying coefficients, time lag and nonlinear behavior. All these topics are treated in this book.

But Tsien went further to deal with optimal control via the variational calculus, optimalizing control and fault-tolerant control systems among other topics! He visualized a theory of guidance and control which would be distinct from, and would support, the practice of these disciplines. This has certainly come to be, and his pioneering effort may be thought of as a major foundation stone of that effort which continues to this day.

Tsien's mental energy continued to dazzle the students, many of whom had no idea what he was going through privately. One time, when working out a lengthy proof on the blackboard, Tsien was interrupted by a student who asked him about another difficult question on a problem unrelated to the topic at hand. Tsien ignored the student at first, continuing to write out the equations until he had filled four blackboards that were each ten feet long and four feet high. "Just keeping that much material in his head was awesome," remembered Frederic Hartwig. "But we were awe-struck when he turned around and gave the answer to the very difficult question which the student had asked. How he managed to solve two gigantic problems simultaneously while putting the answer to one of them on the blackboard I'll never know."

Strangely, despite his own difficulties, or maybe because of them, Tsien became a kinder teacher and mentor. During this period, he was remembered by some students as being unfailingly courteous and supportive when they came in during office hours to ask him questions. "I always enjoyed talking to him," remembered Robert Meghreblian, who was Tsien's first Guggenheim fellow and the first to receive a doctorate under Tsien at Caltech. "Sometimes I would just drop in to say hello. I'd sit in the office—on the windowsill—and he'd pace back and forth, chatting about this subject and that subject. I found him extremely affectionate and congenial and I thought of him as a friend." While Meghreblian acknowledged that Tsien might have been a bad teacher at MIT, "he had certainly evolved far beyond that" by the time he came back to Caltech. Tsien was feared and respected but "certainly not disliked."

Sometimes, rather than simply give a student the answer to a problem, Tsien would ask him crucial questions that would make the student aware of the things he should be investigating and thereby steer him in the right direction. Tsien even changed lives. For example, he sought permission from the Navy for Carl Holmquist, then a promising master's degree student, to write his doctoral thesis—a move that permitted Holmquist to eventually achieve his dream of becoming chief of naval research in 1970.

But as the years drifted by, none bringing any relief, Tsien began to withdraw from the Caltech community. Some faculty members attribute his withdrawal to the actions of other faculty, who began to shun him because they feared that they, too, might become targets of investigation if they became too close to him. Tsien's graduate students remember that his visits on campus became less frequent and that sometimes he practically disappeared. (One of his students, Yusuf Yoler, observed that Tsien appeared "increasingly impatient and irritable" and he wisely switched his thesis topic to study under another professor at Caltech.) By 1954, Tsien seemed more withdrawn and moody than ever. When passing former friends and associates in the hall, he frequently ignored them. People grew concerned for him. "He looked worn out and under stress," remembered Caltech alumnus Franklin Diederich. "I had no idea of his problems with the government and thought he was ill." Theodore von Kármán, who saw Tsien on occasional visits to Pasadena during this time, later wrote: "At times, I actually feared for his mind."

Tsien was not alone in his frustration. His story must be put within the context of a time that we now describe as the McCarthy era. It began when Senator Joseph McCarthy, eager to boost his chances for reelection, gave a speech in Wheeling, West Virginia, in February 1950, in which he condemned the Department of State as being infested with 105 Communists and that he held in his hand a list of their names. This accusation hit the American public at a time when it was already of the verge of panic.

There were a series of world events in 1949 and 1950 that made Americans feel as if they were in the midst of a great Communist conspiracy. In September 1949, the Soviet Union exploded its first atomic bomb, ending the American monopoly on nuclear weapons. Two months later, China fell to the Communists and Alger Hiss, a former State Department official, was charged with being a Communist spy. Three months after that, the British announced that one of their top scientists, Dr. Klaus Fuchs, had given to the Soviets the secrets of the atomic bomb. The United States plunged into a hysterical witch-hunt for Communists that destroyed hundreds of careers in government, academia, and industry.

Government agencies such as the Industrial Employment Review Board began to closely investigate the behavior of military, industrial, and university scientists as they worked on defense contracts. Although exact figures are not available, many scientists who were found to exhibit left-wing tendencies lost not only their security clearances but their jobs. So prevalent was the atmosphere of paranoia that even those not engaged in sensitive research projects suffered. For instance, the Atomic Energy Commission withdrew a sixteen hundred dollar fellowship from a scientist who, they discovered, was a member of the Communist Party, even though he was not working on anything classified.

If the "evidence" that condemned Tsien to deportation was flimsy, what was used to attack other scientists was flimsier still. For instance, Robert Jones, an acquaintance of Tsien's who had developed the famous swept-back wing theory, came under suspicion as a possible Communist or Soviet agent because neighbors accused him of playing Russian language records continuously in his home. Jones, baffled because he had no Russian records besides those of Tchaikovsky, suddenly realized they were referring to the time he had rebuilt his tape recorder so that it would play backwards. (He was eventually cleared of all charges.) Scientists soon realized that mere association with those perceived as left-wing could make one a target. James Bonner, a distinguished professor of biology at Caltech, found himself accused of being a Communist when he mentioned to a fellow rock climber in the Sierra Club that he had known Tsien in the 1930s. (The rock climber was Roy Gorin, the agent who had seized Tsien's luggage, and Gorin relayed his accusation of Bonner's alleged Communist involvement to the FBI.) Bonner, who denied being a Communist, spent two days presenting his case before the FBI and thereafter refused to apply for a security clearance for fear of another false investigation. (Bonner was later given a top-secret security clearance without having to apply for one, however, after being invited to join a presidential science advisory committee.)

Caltech was particularly vulnerable to McCarthyism because it was a university where many foreigners and liberals worked together on high-level defense work. Worse, there was evidence that someone on campus might have leaked information to the Soviets. During the war, the U.S. government had intercepted a Soviet intelligence carrier in Paris who had in his possession papers that originated from the Caltech aeronautics department. Some of them were technical documents that Homer Joe Stewart, a colleague of Tsien's, had written. Hence Stewart began to suspect that there was a spy at Caltech—someone who was near or within the department of aeronautics. "That's the horrifying part," Stewart later said. "That someone close to you might not have been who you had always assumed he was."

While no one has ever been able to confirm or disprove the existence of a spy at Caltech, one thing is certain: the FBI crackdown on Sidney Weinbaum's Communist cell ruined the careers of several promising scientists. Frank

Oppenheimer, brother of the famous J. Robert Oppenheimer, lost his assistant professorship in physics at the University of Minnesota and ended up becoming a cattle rancher in Colorado. (Later in life he would teach high school physics and open a children's science museum in San Francisco, the Exploratorium.) Frank Malina left JPL because he claims he was sick of war work; his first wife, however, remembers he left the United States chiefly to avoid government investigation of his involvement in the Communist Party. (Details of his sudden departure from the United States and activities in the Communist party are still not clear and may be illuminated when his file is released by the FBI.) Sidney Weinbaum fared the worst. Charged with perjury in 1950 for lying on a job application about his Communist membership, he ended up in prison, during which time his wife, unable to handle the pressure, went insane. Weinbaum spent the rest of his life working at a girl's clothing factory and at other menial jobs.

The government investigation of Unit 122 affected not only the individual lives of scientists but entire families. For instance, Weinbaum's daughter, Selina Bendix, remembers vividly that the FBI repeatedly followed her and other family members to the local ice cream parlor in the 1950s because the agents were convinced that the flavors the Weinbaums picked were part of some secret code. Ena Dubnoff, the daughter of John and Belle Dubnoff, remembered that her family suffered socially and professionally for testifying at the Tsien and Weinbaum affair. As a child, Ena did not understand why friends and neighbors shunned them, why teachers gave her strange looks at school, and why her mother, herself a teacher, was unable to get a job in the public school system. Belle Dubnoff ended up starting her own school for delinquent girls and retarded children, but the more prominent she became in her field, the more frequently her past involvement in Unit 122 would resurface in the newspapers. "Every two years, someone would bring it up and there would be some newspaper article," Ena Dubnoff recalled. "It followed her throughout her life. And every time it did come up it was terribly traumatic for her." The elder Dubnoffs tried to protect their children by refusing to discuss the past. But the mystery and tension arising from a multitude of unanswered questions would haunt the Dubnoff children for the rest of their lives. Even when Belle Dubnoff was on her deathbed and Ena begged her to tell her the full story, her mother refused. "She told me: if anyone asks you what happened, I want you to say you didn't know—and mean it."

The investigation would also cast a shadow on Frank Malina's first wife, Liljan, for years. Shortly after she moved from Pasadena to New York and obtained a divorce from Malina, Liljan noticed that she was being followed in the streets. An appointment book and some papers and books disappeared from her car and her mail was routinely opened. A second marriage and motherhood did not make her immune to further investigation. During the summer of 1951, four government agents arrived in an official-looking black car to question her at her home in Glencoe. They demanded information on Tsien, Malina, and other people she had known at Caltech and threatened to give her second husband the names of her former lovers if she refused to cooperate. After she claimed she didn't know anything, the agents interrogated her husband in his office. "They *stormed* in there," Wunderman said. "This was their tactic at the time—to go in and frighten people to death. They figured if they jeopardized my husband's job he would tell them something I had told him. But he was very cool and said, she's told me everything, I know everything about her, I don't know what you want me to do, we're just living out our little lives out in Glencoe! There's nothing we can tell you." That was the last encounter the Wundermans had with the FBI.

It is impossible to say how many scientists left their careers voluntarily during the 1950s because of the crackdown on Communists. Martin Summerfield, for instance, lost his security clearance as a result of his affiliation with the Communist Party and considered switching to a career where security considerations did not play such a great role. Even though he held a prestigious professorship of aeronautical engineering at Princeton University, Summerfield offered to resign. Princeton urged him to stay, but it is unclear how many other scientists had grown disillusioned and left the field altogether.

Of course, scientists were not the only victims. As soon as the Korean War began, the INS and FBI joined forces to root out possible left-wing subversive activities in the Chinese community. Under the approval of FBI Director J. Edgar Hoover, federal agents began tapping phones in Chinatown, subscribing to Chinese newspapers, and monitoring potentially Communist organizations like the Chinese Hand Laundry Association and the Chinese Workers' Mutual Aid Association. They also investigated immigration fraud, as the government feared that some Communist Chinese may have entered the country illegally. Local Chinatown leaders worked out a "confession" program with federal authorities so that many Chinese could work out a new status, but the government often used these "confessions" as an opportunity to get residents to inform on others who they believed were Communists.

Inevitably, attention turned to the scientists who were Chinese nationals studying at American universities. In 1951, there were some thirty-six hundred students from China who were largely doing graduate work at state universities at their own expense. After the Communist revolution and the outbreak of the Korean War, many of these students were stranded in the United States, cut off from funds from their families. University and government funds kept the students afloat, but some taxpayers wrote editorials voicing their resentment of having to foot the bill for the education of Chinese students, especially if they planned to return to the mainland. Others expressed fear: How might the Chinese students' technical skills be put to use when they returned?

According to James Reston of the *New York Times*, the Korean War produced "a whole series of small interdepartmental wars in Washington" over the problem of what to do with the Chinese students. Washington seemed divided between deporting the students and forcing them to stay. In 1950 or 1951, the secretary of state used his authority to prevent the departure of certain Chinese nationals with scientific training. In June 1951, a law was passed to permit these students to take full-time jobs to support themselves, but the Justice Department balked at putting the law into effect. While the State Department urged Justice to act on the law, regional offices of the INS served Chinese students who took full-time employment with warrants of arrest and even warrants of deportation. The situation was further complicated by the fact that many of the students had joined university organizations branded as subversive by the attorney general's office, such as the Chinese Students Christian Organization and the Scientific Workers Association of Engineering and Chemistry, making the students subject to even more harassment from Justice officials.

The confusion in Washington manifested itself at Caltech. When trying to renew their visas, some Chinese students were first told not to leave the United States and then arrested by the INS on charges that they were in the country illegally. The student would have three or four hours to put up a thousand-dollar bond or he would be loaded on a bus with barred windows and taken to the nearest detention center. This happened to several Chinese students, one of whom was Zheng Zheming, a mechanical engineering graduate student at the time and coauthor of a paper with Tsien. Caltech had to pay the thousand dollars for at least two Chinese students who were in this situation.

In retrospect, it is not surprising that Tsien became a target of Cold War hysteria. The 1950s were a dangerous time to be a Communist sympathizer, and also to be a Chinese national or a scientist. If one can imagine these populations as three overlapping circles of a Venn diagram, Tsien would overlap on at least two of the circles and, according to the INS, at the intersection of the three.

By late 1954 there was still no news from the government as to what action they would finally take. If ever Tsien harbored any hopes of clearing his name and going on with work in the United States, perhaps even becoming a citizen, such hopes were buried by 1954 after years of fighting with the INS. The Tsiens kept three small suitcases packed at all times in preparation for the day they could leave.

But what if that day never came? The government had no obligation to deport him. Indeed, the more they were convinced he was a spy, the less reason they had to let him go. What if this purgatory in which he now lived was endless, and he was never incarcerated but never able to do his real work, a cloud of dishonor forever over his head? This was the fate he feared the most. And while the Caltech community, from DuBridge on down, may have tried their hardest to straighten things out, the fact was that they were going on with their lives and he was not.

On December 8, 1954, Tsien's frustration erupted in a letter to Frank Malina:

> Do you expect anyone in the Caltech administration will harm their future (at least they think they will) for the sake of truthfulness to history? Do you believe in history at all, knowing that it is rewritten all the time? Do you think there is justice and honesty in this part of the world? Do you expect to be famous and honored in the USA without being your own public relations man, or without having a public relations man under your employ? Dear friend, let us not believe in fictions! You are now in creative work, so why let such trivial matters bother you? After all, wouldn't it be nice to be able to tell one's conscience during one's last days that he has given more to humanity than he has received from humanity in return?
>
> PS: I am filing your letter with a copy of this letter for future historians.

But neither rage nor frustration nor patience brought any action. January 1955 turned into February, and February into March. Finally one day in June, Tsien and his family briefly eluded the FBI and ducked into a coffee shop. There, he jotted down a note on a piece of cardboard torn from a cigarette box: a message that expressed his wish that the new Communist government assist him in returning to China. He tucked the scrap of cardboard into a letter addressed to Ying's sister in Belgium in which he asked her to pass on the note to Cheng Shutong, a family friend in China. On the way out of the coffee shop,

Tsien quickly dropped the letter into a mailbox. Cheng Shutong, he hoped, would pass his request along to the right Chinese authorities.

A little less than two months later, in mid-August 1955, Tsien's fate would finally be decided and announced. Although the final decision would require the support of INS officials, they would not make the decision alone. Nor would the decision come out of high-level discussions in Los Angeles and Washington. Rather, after memos that would reach the highest level of the government, the office of the president, the decision would be publicly announced a thousand miles away from Washington, in Geneva, Switzerland, over a table set up to negotiate events totally unrelated to Tsien's life, his work, and the accusations against him.

The Wang-Johnson Talks
(1955)

August 1, 1955. Reporters milled about outside the Palais des Nations in Geneva, waiting for news of the historic events taking place within. The subject of their interemst was the negotiations being conducted in the President's Room. On one side of a highly polished table laden with crystal pitchers and glasses sat American ambassador U. Alexis Johnson and his three assistants: Ralph Clough and Edwin Martin, both Chinese language officers of the U.S. Foreign Service, and Colonel Robert Ekvall, a Chinese-language interpreter from the U.S. Army. On the other side was Chinese ambassador Wang Ping-nan and three of his aides. No recording devices, stenographers, or reporters would be permitted in this room; the discussions would remain secret until declassified more than thirty years later. This was the site of the famous Wang-Johnson talks: a series of high-level negotiations to discuss American and Chinese prisoners in both countries captured during the Korean War, which had ended in 1953.

Reaching this stage of negotiation had been difficult. With the Chinese Communist revolution of 1949, the new government had occupied American buildings, arrested consuls, and summoned the U.S. ambassador to court on charges of having abused his Chinese servants. Then, in February 1950, the People's Republic of China signed a thirty-year military alliance with the Soviet Union. This was followed by the Chinese attack on U.S. troops in Korea in November 1950, and two years of bitter fighting, during which prisoners were taken on both sides.

The prisoner-of-war situation had posed numerous problems for the United States during the Korean War. The Americans kept the North Korean and Chinese prisoners on Koje Island, a rocky, barren piece of land surrounded by barbed wire twenty miles southwest of Pusan. The camps had been poorly organized: thousands of prisoners were crowded into compounds designed to hold only one-fifth that number. Adding to the confusion was the fact that the POWs included former Chinese Nationalist Army soldiers who had been taken by Chinese Communists and impressed into Communist forces.

The camps soon split into Communist and anti-Communist factions. Because of the limited number of American guards present (only one for every 188 prisoners), some of the prisoners themselves were given the authority to enforce discipline. Naturally, these were anti-Communist POWs, who rose to positions of power within the camp. Stories soon emerged of their brutality toward the other prisoners. Before long, order deteriorated in the camps until rock slinging, riots, and wholesale killing occurred. In October 1952, on the third anniversary of the founding of the People's Republic of China, guards opened fire on prisoners waving red flags, killing some fifty men and wounding more than a hundred.

The war ended in 1953 with the signing of a truce. When Americans discussed the possibility of a prisoner swap, many Chinese and North Korean POWs expressed unwillingness to return to their home countries. The anti-Communist POWs on Koje Island faced almost certain court-martial, torture, and death if sent back to Communist regimes. Other POWs who had surrendered to United Nations forces after receiving certificates from the United States promising fair treatment feared retaliation if they returned. The Truman administration was torn between their desire to protect these Chinese and Korean POWs from harm and their first responsibility—to secure the release of American POWs. Finally, negotiations code-named Operation Big Switch resulted in the return in 1953 of 75,801 prisoners to North Korea and Communist China and 22,604 to India, where many of them made their way to Taiwan and South Korea. Meanwhile, the Communists handed over 3,326 POWs to the U.S. government.

The returned American prisoners told chilling stories of their confinement. During the early stages of the war, they had been forced to live outdoors or in mud huts in which men were so tightly packed they could not lie down. Most underwent Communist indoctrination: six to eight hours of lectures, interrogations, and self-criticism a day. Those who did not comply were tortured. Some were forced to march barefoot in the snow, others had to stand at attention for twenty-three hours straight, while others were confined in tiny five-by-three-by-two-foot boxes. Newspaper accounts of these atrocities further raised the level of hostility between the United States and China.

After the exchange, there were scores of Americans unaccounted for. Some 155 American civilian prisoners were listed as being detained in China and some 450 Americans were reported missing in action. In 1954 the Chinese, who yearned for admission to the United Nations, agreed to discuss the prisoner situation with the United States. This led to the first talks, in June 1954, between ambassadors Johnson and Wang at Geneva.

During the talks it became clear what each side wanted. The United States hoped for the return of the American prisoners in China, both military and civilian. The People's Republic of China (PRC), on the other hand, wanted the Chinese scientists residing in the United States, many of whom it felt were also prisoners—scientists like Tsien who originally came to the United States to study and were now prevented from leaving because of their technical training. The Chinese understood the value of these scientists in their plan to build up national defense, and during the negotiations Wang used the detention of the Chinese scientists as an excuse not to release the American prisoners. The talks were unproductive and ended after only four meetings.

But the PRC, which desperately wanted international recognition, knew that further talks with the United States would boost its status in the world community, according to Ambassador Johnson. The PRC then embarked on a course that seemed designed to goad the United States back into negotiating. In early 1955, after the United States recommitted itself to defending the Republic of China, now restricted to the island of Taiwan, and some surrounding islands, from attack with a mutual defense treaty, the Communist Chinese seized one of the smaller islands off Taiwan. Secretary of State Dulles soon faced an unpleasant choice: getting entangled in a war with mainland China or engaging the PRC in negotiations that would give the regime more international attention.

The U.S. response was to stall. Over the next few months, the U.S. government evaluated the backgrounds of about one hundred Chinese students and scholars with technical training who had been prevented from leaving the country. On April 1, 1955, Secretary of State Dulles submitted a memo to President Eisenhower suggesting that the release of these scholars would enable the nation to press its case more effectively for the return of the American POWs. That year more than half of the students were told they could leave, and the number kept under restraining orders dwindled to just a few dozen.

In April 1955, the Chinese again tried to resume talks by announcing at a conference at Bandung, Indonesia, that the Chinese did not want war with the United States. In a conciliatory move, they released four American airmen on May 30. On July 11, in order to better deflect international pressure to include China in a four-power summit, the United States informed China through Great Britain that they wanted to resume ambassadorial talks. The talks would focus on the goal of repatriating prisoners on both sides. The day before the talks opened, eleven more American airmen were released from China in a public relations move meant as a gesture of Chinese goodwill.

The talks proved to be a test of nerves. "Despite the background we all brought to Geneva, we felt we were entering a void where past experience had limited value," U. Alexis Johnson wrote later in his memoirs, *The Right Hand of Power*. "We had no real sense of how Peking wanted to resolve the prisoner issue. . . . China and the United States were facing each other across a chasm of ignorance and hostility." A *New York Times* correspondent would later comment: "For sheer endurance there has been no United States diplomatic performance comparable to Mr. Johnson's since Benjamin Franklin's efforts to get financial help from the French Monarchy for the American Revolution."

On the surface was a veneer of formal civility. The two ambassadors followed a strict ritual of negotiation. First each side would open with a statement, usually delivered from a prepared text and translated paragraph by paragraph. Then each side would alternate in giving rebuttals: monologues that sometimes proceeded at such a rapid clip they strained the ability of the interpreters to keep up. While Wang and Johnson spoke or contemplated their next move, their aides would take notes, whisper among themselves, or pass them notes giving advice.

During the first few meetings, Johnson handed Wang a list of the names of forty-one Americans who were being detained in China and asked that they be released immediately. Wang, in turn, demanded a list of the names and addresses of every Chinese in the United States and proposed that the Indian embassy look after their interests. Johnson refused, knowing that this would give the People's Republic of China, not the Republic of China in Taiwan, jurisdiction over the Chinese nationals in the United States. Worse, it could expose those nationals to harassment and intimidation.

On August 8, 1955, to the surprise of all, Ambassador Wang opened with a statement about Tsien Hsue-shen—the first and perhaps the only Chinese national in the United States mentioned specifically by name during the entire course of the talks. Wang announced that his government had received a letter from Tsien that indicated his desire to return to China. The letter, Wang asserted, was vivid proof that many Chinese scientists in the United States who wanted to return to the mainland were unable to do so. Again, Wang demanded a list of all Chinese nationals still residing in the United States.

As it turned out, the U.S. government had already spent considerable time in trying to decide what to do about Tsien. In June 1955, the secretary of defense presented President Eisenhower with a memo regarding the question of Chinese scientists educated in America who now wanted to return to China. It was determined that of the more than five thousand Chinese students who came to America since World War II, only 110 possessed technical knowledge that could endanger national security. Of the 110, the memo noted, all but two Chinese scientists had been given permission to leave the country. The Department of Defense remained dubious about these two Chinese scientists because of their backgrounds working with highly classified information. One was David Wang, who had worked on the Nike missile. The other was Tsien Hsue-shen.

The Department of Defense had doubts about the merits of letting Tsien go. "Dr. Tsien would, if he were released, take back with him high competence in his professional field, much background information on jet propulsion as applied to weapons, and unusual ability to interpret technological progress in the U.S.," the memo read. Defense officials acknowledged, however, that it was "quite probable that any classified information which he [Tsien] possessed at that time is by now outdated by later research and is common knowledge in the Soviet Bloc."

Ultimately, it was Eisenhower who would make the final decision about Tsien. (Years later, Lee DuBridge, who became the chief science advisor to President Nixon, said that Eisenhower had probably never heard of Tsien. DuBridge was wrong.) On June 12, 1955, a government memo from Dulles's secretary Mildred Asbjornson indicated that the President was in a "give them all back" mood. The next day, June 13, 1955, Eisenhower decided that Tsien and Wang would be released. By August 3, the Department of Defense had withdrawn all objections to the release of Wang and Tsien, and the U.S. government began making preparations to send Tsien back to China. In a letter dated August 4, 1955, the INS informed Tsien that he was free to leave. Johnson knew this as he sat at the negotiating table.

After numerous meetings between August and September 1955, the United States and China worked out a formal agreement to repatriate citizens on both sides. The government decided that Wang's demand for a list of all Chinese nationals would not be honored, but that any Chinese student who wished to return to the People's Republic could make the proper arrangements through the Indian embassy in Washington. As it later turned out, it would be the only formal agreement between the two countries until 1972, when President Richard Nixon and Premier Zhou Enlai signed the Shanghai Communiqué.

Once the news of Tsien's deportation was made public, there was much discussion in the press as to whether a swap had taken place. In September, while the People's Republic of China announced the release of twelve more American prisoners, the United States put Tsien's deportation order into effect. Newspapers speculated on the front pages that the entire thing had been a swap, which the State Department, naturally, officially denied. "I note the newspaper suggestion that Dr. Tsien was 'swapped' as part of a deal for the release of the eleven airmen," noted a State Department spokesperson in the *Los Angeles Times.* "It would be totally contrary to our principles to be swapping or exchanging individuals or groups of individuals. The United States has not been engaged and is not now engaged in swapping human beings."

But decades later, U. Alexis Johnson himself confirmed, during a telephone interview with the author, that Tsien's departure had been part of a trade. Tsien and other Chinese scientists, he said, had served as "prime swapping goods" for the Americans still in captivity in China.

As Tsien prepared to leave, there appeared to be some more maneuvering behind the scenes to keep him here. According to R. B. Pearce, who was given this information by Kármán, Tsien was told he could remain in the United States if he would testify before congressional committees. "But that was too much 'loss of face' for Tsien and he refused," Pearce wrote.

In the end, history will record, both sides got what they wanted. Decades later, when writing his memoirs, Johnson noted: "Of seventy-six American prisoners in China in August 1955, forty-one civilian and thirty-five military, all but thirteen had been returned by September 1957." The Chinese eventually got some ninety-four Chinese scientists educated in the United States. That would make up nearly half of the key scientists who would take China into the nuclear age. And years later, when Premier Zhou Enlai spoke triumphantly of the Wang-Johnson talks, he said: "We had won back Tsien Hsue-shen. That alone made the talks worthwhile."

"One of the Tragedies of This Century"

Tsien was observed in his office writing long letters in Chinese, or talking about plans for his life in China. "I remember Tsien talking about his imminent departure with regrets and, occasionally, with bravado, which is normal for almost any person in that position," remembered S. S. Penner, then an assistant professor in jet propulsion at Caltech. "He certainly had mixed feelings about going, and fundamentally he was not at all happy about going."

Almost everyone who knew him felt that Tsien's departure would be a tremendous loss to the country. And indeed, forty years later, the U.S. government would be faced with one of the supreme ironies of Cold War history. It had accused Tsien of being a member of the Communist Party when there was no clear evidence that he had ever been a member, and to punish him had deported him to Communist China . . . where Tsien was credited with revolutionizing the Chinese ballistic missile program. Who was accountable for this situation?

The largest share of the blame rests on the U.S. government at the time. Initially, it had every right to be suspicious of Tsien—to cancel his clearance, to seize his baggage. But further investigation should have exonerated Tsien when no concrete evidence, either documentary or testimonial, could be found about his alleged membership in the Communist Party. The INS never produced a list with Tsien's name on it or a card or anything on Communist Party stationery proving he was a member, and not a single witness among the many they brought in could testify with certainty that Tsien was a member of the Communist Party. In addition, an independent investigation conducted by the author revealed that it was unlikely that Tsien had ever joined the party. Liljan Malina Wunderman, the first wife of Frank Malina, does not remember seeing Tsien at a single official meeting of the cell. Betty Weinbaum, the second wife of Sidney Weinbaum, also insists that Tsien was never a member.

The government also failed to put Tsien's involvement with Weinbaum and his group within the context of the times. Remember, this was a very short period in Tsien's life, 1938–39, when he was unmarried and free of administrative duties and had the time to go to social events. Had Tsien known in the 1930s that his entire career would be in jeopardy fifteen years later because of his involvement in the cell, he probably would never have attended Weinbaum's parties. But at the time, Tsien was still on a student visa and did not even know that he had the option of staying in the United States for more than a few years. He had no idea that the war would escalate so rapidly worldwide that his skills would be sought by the highest levels of government. It is likely he never dreamed that he would be given top-secret clearance and later the most coveted professorship in aeronautics in the country, and that Communism, then a chic concept among young intellectuals during the Great Depression, would be condemned.

It is possible that Tsien didn't even know that Weinbaum was running a Communist cell. Some former Communists of Unit 122 said that the members of the party sometimes kept the nonmembers and potential recruits entirely ignorant of the situation. Andrew Fejer, one of Tsien's closest friends in the 1930s, told the author that both he and Tsien were duped into thinking Unit 122 was a music group and not a Communist group. Homer Joe Stewart speculated that someone might even have put Tsien's name on a membership list to play "a dirty trick on him." Whether he knew Unit 122 was Communist or not, Tsien, Liljan Malina Wunderman claimed, had been "royally railroaded" by the whole affair.

When Grant B. Cooper, Tsien's lawyer, was asked by *60 Minutes* if he thought Tsien was a Communist, he said:

While I don't rule out the possibility that he might have been, I don't think he was for several reasons. First, he had filed an application to be a citizen of the United States. Secondly, he was married to a charming Chinese lady who was the daughter of one of Chiang Kai-shek's generals. Hardly a Communist. Third, he made outstanding contributions to this government scientifically and through war work, and had been cited by the government for his outstanding scientific contributions. And fourth, and most importantly, was his sincerity. You know, when you face a man, face to face, you either believe him or you don't. Tsien, in my opinion, was telling me the truth.

Actually, the issue of whether Tsien was Communist or not should have had no bearing on whether he would be permitted to return to China. INS officers failed to see the irony of deporting a scientist accused of Communist leanings to a Communist country—especially when this scientist was a world-renowned expert in ballistic missile design.

The Defense and INS officials should have coordinated efforts toward the single goal of keeping Tsien here. To this day, it is still unclear exactly what the INS had hoped to accomplish by deporting Tsien. Even if Tsien were proven to be a Communist, or worse yet, a spy, interests for national security should have kept him in the United States permanently—in prison, if necessary. But bureaucracy and disorganization kept Tsien in limbo while government officials tried to decide what to do about him—a move that both embittered Tsien against the United States and made him an easy pawn for the trade that brought home more American POWs.

Did Caltech do everything in its power to help Tsien?

While there is no question that Lee DuBridge fought long and hard to try to keep Tsien at Caltech, it is unlikely that the rest of the aeronautics department—including Theodore von Kármán—made an equally heroic effort. If they had, they were ineffectual in getting the national press and the academic community to rally around Tsien in protest.

One thing that remains somewhat controversial is Theodore von Kármán's role in the Tsien affair. Half a century later, some academics believe that Kármán did not intercede enough on Tsien's behalf to keep him in the United States. Why, they ask, didn't Kármán write more letters from Europe to the U.S. government protesting the treatment of his star student? Why didn't he give angry interviews to the press in protest? Kármán, after all, knew some of the most powerful men in the Air Force and other government agencies. But the

U.S. military, as some people pointed out, was unusually important to Kármán, for it filled an emotional void in his life created by his lack of a wife and family. Is it possible, they wonder, that Kármán did not fight hard enough for Tsien—in order to not jeopardize his relationship with the military?

Years later, Kármán himself acknowledged that Tsien may have believed that his old professor did not do enough for him during this time of crisis. In his autobiography, *The Wind and Beyond*, Kármán wrote: "I think . . . Tsien thought that because of my strong ties with Washington I could have done more for him than I had done. The sad truth is that in this time of unreason one could do little once these situations started, even with the strongest of auspices." Michael Gorn, an Air Force historian who wrote *The Universal Man*, a biography of Kármán, claims there was probably nothing Kármán could have done for Tsien, as he was under attack himself. In 1951 the FBI had questioned Kármán about his role as minister under the Béla Kun regime in Hungary; later, in 1952, the Atomic Energy Commission refused to renew Kármán's security clearance until he submitted written statements testifying to his loyalty to the United States and explaining the political activities of Martin Summerfield, Frank Malina, Tsien, and even Kármán's sister, Pipö.

On several occasions, Kármán wrote letters attesting to Tsien's loyalty. In an affidavit to the AEC in 1954, Kármán said he had no reason to doubt the loyalty of Tsien, Malina, and Summerfield and considered questions regarding his sister's Communist sympathies "an inexcusable insult." Earlier, on November 14, 1950, Kármán also gave a notarized statement defending Tsien by stressing that Tsien was one of his best students and that he had high esteem for Tsien's integrity. "I had the impression that he was always conscious of the obligations connected with any work he took on," Kármán wrote, ". . . both in matters scientific and in matters of life he always said what he believed to be the truth."

There is some evidence that Kármán tried to help Tsien during the 1950s but that Tsien was reluctant to cooperate. For some reason, Tsien did not appear too enthusiastic about corresponding with Kármán about the difficulties he was having in Pasadena, forcing Kármán to get news about Tsien through other sources. In a letter from Kármán to Marble on October 22, 1951, he wrote: "How is Tsien? Anything happened in the matter?" A month later, on November 11, 1951, Kármán wrote again to Marble: "Maybe he can write me a letter sometime and tell me what he is doing." In addition, Rene Miller, Tsien's best friend at MIT, claims that Kármán was willing to do everything he could to help Tsien but that Tsien refused to let him.

"Tsien had broken communication and wouldn't cooperate with [Kármán's efforts] to help," Miller recalled. "Kármán was chair of the Air Force's advisory board and from that position could talk to anyone he wanted to. He told me he felt very badly that Tsien had cut himself off from us, but Tsien was very bitter."

Because Kármán is no longer alive to confirm or deny this, and because his personal records contain almost nothing on this subject, the truth about his relationship to Tsien during the McCarthy period cannot be known.

Other Caltech faculty probably balked at getting too involved in the Tsien case because they feared investigation themselves. One professor recalls that several colleagues in the aeronautics department shunned Tsien for precisely this reason. It reflects the perniciousness of McCarthyism at the time: guilt by mere association. People could see clearly what could happen when one came too close to a suspected Communist. It was, after all, Tsien's friendship with Weinbaum—and later his refusal to testify against him—that tainted his reputation in the eyes of the government.

Another possible reason for apathy concerning Tsien's plight was that few of his friends, acquaintances, and associates could agree on exactly where Tsien's true loyalties lay. Some people—both within and outside the department of aeronautics—were convinced that he wanted to remain at Caltech. Martin Summerfield, for instance, adamantly believed that Tsien wanted to stay, and Shao-wen Yuan, Tsien's former roommate at Caltech, insisted: "Tsien had no intention of going back to China—Period! Ever! There were no facilities there with which to do research! He went back to China because he was *forced* to go!" Others, however, were equally convinced that Tsien had always wanted to go back to China. "You don't ship 1,700 pounds of books if you want to go for a little ride," asserted Hans Liepmann, a Caltech aeronautics professor who had been both friend and rival to Tsien in the department. "I think he wanted to go home, and to a certain extent I think Tsien always had a mind to help China." One professor, Fritz Zwicky, pointed out that if Tsien was truly determined not to go back to China he could have made plans to resettle in Europe or some other part of the western world. In addition, Tsien's former student Chester Hasert recalled, "It was very clear that he felt a certain duty to go back to his homeland to help them recover from the war damage. I got the impression that he intended to eventually go home to China and to do what he could to help them back there. So when the big furor came up later on about his wanting to go back, I was not surprised that he was going back."

Ultimately, a great burden of responsibility rests on Tsien himself. Had he spent less time vacillating between the desire to stay at Caltech and the desire to return to China, perhaps fewer people would have doubted his loyalty to the United States. Had he been less abrasive and arrogant, the aeronautics community might have been more willing to rally behind him—organizing demonstrations, writing letters to the newspapers, circulating petitions in the kind of public outcry that a more popular professor such as Kármán would have inspired had he been at risk for deportation. "We used to talk about that in the early days," Thomas Adamson remembered, "wondering how many times people who were angry at him therefore didn't jump to his defense as quickly as they might have."

But the flaw that brought about Tsien's downfall in the United States was his pride. Tsien, after all, was not the only one accused of Communist membership at Caltech: a number of Chinese students from Caltech were detained by the INS, and other colleagues lost their clearances and even their livelihoods. Tsien was not in as precarious a situation as other scientists such as Weinbaum, who faced the humiliation of unemployment and the prospect of prison. Most likely, if Tsien had kept a low profile during the McCarthy era, he, like Summerfield, would have suffered a decade of lost clearance, retained his professorship at a top university, and reclaimed his clearance at a later date. But impulsively, Tsien decided to leave the United States—a decision that made him starkly conspicuous to American investigators, who saw this action as confirmation of their worst suspicions.

Packing up and trying to leave is consistent with Tsien's behavior in the past. He had a habit of reacting to unpleasant situations by refusing to deal with them at all. He left MIT twice after brief stays because the environment was not entirely to his liking. He never learned the techniques of compromise and negotiation, because pride made him incapable of it, and he was also completely inept at manipulating people politically. He was too brutally honest, too impatient, too impulsive, and too direct. When the FBI visited him in 1950, Tsien saw only two options: staying in the United States with his clearance and pride intact, or leaving the United States forever.

It didn't have to be that way. He could have waited things out, worked on nonclassified research, bided his time until McCarthyism was but a memory. If a single small link in the chain of events had fallen out, perhaps Tsien's story—and thereby world history—would have been different. What if the Bekins packing company waited to inform the Customs officials about Tsien's shipment, giving him more time to withdraw it when he returned from Washington? What if Kimball had never made his phone call to the Justice Department? What if Gorin had contacted Tsien first about the shipment of books and resolved the matter quietly instead of going to the media? What if the INS had a less zealous hearing officer than Albert Del Guercio and had opted to drop the whole matter? So many details of the story seem to be accidents of bad luck and circumstance.

Could something like this happen today? Theoretically, yes. According to Marian Smith, historian of the Immigration and Naturalization Service, the federal government has the power to prevent anyone (alien or citizen) from leaving the country for any reason deemed to serve national security. But while it is possible that a repetition of Tsien's case could occur, it isn't likely. His situation was unique. According to the INS, "Tsien was probably the only Chinese rocket scientist to be deported and prevented from departing at the same time."

Today we have a press that is much more adversarial toward the government. In an age in which people are eager to denounce the media as too aggressive, Tsien's case stands as a reminder of what can happen in a country with a docile media. The newspapers of the 1950s failed to do their own research on the Tsien story: no follow-up stories ever appeared about Tsien's life being shadowed by the INS and FBI; no in-depth interviews were done with Tsien at his home in Pasadena; there was no digging into Tsien's actual activities with Unit 122 or the possible technological consequences of his return to the PRC. Rather, in story after story, journalists merely quoted the government's assessment of the situation. The trust the press had in government was reflected in its coverage of the seizure of Tsien's papers. Los Angeles headlines announced that "secret data" had been discovered in Tsien's crates, but the follow-up story—that Tsien had not violated any security regulations—was buried in the back of the papers. For that reason, Wiliam Zisch said in the 1960s, "my mother-in-law to this day still believes he was escaping with the crown jewels."

Americans are also more sensitive to issues of race and ethnicity in the 1990s than they were in the 1950s. If a case such as Tsien's occurred today, Asian groups and civil libertarians would be lobbying the government, writing to their congressional representatives, debating the issue over the airwaves. It is likely, too, that the deportee would also be more sophisticated in his or her dealings with the government. "[H]igh profile cases can and do frustrate INS efforts to carry out lawful deportations because the deportee and his/her supporters are so successful in their use of modern lobbying techniques," wrote Smith, based on her discussions with a senior deportation officer. "Their advocates are able to marshal support using advertisements or direct mailings to specific markets or groups. With this support the subject can hire attorneys to use every delaying tactic and exhaust every appeal. Furthermore, they are able to prevail upon officials with discretionary authority (i.e., judges, Congressmen, the Attorney General) to stay the deportation."

But during Tsien's twenty-year stay in the United States, there was more naivete and faith in government as well as blatant discrimination, both socially and legally, against Asians. The Chinese could and would be routinely turned away from restaurants, barbershops, and hotels; their children were often segregated in the public schools; they and their children might be segregated in movie theatres; intermarriage between whites and Chinese was illegal in thirty states. Hollywood picked up on existing stereotypes and depicted Asians either as Charlie Chan types, after the apologetic little detective who spoke comically broken English, or the more sinister Dr. Fu Manchu, depicted with long, claw-like fingernails, a pointed beard, and a hissing voice. This, combined with general political apathy among the Chinese-American community, resulted in little protest when Tsien was deported to China.

What exactly did the United States lose when it deported Tsien? According to my interviews with the scientists who knew him, it lost a first-tier scientist in the areas of applied mathematics, high-speed fluid dynamics, structures, and jet propulsion, with the ability to work in many different fields and to organize and master vast quantities of technical material. He was a rigorous theoretician who was also capable of planning the next step for research and development in defense, as evidenced by his works for the book *Jet Propulsion* and the *Towards New Horizons* series. While we can never say with certainty what Tsien might have achieved had he remained at Caltech, it is likely that he would have made significant contributions to the lunar and planetary scientific missions developed and managed by JPL, and to various other space flight programs. Most of his former colleagues and students moved years later into powerful industrial or advisory positions to the government in matters of space development, such as Homer Joe Stewart, who headed an important defense committee for earth satellites, or Allen Puckett, who became the chief executive officer for Hughes Aircraft, or Joseph Charyk, who served as undersecretary for the Air Force and chairman of the board for Communications Satellite Corporation.

Grant B. Cooper put it most dramatically: "That the government permitted this genius, this scientific genius, to be sent to Communist China to pick his brains is one of the tragedies of this century."

23

A Hero's Welcome (1955)

$\mathbf{F}$rom the moment of the announcement of his deportation to September 17, 1955, the day Tsien and his family gathered at the Los Angeles harbor armed with third-class tickets, waiting to board the *President Cleveland*, the atmosphere had been solemn at Caltech's Jet Propulsion Center.

The dock was so crowded with reporters that some friends of Tsien could not even get close to him for a final farewell. To the newspapermen he said: "I do not plan to come back. I have no reason to come back. I have thought about it for a long time. I plan to do my best to help the Chinese people build up their nation to where they can live with dignity and happiness. I have been artificially delayed in this country from returning to my country. I suggest you ask your State Department why. Of your State Department and myself, I am the least embarrassed in this situation. I have no bitterness against the American people. My objective is the pursuit of peace and happiness."

On board the ship, Tsien and his family posed for numerous news photographs. There was Tsien, faintly smiling in suit and tie, with a tiny curl of hair sticking up from the top of his head. Standing to his right was his wife, wearing a dark dress with a corsage over a floral print blouse. In front were his two children: his seven-year-old son, Yucon, broadly grinning and sporting a crew-cut, shorts, and striped shirt under a bowtie and white jacket; and four-year-old Yung-jen in blunt-cut bangs and bob, wearing a white petticoat dress and clutching a small doll. They looked freshly scrubbed, healthy, and very American. If one ignored the Chinese features, one could almost say that they embodied the image of the all-American family of the 1950s.

Then, at 4:00 P.M., Tsien and his family sailed for China. He left behind dozens of people who were as stunned as he was that as a result of the flimsiest of allegations, he was now on his way back to China, giving China something it could not have gotten without U.S. cooperation—a man not only enormously knowledgeable about the science of rocketry but also capable of organizing the effort needed to move his country far ahead scientifically.

In many quarters, the reaction to Tsien's departure was rage. "I'd rather shoot Tsien than let him leave this country," Dan Kimball said to his friends during the 1950s. "He knows too much that is valuable to us. He's worth five divisions anywhere." When questioned about the event years later, he said: "It was the stupidest thing this country ever did. He was no more a Communist than I was—and we forced him to go."

Grant B. Cooper, Tsien's lawyer, was equally angry. The INS had never informed Cooper of the change in Tsien's status, nor did they bother to notify him of Tsien's departure. It had long been Cooper's plan to file a petition for a writ of habeas corpus when the INS decided to deport Tsien so that the proceedings could be challenged in a court of law. But Cooper was attending a bar convention in San Francisco when Tsien sailed for China and he learned of the news only when his secretary read about it in the *Los Angeles Times*. "To say that I was shocked," wrote Cooper to the INS on September 29, 1955, "is to put it mildly. If I, as an attorney in any proceeding, were to deal with a client in the opposing side without notice to or consent of the counsel for the opposing side, I would be subject to severe discipline and probably even disbarment—and I do not feel that an agency of the government is in a different position."

Between late September and the first week of October the *President Cleveland* stopped at ports in Hawaii, Japan, and the Philippines. But while other passengers stepped ashore to go sightseeing, Tsien and his family remained on board. This, he wrote later, was chiefly because the U.S. government would not be responsible for his safety if he ventured off-ship.

To break the monotony of the voyage, Tsien and his family befriended some of the other Chinese passengers. On board there were twenty or thirty American-educated Chinese scholars who had chosen to return to their home country with their families. On October 1 the group celebrated the sixth anniversary of the founding of the People's Republic of China. Tsien played his bamboo flute while his wife and children sang the Chinese national anthem. Under Tsien's urging, they formed a small club that they named the President Cleveland Association, with members ranging from distinguished scientists to a seven-month-old baby.

One man with whom Tsien particularly enjoyed talking was Xu Guozhi, a young doctor of engineering who had worked at the University of Chicago and the University of Maryland. Xu was impressed by Tsien's mind and range of scholarly interests. He observed Tsien spending long hours on the ship reading technical books and articles and yet conversing on numerous topics outside of science. Tsien, meanwhile, took note of Xu's mathematical skills and discussed with him theoretical problems that they could work on together in the field of engineering cybernetics. As the two of them talked about the announcements they would make at a future press conference in China, Tsien seemed excited by the challenge of helping his homeland develop its science and technology infrastructure. He did, however, express his concerns about the lack of computers in China, which would hinder the speed of engineering research.

In the early dawn on October 8 the *President Cleveland* approached Hong Kong harbor. With his head pressed against the glass porthole of his cabin, Tsien gradually began to make out the silhouettes of large rocks jutting out of the sea and the coastline of China. "I was looking out so eagerly because this was my homecoming after twenty years in the United States," Tsien later wrote. It was daylight when the steamship was at the dock. Tsien and all the other Chinese passengers boarded a small boat that took them to a railway station in Kowloon. There they were ushered past a crowd of reporters into a large room guarded by a phalanx of police officers, with two policemen for every door. For two hours the police kept out the press, but eventually capitulated to their demands.

Then, as Tsien said later, a "flood" of newspapermen broke into the room. Every Chinese scientist was suddenly surrounded by four or five reporters and bombarded with questions like "Will you be working for an armament plant?" "Will you be working on atomic rockets?" "Were you exchanged for the American flyers?" "Do you hate the United States?" A transcript of the interview reveals that Tsien was in no mood to talk to any of them.

REPORTER: What about the confiscation of your papers?
TSIEN: At the moment, I cannot talk about this.

REPORTER: Are all the Chinese students in the United States willing to return to China?

TSIEN: I cannot talk about this.

REPORTER: Before you went to the U.S.A, where were you educated? Can you tell us about that?

TSIEN: That is not important at all. I don't think I have to answer you.

REPORTER: Can you tell us of any of your friends who have not been released?

TSIEN: We Chinese do not have the freedom of speaking in the United States, and I have no intention of answering for them.

REPORTER: Did the American Consulate send someone to question you?

TSIEN: Why don't you go and ask the American Consulate yourself?

REPORTER: After returning to China, where would be your destination? Shanghai, Peking, or Tientsin?

[*No answer.*]

REPORTER: Was your daughter born in the States?

TSIEN [*after a moment of reflection*]: Yes.

REPORTER: If your daughter was born in the States, is she still a U.S. citizen?

TSIEN: You can check this yourself.

REPORTER: Does your daughter speak Chinese?

TSIEN: This is my private affair. I refuse to answer the question.

REPORTER: When did you marry your wife?

TSIEN: It is beside the point.

REPORTER: Were you under observation?

TSIEN: I don't know.

REPORTER: Do you have all your books and luggage?

TSIEN: Majority of them.

REPORTER: Was part of it confiscated?

TSIEN: Yes.

REPORTER: What was confiscated?

TSIEN: According to American export rules, you are not allowed to bring out anything which you cannot get from a shop.

REPORTER: Do you mean that your aeronautical engineering notes were confiscated?

TSIEN: All the notes, such as a diary.

[*A Chinese reporter asks a question in English.*]

TSIEN [*smiling*]: I think every Chinese should speak the Chinese language.

REPORTER: I speak only Cantonese and English.

TSIEN: I think Mandarin is widely used in China, and you are a Chinese. You should learn Mandarin.

[*General laughter.*]

"Same questions, same mentality as the reporters I had met on the day of my sailing from Los Angeles harbor!" wrote Tsien of his vexation. "We said nothing to these men. When the disappointed sensation-hunters finally melted away, we were able to get on our way."

The train took Tsien and his family to the town of Shenzhen, the last British checkpoint before the border of Communist China. It was there that someone spied the five-star Chinese flag. "Yes, it was our flag!" Tsien remembered. "So bright, so shiny under the noonday sun! All of us were suddenly silent, many had tears in their eyes. We walked across a little bridge. We were in our country, our homeland, our proud homeland—a land with four thousand years of unbroken civilization!"

Then they heard a voice addressing them over a loudspeaker. "Welcome countrymen!" it said. "The whole country welcomes you! We are now in the third year of our first Five-Year Plan. We need you! Let us work together. Let us strive for a better, more prosperous life!" The official welcome for Tsien had began. In Shenzhen, representatives from the Chinese Academy of Sciences and other governmental science associations were there to greet him. "What a difference!" Tsien recalled of that moment. "What brotherly warmth! No sensation-seeking reporter, no lurching FBI man, no vulgar advertising poster! We breathed pure, clean, healthy air!"

Thus began weeks of homecoming festivities and sightseeing for Tsien. After a brief customs check in Shenzhen, Tsien boarded a train for Guangzhou: a southern metropolis on the banks of the Pearl River, located barely a hundred miles away. The train stopped at every tiny village on the way to the city, and Tsien described each as clean and orderly. "Outside, on the stations, no trash was visible. No newspapers on the ground, no comics, no cigarette butts," he wrote. "I was glad to have this first acquaintance with the spirit of my reborn homeland."

In Guangzhou, Tsien was treated like a celebrity. Members of the local elite greeted him at the station and invited him to a banquet hosted by the Guangzhou branch of the Chinese National Union of Natural Sciences. Tsien also embarked on a well-publicized tour of the city. There was much to see.

Thousands of boats arrived and departed from the city with cargo every day, and entire communities of people lived on the river in boats. During the day, the hot, crowded streets offered a profusion of silk, artwork, bamboo crafts, and *dimsum* restaurants; at night, the river glowed with thousands of red lamps and candles.

But what proved to be most memorable for Tsien was a visit to two museums of Communist history. One showcased the achievements of Soviet economy and culture, another the humble beginnings of the Chinese revolution. The latter, housed in a Confucian temple, had served as the first school for the organizers of the Chinese peasants. Mao Tse-tung had worked there as head lecturer. Tsien was struck by the poverty of the revolutionary school: the lecture hall furnished with only crude narrow benches, the bed in Mao's room that was little more than a few planks laid over two benches. He began to express more interest in Communism and read Mao's selected works, the new Chinese constitution, and books on the five-year economic and scientific plan.

A few days later, on October 13, Tsien arrived in Shanghai. His father, now 74, met him at the train station. Knowing well his son's love for art, the elderly Tsien greeted him with a set of famous Chinese paintings. For the first time, Tsien Chia-chih had the opportunity to meet his two grandchildren. Communication must have been a little difficult at first, as Yucon and Yung-jen understood barely a word of Chinese. Nevertheless, it was a joyful occasion. The day of the reunion coincided with Yucon's birthday, and the family celebrated by eating a meal of noodles in keeping with the Chinese tradition, the length of the noodles symbolizing longevity.

Tsien stayed in the Shanghai region for two weeks, seeing old friends and reliving old memories. He visited his alma mater, Jiaotong University, which had grown from an enrollment of around seven hundred students in the 1930s to almost six thousand. He also made a side trip to his childhood home of Hangzhou to pay his respects to his deceased mother and to sweep her grave.

He marveled at the changes that had taken place since his visit eight years earlier, or at least publicly spoke as if he did. "[Shanghai] was no longer familiar to me," Tsien wrote. "The streets were so clean, and there were no pickpockets, no thieves, no crowding peddlers, no high-and-mighty foreigners. In their stead, there were eager men and women in ubiquitous dark-blue cotton jackets and happy children in red scarves, the Young Pioneers. And in the stores, we found the price was the same for the whole city: no haggling was required. Well, for Shanghai, this was really something new!"

In late October, Tsien and his family boarded a train and headed for their

Handwriting specimen of Sidney Weinbaum during his trial, September 1950. Earlier in 1950 Weinbaum had been arrested on charges of perjury for lying on a security application to JPL that he had never been a member of the Communist Party. Tsien, who had recommended Weinbaum for the job and refused to testify at Weinbaum's trial, also fell under FBI suspicion of possible Communist membership. *(photo: Hearst Newspaper Collection, Special Collections, University of Southern California Library)*

Convicted on charges of perjury and fraud and sentenced to four years of prison, Weinbaum is led out of the courtroom in handcuffs, September 1950. *(photo: Hearst Newspaper Collection, Special Collections, University of Southern California Library)*

In August 1950, Roy Gorin, head of the exports division (*left*), and Alfred Gonzales, a Customs inspector, seized eight cases of Tsien's scientific papers about to be shipped to Shanghai. The headlines in Los Angeles papers proclaimed that Tsien was trying to ship out "secret data" and "codebooks" to China. The "codebooks" later turned out to be logarithmic tables. (*photo: Los Angeles Times Photographic Archive, Department of Special Collections, University Research Library, UCLA*)

Terminal Island. Tsien was held on the island's immigration detention center for two weeks in September 1950 following his arrest by immigration authorities. (*photo: Courtesy of the San Pedro Bay Historical Society Archives—NewsPilot photo by Tom Coulter*)

A recent photograph of the Immigration and Naturalization Service detention facility on Terminal Island, San Pedro, California.

The INS pushed for Tsien's deportation on charges that Tsien had been a member of the Communist Party prior to his entry into the United States in 1947. Here, Tsien attends an immigration hearing in downtown Los Angeles, November 1950. *left to right:* Tsien's attorney, Grant B. Cooper; Tsien; a hearing reporter; examning officer Albert Del Guercio; and hearing officer Roy Waddell. (*photo: Los Angeles Times Photographic Archive, Department of Special Collections, University Research Library, UCLA*)

The famous Wang-Johnson talks in the Palace of Nations in Geneva, August 1955. PRC representatives wanted to secure the return of Tsien and other Chinese scientists detained in the United States; U.S. diplomats demanded the release of American prisoners of war captured by the Chinese during the Korean War. Wang Ping-nan (*third from left*) leads the Communist Chinese diplomats in discussions; Ambassador U. Alexis Johnson (*second from right*) sits with one of his aides, Ralph Clough (*far right*). Of all the Chinese scientists in the United States, Tsien Hsue-shen was singled out for discussion. (*photo: UPI/Bettmann*)

Five of the eleven American airmen freed from Communist China as a result of the Wang-Johnson negotiations make a stop in Honolulu, Hawaii, August 1955. *left to right:* Capt. John Buck, Tech. Sgt. Howard Brown, beauty queen Mary Snively, Mjr. William Baumer (with crutches), beauty queen Audrey Garcia, Daniel Schmidt, and Steve Kiba. (*photo: UPI/Bettmann*)

On November 17, 1955, Tsien and his family boarded the *President Cleveland* to return to China. "I do not plan to come back," Tsien told reporters. "I plan to do my best to help the Chinese people build up the nation to where they can live with dignity and happiness." *(photo: Los Angeles Times Photographic Archive, Department of Special Collections, University Research Library, UCLA)*

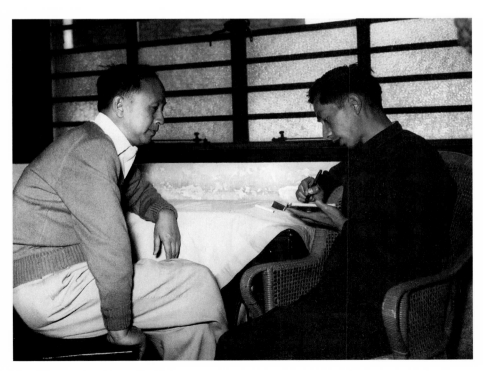

Tsien with a Xin Hua News Agency reporter in China. *(photo: Xin Hua News Agency)*

In October 1955, Tsien stopped in Shanghai on his way to Beijing to meet with Lou Zhongluo, director of the Institute of Physiology and Botany of the Chinese Academy of Sciences (*left*), and Yin Hongzhang, a deputy director and former classmate of Tsien's (*middle*). (*photo: Xin Hua News Agency*)

Tsien (*right*) shaking hands with Premier Zhou Enlai.

Tsien with Mao Zedong. *(photo:* People's Daily *Overseas Edition)*

A recent photo of Tsien's home in Beijing during the 1950s and early 1960s.

气是很好的点灯燃料和煤气机的燃料。如果把沼气压缩在高压气瓶里，那就是拖拉机、联合收割机、运输汽车等的燃料。燃烧沼气的发热量比汽油还高，估计6,000匹马力小时所需要的沼气量只有1.2亿吨，再加上点灯等的需要，每年沼气的用量大约有2亿吨。2亿吨沼气需要多少酸醇原料呢？这我们还没有确切的资料，但估计還料大约有100亿吨。酸醇后的利渣大约有90亿吨，又是很好的有机肥料。以16亿多亩树地来计算，每亩可以施酸醇后的泥法1万多斤，再加上化学肥料和其它杂肥，就能保证农作物的丰产。

那么，我们每年能不能聚集約100亿吨的酸醇原料呢？因为人口会不断增加，我们姑且假定以7亿人口計算，其中5亿为农村人口，2亿为城市人口。每一农村人口如果能养一头猪，那就是5亿头猪，而每头猪每年产粪尿2.5吨，所以每年猪粪尿共有12.5亿吨。如果有一部分城市人粪尿可以少多，以6亿人的粪尿計算，每人每年产

0.85吨，每年共有2.1亿吨。因此光是人和猪的粪尿每年就有15亿吨，再加上其他牲畜的粪尿，想来可以有20亿吨以上。如果全国16亿多亩地，除了栽培飼料之外，每亩每年还能产2,500斤茎叶，秸秆，每年就有约20亿吨。树叶，杂草假定是30亿吨。三样加起来一共有70亿吨，所以再找补上30亿吨的垃圾和污水，就可以满足沼气原料的全部要求。因此看来，我们的沼气計划是可以实现的。

我们也应当注意到沼气机和其他农业机械的润滑问题。每年生产6,000马力小时的农业机器所需要的润滑油是不少的，估计每年润滑油总是在200万吨以上。这对我国石油工业的规模来讲将是一个不小的数字，在我们规划我国石油工业的时候必需作适当的安排。

养殖小球藻成了农村的新副业

在利用沼气的过程中，因沼气体里含有大量的二氧

A page from Tsien's controversial article that appeared in the June 1958 issue of *Kexue Dazhong* magazine, in which Tsien claimed, "We only need necessary water conservation, Manure, and labor for the yield of the fields to rise ceaselessly." Tsien'd critics later blamed this article as one of the instigators of Mao's Great Leap Forward, which resulted in a famine that killed millions of people.

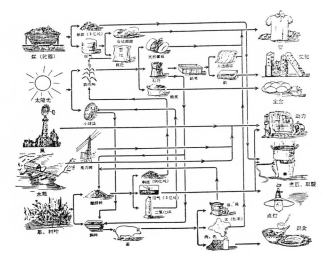

Soviet and Chinese scientists and families gather at the Beijing train station in 1960 for a final farewell after the Mao-Krushchev split. Tsien is in the first row, third from left, with his head turned toward the center. (*photo: Collection of Lin Jin*)

Tsien *(third from right)* and others during an inspection of a Chinese sounding rocket.

Tsien with colleagues Liu Yalou *(center)* and Wang Bingzhang *(right)* at a rocket test station.

Tsien *(right)* and Zhu De *(left)*.

Tsien in a group picture with Marshall Nie Rongzhen at the Shuangchengzhi missile test site in the Ganzu province in October 1960, before the November 5, 1960, launching of the Chinese-made R-2 rocket.

An intermediate-range ballistic missile at a Chinese launch site.

On October 27, 1966, China successfully launched a ballistic missile tipped with a nuclear warhead—one of the most dangerous nuclear experiments in world history. (No other country has ever admitted to testing a nuclear warhead and a missile simultaneously.) Here the missile takes off from the Shuangchengzi base in Gansu.

The launching of a medium-range Chinese missile.

A strategic missile unit of the People's Liberation Army passes through Tiananmen Square in Beijing during a parade in 1984 to mark the thirty-fifth anniversary of the founding of the People's Republic of China. (*photo: Glofa Enterprises, Inc.*)

The assembly and testing of the Dongfanghong-1, China's first manmade satellite. Tsien was key to its development, spending years cultivating scientists for the project while lobbying the Chinese government for support.

Tsien talks to reporters about new Soviet rockets. (*photo: Archives Photos Stock Photo Library*)

A Silkworm missile. *(photo: China National Precision Machinery Import & Export Co.)*

Tsien with Li Peng, then Premier of the State Council, and Jiang Zeming, then General Secretary of the CCP Central Committee, in the Ziguang Hall in Zhongnanhai, Beijing, August 7, 1989. Jiang said, "Not only should we emulate Comrade Tsien Hsue-shen's academic achievements but also his political character, which is even more important."

On October 16, 1991, Tsien was given the honor of "State Scientist of Outstanding Contribution," the highest honor a scientist can achieve in the People's Republic of Chhina, during a ceremony in the People's Hall in Beijing. Students from Tsien's high school alma mater present flowers to Tsien *(second from left)* and his wife *(far left)*. *(photo: Xin Hua News Agency)*

The aging Tsien. *(photo: Xin Hua News Agency)*

final destination: Beijing. They arrived at the capital on October 28, where a delegation of twenty distinguished scientists greeted them at the station. The group included three of Tsien's acquaintances from Caltech: physicists Zhou Peiyuan and Zhao Zhongyao and aeronautical engineer Qian Weichang (or Chien Wei-zhang), who had worked under Tsien at JPL. For the next few weeks, Tsien reigned over the city as a hero. He was regally wined and dined as journalists, scientists, and officials all clamored for the chance to talk with him. The scientific elite of China toasted him at lavish banquets hosted by Guo Moruo, the president of the Chinese Academy of Sciences, and Ma Yinchu, the president of Beijing University. Meanwhile, Tsien had the opportunity to meet with vice premier Chen Yi and other members of China's top leadership.

Despite the protocol of official welcome, Tsien found the opportunity to explore the city of his childhood. Officials on the train had boasted over the public address system that Beijing now possessed districts devoted to education and culture, and new parks and museums. Now, as Tsien observed, conditions in Beijing had indeed changed for the better.

The contrast between what Beijing had been when he was a boy and what it had now become was astounding. Gone were some of the obvious signs of filth and misery. During the Nationalist era, some working-class homes in Beijing were nothing more than mud kennels into which overflowing sewers poured rot and garbage during heavy rain. Now, new housing projects of concrete emerged, along with paved roads, telephone poles, and citywide plumbing. Wrote Simone de Beauvoir in her 1955 memoir, *The Long March*: "Today there are no more prostitutes, no more smell of opium in these streets, only opera airs coming over radios, red-and-black shop signs swaying above shops. . . . In the past one went long distances to fetch polluted water; today pure water runs from a tap placed at every intersection in Peking. . . . You no longer hear any screaming or shouting when two bicycles or pedicabs collide." One Beijing native, nostalgic for the colorful brawls, flies, and poverty of the city, was said to have exclaimed: "No more beggars! Why, this isn't Peking anymore!"

Poverty was being replaced by drab but clean uniformity. Earlier efforts by the Communists to erect pagoda-style government architecture had been criticized for wasting materials on decorative facades; the new skyscrapers and compounds that shot up in the city were stark, functional, and boxy. In the streets men and women wore simple blue cotton suits to maintain the appearance of an ideal classless society.

Two days after their arrival in Beijing, Tsien and his family visited Tienanmen Square, the Chinese equivalent of Moscow's Red Square. Here, in contrast to the ubiquitous concrete, were marble bridges, sculptured balustrades, and the famous Gate of Heavenly Peace, which for hundreds of years had served as the main entrance to the ancient Forbidden City. Painted in imperial violet, the gate of Tienanmen Square towered over speaker's platforms and grandstands set up for mass rallies and demonstrations. So powerful was the gate as a symbol of righteous government it appeared in the PRC national emblem: a gate between five stars and a cogwheel within a larger circle of wheat. "The vista was simply breathtaking," Tsien recalled. "To me, no city in the world, however famed, can compare with Beijing, and in Beijing nothing can compare with Tienanmen."

Tsien's demeanor with reporters—his blatant anger toward the United States, his effusive praise for the new China he found on his return—gave no hint of the inner turmoil that must have arisen as the reality of his new life, most probably the rest of his life, took form. True, the United States had given Tsien few options—at least few options his pride would let him exercise—but this did not mean that Tsien was returning to China without severe reservations about his future in a new China he really did not know.

At a minimum, he knew that his period making original contributions to cutting-edge space-age science had come to an end. He might do brilliant work in China, but it would be work of a different sort—it meant applying the insights and information developed by others. He may have told himself that if he could no longer be the theoretician he might have been, he could be something just as good—the Kármán of China, someone who helped the military put new scientific insights to practical use, someone who advised a new generation of brilliant young scientists about the problems they would have to solve to go beyond the work he had done. But Tsien knew he was not returning to a university setting and those he would advise would not be doing pure research. Wernher von Braun was a more appropriate subject of comparison. Tsien was returning to develop for China the weapons the country felt it needed to defend itself: long-range missiles, perhaps nuclear missiles, and perhaps satellites. It would not be surprising that he saw nothing immoral in this, nothing threatening to world peace and security. The China of recent history had never been an aggressor nation; it had always been the target of colonialist exploitation, and its own national integrity had repeatedly been violated. It was China who had been carved up by foreign exploiters, bombed by the Japanese, had its territory taken from it by the Japanese and the Russians, been invaded by sea by the French, the Germans, the British, and finally the Americans. Still, the work he would now be doing was not the same kind of scientific work he had been doing.

And then there was the bigger problem. Tsien had seen firsthand what happened when a bureaucracy seized control of a life. For all his moral rightness, all the influence of Caltech, all the contributions Tsien had made as a scientist working in the United States, he was unable to clear his name of a charge whose very vagueness made it difficult to refute. Now he was returning to a country where the government was a huge bureaucracy that saw as its responsibility the direct and continuous control of the actions of all its people. Tsien was certainly capable of seeing the similarities in process despite the differences in ends. Arbitrary power had flexed its muscle to force his deportation; now arbitrary power would be controlling every aspect of his life and his work. What if he found himself in conflict with those at the highest levels of the Chinese Communist government? What then? Where would he go? What would he do? What would his family do?

In past times, his salvation had lain in work. Now not only his salvation but his very survival depended on work—work the Chinese government believed he alone could perform. What would the government ask of him scientifically? Did they really believe that one man could bring back to China the entire body of American scientific knowledge regarding missiles and space flight? Did they think that he would be able to perform miracles? Would they give him the resources and the freedom to do what he could? And could he do it even with those resources? Would they understand temporary setbacks? Did he have within him whatever it would take to do whatever it was they asked of him?

Years later, he would supply the answer to only one of these questions—the last. Years later, he observed that he had been terribly unsure that he would be able to help China build its first missile capable of hitting Japan.

Missiles of the East Wind

The Chinese quickly figured out—probably with Tsien's help—that even with his extensive knowledge and experience, they would not be able to create a missile program immediately. There were no factories in China that could easily produce the complex materials they would need. There were no major wind tunnels, no engine test sites or launch sites, no university research institutes devoted to jet propulsion. There were not even indigenous textbooks on the subject.

When Tsien grasped the full extent of the backwardness that existed in Chinese science, education, and national defense, even his enthusiasm waned. "[We] had no research team personnel and metalworking workshops," Tsien wrote. ". . . At that time my thoughts completely changed, from optimism to pessimism. I really felt that in scientific research it would be difficult to progress even an inch, and I was worried to death about it. . . . I didn't know how to struggle in a difficult environment . . . how to start from scratch."

Four decades later, China possesses the third largest nuclear missile arsenal in the world and has become one of the greatest arms suppliers to the Third World. It is the only country other than Russia to have intercontinental ballistic missiles pointed at the United States. Chinese newspaper and magazine articles credited Tsien for pulling off a miracle: revolutionizing space development in China when the country in 1955 had barely enough technology to build a decent car or bicycle.

How did he do it? What, exactly, was Tsien's role in this incredible evolution?

It is not easy to answer this question, because until recently almost all past missile-related activities were secret. The disclosure of such activities was punishable by death. Only in the past couple of years have the Chinese began to release information in official histories or monographs with limited circulation, or in vague outline to the news media. Currently the Chinese are engaged in a public relations effort to publicize their missiles in order to enhance their prestige on the world market. Yet even with this increase in openness many questions remain unanswered. It is difficult even to obtain copies of photographs of missile institutes or historic launches.

Tsien's own refusal to talk to most reporters in China—or even to permit an official biography to be written while he is still alive—has compounded the problem of getting information. There are other difficulties inherent in trying to construct a narrative of events: a tendency of the Chinese government to change the names of institutes and missiles; the different Chinese, American, and Soviet designations for the same missile; faulty translations that render historical descriptions vague and confusing.

Although the information available is sparse, three years of research has pieced together an obscure journal article here, an official history there. The best source of information was the many discussions the author had with top missile scientists in China who found the courage to talk with her during the World Space Congress in Washington in 1992 or during her visit to Shanghai and Beijing in the summer of 1993.

It appears that over the past four decades, Tsien has given the People's Republic of China four general contributions.

First of all, Tsien gave the government the most important thing—confidence. After all, Tsien was not the first scientist to urge that the Chinese government develop missiles—Russian advisors had proposed it before Tsien had. But it appears that he was the first scientist to whom they seriously listened and on whose words they took concrete action. "Tsien's role was symbolic," Lin Jin, a missile scientist who worked under Tsien, told me in 1993. "He did not make any specific contribution or specific missiles, but it was his overall vision and organization that mattered. He was the one who made the proposals and gave advice to Mao and Zhou Enlai. They listened to him. He got us the funding. If there had been no Tsien, they might not have made the final decision to start the missile program at all."

Second, time and time again it was Tsien's ability to inspire those below him that made the difference. A couple of rungs below Tsien on the missile hierarchy is a small group of Chinese rocket scientists a generation younger who are credited today for being the backbone of the Chinese space program. Mostly educated by Soviet universities, these men later assumed powerful positions within the system, commanding teams of scientists. Many of these men Tsien coached, advised, and goaded to bring out the best in them. And many worked continuously to please him, even cutting back on food and sleep.

This is not to say, however, that Tsien abandoned his old ways of intimidation and arrogance, his utter lack of charity toward those whom he believed were not as mentally quick as he. If anything, he was more aloof and awesome a figure in China than he had ever been at MIT and Caltech. "Tsien's reputation was very high," one scientist who worked under him remembered. "He was like a demigod." Years later, some of the men who served as Tsien's high-level assistants—even the ones Tsien respected—told me they never remembered having a relaxed personal conversation with him. All of this seems to indicate that once Tsien settled in China, he appeared to have trusted no one—not even members of his own family—with his most intimate thoughts.

His third important contribution was harping on the fact that the answers to many of the questions the Chinese had were to be found in books. His office was always stocked with books, he was always seen reading books, and one of his secretaries initiated a secret program to lend out his books to other scientists. When advising young scientists, Tsien would repeatedly stress the need to "study and analyze the reference materials." Much of the information that Tsien deemed important could be found in American publications, and Tsien taught others how to scrutinize American research journals and even airline advertisements in commercial magazines to estimate the levels of aeronautical development in the U.S. defense industry. In China, Tsien may have even changed educational philosophy in military circles; one of the biggest problems he came across was that young Chinese scientists were being trained to read Russian while all the best space literature was in English.

And finally, he proved himself capable of creating an organization. One does not build a missile with a couple of scientists. Teams of scientists and technicians must be brought together and a structure for work established. Some of Tsien's greatest achievements in China involve his contributions to a twelve-year national plan for Chinese scientific development and his leadership in establishing the Fifth Academy of National Defense, China's first institute for missile design. It was he who initiated and oversaw programs to develop some of China's earliest missiles, the first Chinese satellite, missile tracking and control telemetry systems, and the infamous Silkworm missile. And it was he who helped turn systems engineering into a science in China, by working out a management structure that would facilitate communication between tiers of engineers with a minimum of confusion and bureaucracy.

Here is, in part, the story of what Tsien did for China scientifically and organizationally.

During his first year in China, Tsien was preoccupied with building an institute devoted to the study of applied mechanics and the development of high-speed aerodynamics for defense purposes. On January 5, 1956, the government officially founded the Institute of Mechanics in Beijing, with Tsien as its director.

Facilities at the fledging institute were primitive at first. There was only one telephone in the entire building, which rang incessantly for Tsien. His office was on the fourth floor and the phone was on the ground floor and he had to run up and down those stairs to answer the phone. There was also little usable equipment. The institute purchased some desk calculators that had to be wound up by hand, though Tsien dreamed of the day when the Chinese could build small electronic computers.

Tsien divided his time between meetings to refine organizational research goals and seminars given by the different branches of the institute. A small team of American-trained scientists assisted Tsien. The deputy director was Guo Yonghuai, Tsien's former student and protégé, who had received his doctorate in aeronautical engineering from Caltech in 1946. The director of operations research was Xu Guozhi, the systems scientist whom Tsien had met on the ship. During the first year of his time in China, Tsien held weekly workshops at the Institute of Mechanics, attended by scientists from local universities and also from the nearby city of Tianjin.

Tsien shaped a plan for Chinese scientific research that bore a haunting resemblance to the monumental blueprint for U.S. air defense he had worked on only ten years earlier: *Toward New Horizons*. In March, hundreds of Chinese and Soviet scientists arrived in Beijing to formulate a twelve-year plan for scientific development that focused on fifty-six major areas. Tsien, a powerful member of the State Council for Scientific Planning, "played a decisive role," according to He Zuoxiu in his article "The Twelve-Year Development Plan for Science and Technology." Tsien ensured that special emphasis would be placed on subjects vital to defense: atomic energy, missiles, computer science, semi-conductors, electronics, and automation technology. He was particularly instrumental in pointing out to the Chinese government that the benefits of missile development outweighed aircraft development, because missiles had a higher Mach number than planes and could carry weapons intercontinentally.

Self-discipline distinguished Tsien from the others. He was one of the most industrious people in the institute, arriving at the office at 7:30 A.M., usually half an hour before the others. He spent hours writing at his desk and reading technical books, breaking only for lunch and dinner. Often, he would return to his office from seven to ten in the evening. He worked in a room simply furnished with a large wooden desk and chairs, a small round table with a telephone, a metal chest for papers, a typewriter, and two large bookshelves that stretched from floor to ceiling, filled with volumes ranging from biographies of scientists to math, physics, mechanics, biology, and music.

Though Tsien did not know it, these books educated an entire institute. The other researchers yearned to read his books, starved as they were for news from the world of science outside China and for an understanding of what made a scientist such as Tsien great. But there was that aura of reclusiveness and detachment about Tsien that most people found impossible to penetrate. Zhang Kewen, his secretary at that time, solved the problem by secretly loaning his books out. "My philosophy was that if ten people could read a book as easily as a single person, why not increase the sum of knowledge?" she said. "So I told the others they could read Tsien's books but that they couldn't take them out of the institute. These books had to be retrievable if Tsien needed them right away. I kept a log of which books were loaned to whom. Tsien Hsue-shen never knew."

He was also involved in the founding of the Fifth Academy, which marked the official beginning of Chinese missile development.

Within a few months of his return, Tsien was one of the most powerful scientists in the country, serving as a liaison between the scientific community and the highest reaches of government. Minister of Defense Peng Dehuai discussed short-range guided missiles with Tsien when he met him and later sent a general staff official to make a detailed analysis with Tsien of advantageous and problematical conditions for guided missile research and development. Tsien also talked with other high-level military officials, urging them to make satellite and launching vehicle development a national priority. On February 17, 1956, Tsien submitted to the party leadership a secret proposal to establish research facilities for aeronautics and missile development. Premier Zhou Enlai held a special conference to consider Tsien's proposal, and on October 8 the Chinese leadership established the Fifth Academy of the Ministry of National Defense. Tsien was appointed its first director.

The beginnings of the academy were humble indeed. An old hospital and two former sanitariums were converted into office space for the academy. Missile scientists remember that the staff consisted of only a hundred workers with secondary school educations and one to two hundred Chinese college graduates who were eager to begin their apprenticeship with Tsien, the sole rocket expert in the institute. Immediately recognizing the importance of having a good training program, he held informal engineering courses at the institute, teaching a class entitled "Introduction to Rocket Technology" while his former Caltech student Zhuang Fenggan taught courses in aerodynamics. Many of their students became the backbone of the Chinese space program. In an article that Tsien wrote for *People's China*, he described how universities and research institutes throughout China were coping with the lack of educated manpower:

> First we recognized that the pressing problem was to teach, not immediately to do independent research. Since we lacked professors, college graduates were drawn into the teaching service. The recruits within each institute were divided into groups, each concentrating on one subject. They wrote the class notes and discussed the plan and the method of presentation. Each group was led by a professor. When the subject was new and there was no professor, a Soviet specialist on the topic was invited to China to lead the group.
>
> Naturally such green instructors sometimes had trouble in facing the students. Here is how the problem was solved. When a teacher was confronted with a difficult question from the students, he generally did not answer it immediately. During the evening meeting of the group of teachers on that subject, the question was thoroughly discussed and the correct answer formulated. Thus not only was it possible to obtain the huge number of instructors, but when they appeared before the class, they were able to discharge their duties admirably.

During the first few years Tsien also invited the section leaders of the burgeoning Fifth Academy to his house every Sunday afternoon for brainstorming sessions. "To tell the truth," Tsien said decades later, "I was not sure I could carry out the mission assigned to me by the Party and the State. When I was in the United States, I had done some work on guided missiles and satellites. I was, however, never involved in any of the launches. I had no option but to consult my colleagues."

It was obvious from the beginning that the Chinese would need outside help—logically, Russian help. Ideally, what the Chinese needed were actual missiles, no matter how outdated, to use as models in designing their own.

In the 1950s the Chinese began to negotiate with the Soviet Union in order to purchase missile technology. On September 13, 1956, officials in Moscow agreed to sell China two R-1 missiles, which upon delivery in December that year turned out to be nothing more than replicas of the German V-2. The Chinese government demanded something more sophisticated. Moscow did not comply until the following year, when Khrushchev desperately needed Mao's support against enemies in Eastern Europe and the Kremlin.

Tsien went to the Soviet Union in the summer of 1957 as part of a Chinese military delegation. While the details of his mission remain secret, it would be reasonable to assume that Tsien was either directly or tangentially involved in negotiations with the Soviets to obtain better missiles for China. On October 15 Chinese officials met with the Russians to sign the Sino-Soviet New Defense Technical Accord—an agreement specifying that the Soviet Union provide China with missile models, technical documents, engineering designs for research and development, launch sites, and technical specialists to help with licensed copy production and to train Chinese students in rocketry.

After months of talks, the Soviets sold China two R-2 missiles, which were improved versions of the R-1. On December 24 a Soviet army missile battalion arrived in Beijing with two R-2 missiles and associated launching equipment. Some scientists remember that the missile parts were delivered secretly to the Fifth Academy by train in the middle of the night. The acquisition of the R-2s, experts later attested, gave the Chinese the opportunity to work with an existing rocket system and marked the true beginning of Chinese missile development.

That year, more Soviet technical assistance arrived to help China's space effort. During the second half of 1958, the Russians delivered a total of 10,151 volumes of blueprints and technical documents for manufacturing, testing, and launching the R-2 missiles. The Fifth Academy also purchased twelve more R-2s. Meanwhile, approximately one hundred Soviets arrived in Beijing to serve as scientific advisors.

Between April 1958 and April 1959, the Chinese People's Liberation Army (PLA) transferred more than three thousand technical professionals and cadres from other agencies to assist Tsien in the Fifth Academy, along with three hundred engineering experts from industry. Meanwhile, the government sent more students to Soviet universities to learn rocket technology. The first wave of Soviet-trained Chinese graduates of aeronautical engineering programs in Russia rapidly filled the first tiers of a growing missile hierarchy.

The typical recruit for the Fifth Academy was an unmarried man in his twenties who had studied engineering at a Soviet university. The moment a recruit received his assignment, his life was enveloped in secrecy. The young engineer was assigned and bused to the Fifth Academy before being told what his position or responsibilities would be. Forbidden to tell friends or family where he lived or worked, he would specify his address on correspondence only by post office box number. In 1958, the institute operated in such secrecy that even the name of the Fifth Academy was classified. To disclose its existence was punishable by death.

Unbelievably, the Chinese at first refused to copy or seriously study the R-2 missiles they had purchased from the Soviet Union. In an atmosphere of overconfidence and even self-delusion, many Chinese scientists and officials believed that they could build their own missiles without any Soviet help at all. Soon it was beginning to dawn on senior scientists that the initial goal of launching a satellite by 1959 would be impossible to achieve. While the R-2 had too short a range to hit American bases in Japan and too little payload to carry the Chinese atomic bomb under development, the R-2 rocket could provide the scientists with enough technology to jumpstart its missile program.

By early 1959, Nie Rongzhen, a senior commander in the revolution who had assumed responsibility for the strategic weapons program, announced that the Fifth Academy would copy the R-2 missiles China had purchased from the Soviet Union. The Fifth Academy embarked on copy production of the R-2 missiles under the code name "1059," which was also the Chinese name for the missile. In January 1959, the Soviets arrived to assist the Chinese in production. In April and July, a Chinese delegation traveled to the Soviet Union to negotiate the delivery of more machines and facilities for the task.

Although the Chinese boasted they could develop missiles without Soviet help, they soon found that just *copying* an existing missile was arduous and the scale of the project intimidating. Fourteen manufacturers and fourteen hundred work units took part in the copy work of the R-2 engines. Tsien soon realized that the country was woefully deficient in just about every area of missile production. The Fifth Academy encountered difficulty in getting basic materials—rubber, stainless steel pipes, and aluminum plates. They tried to solve the problem by importing some materials and producing others, or substituting similar material for rare items. The percentage of substitute material used in Chinese copies of the Russian rocket eventually reached 40 percent, but defects marred some of the substitutes.

Another problem Tsien confronted immediately was the lack of tools. The scientists needed large punching presses, lathes, welding equipment, and assembly machines to produce the missiles. Because no machinery was available to do the job, workers forged the large circular frame of the first Chinese-made rocket by pushing against it manually. The Fifth Academy also needed trained technicians and welders to put the missile together. To ameliorate the situation, Soviets started a program in welding that helped the Chinese master the intricacies of inert gas arc welding as well as other necessary techniques.

While new Soviet-style buildings for the space program were being constructed, the early staff of the missile academy labored under makeshift conditions. It was not unusual to see engineers laboring at night in crowded corridors lit by a single bulb. Administrative offices were temporarily placed in a former hospital and a former military school, a rocket assembly plant took shape in an old airplane repair factory. Some scientists toiled within a windowless airplane hangar with brick walls and a sheet-iron roof, where overhead electric lights burned day and night. In the summer the heat was almost unbearable, forcing the young men to work out equations barechested or in underwear, cranking away at their mechanical hand calculators. Housing was so limited at the time that some scientists actually lived where they worked, or slept in tents or military barracks. Most, however, lived in dormitories nearby, which were standard brick five- and six-story buildings. The young engineers practiced a Spartan lifestyle: eating in communal mess halls, washing their own clothes and hanging them up to dry in their rooms.

Such conditions prevailed during the development of China's first liquid-propellent sounding rocket in the 1950s. The scientists who worked on the project labored in conditions reminiscent of Tsien's Suicide Squad days of the 1930s: primitive desk calculators, shoddy equipment, small-scale rockets. In 1960, engineers at the Shanghai Design Institute completed a two-stage uncontrolled rocket with a liquid-propellent main section and a solid-propellent booster, as an experimental model for the T-7 sounding rocket then under development. In early 1960, the model was moved near Laogang, a coastal town outside of Shanghai, for its first launch.

If Tsien visited the launch site, he might have wept at the crudeness of it all. On a flat, desolate area next to a river stood the power source for the post: a fifty-kilowatt generator in a hut with reed walls and an oilcloth roof. A bike pump was used to fill the rocket with propellent. Across the river stood the "command post": a small heap of sandbags to protect the observer during a launch. Since no walkie-talkies, loudspeakers, or phones were available, the launch commander had to shout across the river to his henchmen on the other side. But despite the primitive equipment, the launch was successful, and the model of the T-7 rose eight kilometers into the sky. A few months later, on the rainy evening of April 18, 1960, Tsien arrived at the Shanghai Jiangwan airport to inspect the main engine of the T-7 and to watch several tests of the engine conducted on a simple test stand. That September, the T-7 was launched successfully and became the pioneer of China's first generation of sounding rockets.

The Soviets, meanwhile, appeared to be under orders to keep the Chinese ignorant of other methods crucial to rocket development. Frequently the Soviet scientists would limit the documents they showed the Chinese and go back to the embassy to check their own reference materials. The Chinese were particularly frustrated by the secretive nature of two Soviet atomic specialists, whom they derided as "the mute monks who would read but not speak."

The Sino-Soviet collaboration, originally designed to last thirty years, ended within three years after the signing of the New Defense Technical Accord. One reason for the break was the Soviet decision to renege on the agreement to help China with its nuclear bomb program. But undoubtedly the greatest reason for the failure of the Sino-Soviet collaboration was the Great Leap Forward and Mao's megalomania. Khrushchev became convinced that Mao was not only a tyrant but a madman, a Chinese version of Stalin who would blow up the world if he had the means. In 1957, Mao had shocked the Russians by welcoming the prospect of nuclear war: "We may lose more than three hundred million people. So what? War is war. The years will pass and we'll get to work producing more babies than ever before."

In August 1960, the Soviets abruptly withdrew all of their scientists from China. Some 1,390 Soviet experts and advisors were summoned home, and at least 343 contracts and 257 technical projects were canceled. The Chinese denounced this action as "treachery by the socialist imperialists who overnight tore up their contracts and recalled their technicians." That summer and fall, Soviets packed their belongings, taking with them blueprints and papers. They systematically shredded documents they could not bring back with them to the Soviet Union. On August 12, Tsien, snug in an overcoat and cap, gathered with Chinese scientists at the Beijing train station to see the Soviet specialists from the Fifth Academy off. After a round of farewells and picture-taking, the Soviet scientists and their families boarded the train and departed, most of them never to return.

The Sino-Soviet rift, coupled with the technological advances of both the Soviet Union and the United States, had left China increasingly insecure about its political future. The building blocks for the Chinese missile program consisted of little more than the Soviet R-2 missiles and some quasi espionage committed by Chinese students in Russia. (In the late 1950s Chinese students majoring in rocketry at the Moscow Aviation Institute had stolen books from the university libraries, quizzed talkative professors, and copied formulas from restricted notes to secure crucial information about state-of-the-art Soviet missiles.) To counteract the perceived threats to national security, the Chinese leadership allocated enormous financial and labor resources for the nuclear weapons industry, causing a tremendous explosion of growth at the Fifth Academy.

Even though the Soviets were gone, they had helped plant seeds for the development of China's first short-range rockets. In September 1960, the Chinese launched a Soviet-made R-2 rocket fueled with homemade propellents in preparation for the launching of its Chinese-made replica. Soon after, Tsien, PLA colonel-general Zhang Aiping, and electronics expert Wang Zhen headed a committee to organize the first flight test of a Chinese-built R-2.

The missile was moved by train to the Shuangchengzi base, hidden away in the Gobi Desert in remote Gansu province. The Chinese often refer to the Shuangchengzi base as the Jiuquan launch site because of its proximity to the town bearing that name. Jiuquan's history dates back to the Han dynasty, when Chinese troops were stationed there to guard the frontier against the Huns. For centuries the landscape remained virtually the same: a lonely country of mountain and desert, with winter winds strong enough to hurtle basketball-sized rocks across the horizon.

In 1958, the Chinese government dispatched the 20th Corps of the PLA to the Gobi Desert to construct the rocket site and living facilities. The troops dug wells, planted willow and poplar trees, and built roads and houses. Water had to be transported to the soldiers by truck from distant cities. They also laid railroad track. By 1960 a railway connected Jiuquan to Beijing, threading through the capital into the secret confines of the Fifth Academy itself. Engineers who journeyed to Jiuquan by train typically arrived there in four to five days. It is difficult to say how many missile scientists, military personnel, and their families were living in Jiuquan in 1960, but today the town holds a total of perhaps ten to fifty thousand residents.

In late October 1960, Tsien journeyed to the base to supervise the first launching of the Chinese version of the R-2. When he arrived, he could see asphalt roads stretching far into the desert, a concrete launch pad and a few buildings two or more stories high. During the first few days after its arrival, the missile was tested in a building, loaded onto a transport trailer, and moved to the forwarding station of the launch site. There, a crane hoisted the missile onto an erecting bracket that slowly lifted the R-2 into a vertical position on the launch pad. Operators on three different floors of scaffolding examined the rocket while filling it with propellent and installing batteries and other equipment.

At 9:00 A.M. on November 5, 1960, the Chinese successfully launched the Chinese R-2 from the Jiuquan launch base using radio lateral control. Marshall Nie Rongzhen, who was present during the test, toasted the scientists in the celebration banquet that followed. This, Nie said, was the first Chinese missile to fly over the horizon of his motherland, marking a turning point in its history. It had taken Tsien and his colleagues almost three years from the moment the R-2s first arrived in Beijing to achieve this historic flight.

The next few years, however, were beset with failure for Tsien.

Even though in December 1960 the Chinese successfully launched two other homemade R-2 missiles, they initially seemed incapable of designing a more advanced missile that actually worked. As head of the Fifth Academy, Tsien presided over the development of China's first generation of land-based ballistic missiles, the Dongfeng ("East Wind") missiles. In March 1962, the Dongfeng-2, or DF2, was moved to the Jiuquan base for its first test. On March 21, 1962, it lost stability, fell to the ground after 69 seconds, and exploded. Tsien flew to the base to examine the remains and to direct the analysis of what went wrong. "After the launch, Tsien pointed out correctly what caused the crash," Tsien's protégé Zhuang Fenggan said. The scientists did not take the flexible vibration of the rocket into account when designing the guidance control systems. The link between the engine and rocket structure had been weak, and the position of the gyroscope was not correct.

To make matters worse, a pet project of Tsien's was deemed too ambitious and abandoned. On November 14, 1961, Tsien had appointed himself chief designer of the DF3, originally intended to be a 10,000 kilometer range intercontinental ballistic missile fueled by liquid oxygen and kerosene, like the U.S. Atlas missile. But these goals were simply too advanced for the Chinese infrastructure to handle, and after numerous economic and technical setbacks it was canceled. Hua Di, a former Chinese rocket scientist who had worked under

Tsien and who is now a research associate at the Center for International Security and Arms Control at Stanford University, said that while Tsien continued to guide other projects, this was the last time he appointed himself chief designer on any specific missile. For the perfectionist that Tsien was, a scientist more theoretical than pragmatic, it was tempting but mistaken to pursue a vision far more grand and complex than Chinese technology at that time was able to sustain.

The failures seemed to only stiffen Tsien's resolve to work harder. During the 1960s and 1970s he made four significant contributions to the Chinese missile program: he inspired and goaded his underlings to produce, introduced them to key theoretical formulas which they adapted for practical purposes, developed a systems management technique to minimize bureaucracy, and shaped the organizational and technical direction for China's first generation of missiles.

Tsien's tactics for inspiring others involved a combination of pep talk and fear. After the failure of the DF2, Tsien presided over technical meetings for its redesign, where he would listen to the ideas of top rocket scientists and urge them to "sha xuelu"—literally, "kill a bloody path"—through the hordes of mathematical problems. His protégé Zhuang Fenggan remembered that Tsien would often say, "What the Americans can do, we Chinese can do also! The Chinese are no worse than the Americans!" However, if students had been "scared stiff" of Tsien at Caltech, they were absolutely petrified of him in China. When he strode into the room, everyone jumped to his feet, even scientists several years his senior. Then, enthroned in a chair, Tsien would listen to their presentations. Sometimes he sat still, musing over a proposal. Sometimes he would pace the room with his hands clutched behind his back—a sign of impatience and contempt for the speaker's intelligence. This inspired terror among the junior engineers, for Tsien's criticism could be so sharp the victim often turned into a quivering lump of jelly. Recalled Lin Jin, who had worked with Tsien closely on the DF2 project: "We all felt like small pupils facing a stern teacher." No one wanted to disappoint Tsien, or endure humiliation before him, and so they worked harder in hopes of pleasing him.

Tsien also referred his engineers to the appropriate reference materials. His book *Engineering Cybernetics*, for instance, was a valuable guide for engineers in redefining the DF2. Particularly important was Tsien's interpretation of the Bliss formula, based on the work of a mathematician best known for his work on the calculus of variations and for the application of calculus to the field of ballistics during World War II. In *Engineering Cybernetics*, Tsien applied Bliss's work to the science of guided missiles, and his application was later exploited by Chinese engineers to design a simple guidance system for the new, improved DF2. Chinese aeronautical engineers claim today that Tsien's book helped provide a sound theoretical foundation for an entire generation of Chinese rockets: from the DF2 to the DF5.

During this period, Tsien introduced to China a systems engineering management plan modeled after an existing program in the United States. The sheer complexity of designing a missile requires a system to minimize bureaucracy and confusion. Administrators faced with missile projects involving thousands of engineers and technicians typically organize them into several tiers: the top levels governing the structure of the entire system, the middle ones the engine and guidance systems, the bottom ones the individual components. In 1962, Tsien created a plan to facilitate communication between the tiers—a plan similar to the Program Evaluation and Review Technique (PERT) system. Developed by the United States Navy for the Polaris missile, PERT permits computerized analysis of variables in projects involving systematic completions of tasks. In essence, it displays a flow chart of all the processes involved in a large engineering task: the division of labor, time requirements, interaction between departments, the different stages of progress. Tsien's version of the PERT plan was later used on computers to design the guidance systems for long-range Chinese rockets.

In 1964, Tsien was an active participant in several important meetings held by the Chinese leadership to determine the direction of missile development. During these meetings, the scientists and the leadership revised their goals for the Dongfeng program. There soon emerged a logical sequence for the goals of the DF missiles: the Chinese R-2 missile was renamed the DF1, the DF2 would be capable of hitting Japan, the DF3 the Philippines, the DF4 Guam, and the DF5 the continental United States.

During the meetings there arose a clash of philosophy involving the concept plans for missile guidance and control, which boiled down to a fundamental difference between the American and Soviet schools of thought. Tsien championed the U.S. approach, which placed heavy reliance on individual, sophisticated pieces of equipment. Soviet-educated Chinese engineers advocated the Soviet approach of being less concerned about the quality of individual components than the ease with which a rocket worked as a system.

The Chinese leadership decided to compromise. The shorter-range missiles—the DF2 to DF4—would use primitive strap-on accelerometers, while the DF5, China's first ICBM, would use advanced gyroscopes with stabilized platforms and gimbals installed inside the rockets. Tsien was adamant on that point. "We should not," he declared, "be satisfied with a primitive intercontinental ballistic missile."

Within a few years, Tsien began to see results. Several successful launches were accomplished in 1964. On May 29, three medium short-range surface-to-surface missiles lifted off without mishap. Later, on June 29, the Chinese launched the DF2 after a major redesign that reduced the engine's liftoff thrust and shortened the range to 1,050 kilometers—barely enough to reach Japan. In July, the missile designers started a renewed effort to increase the range of the DF2 and to make additional improvements.

One major triumph for the Chinese was the November 1965 launching of the DF2A, the improved version of the failed DF2 and the country's first inertially guided missile. The payload of the DF2A remained the same as the DF2 (1,500 kilograms, which was too light to sustain the 1,550 kilogram weight of the first Chinese atomic bomb), but the range increased by about 20 percent. The DF2A also replaced radar control with a small, primitive computer placed inside the body of the rocket, thus making it impossible for an enemy power to change the direction of the rocket by interfering with its radio signals.

The DF2A was soon used in what was possibly the most dangerous nuclear experiment in history. On October 27, 1966, the Chinese recklessly tested an atomic bomb with a nuclear missile *simultaneously*—the first and only country ever to do so. The test was conducted at the Shuangchengzi base in Gansu using a DF2A rocket tipped with a smaller, 1,290 kilogram nuclear device, specially designed so that it would be light enough for the missile to carry. The missile crew launched the DF2A rocket eight hundred kilometers west of the base into the deserts of Xinjiang province, where upon landing the bomb exploded with twelve kilotons of force. Nie Rongzhen later said of the test: "If by any chance the nuclear warhead exploded prematurely, or fell after it was launched, or went beyond the designated target area, the consequences would have been too ghastly to contemplate."

The test unloosed an avalanche of publicity for Tsien. The *New York Times* credited the advances in Chinese missile weaponry to Tsien, putting him in a front-page story as well as in the "Man in the News" section. "The irony of cold war history," the *New York Times* reported on October 28, 1966, "is that the man believed responsible for putting Communist China's first atomic bomb on the nose of a missile was trained, nurtured, encouraged, lionized, paid, and trusted for 15 years in the United States." More publicity followed when the writer Milton Viorst saw Tsien's story in the *Times* and wrote a profile about him for *Esquire* magazine.

"Properly speaking, this piece on Tsien Hsue-shen is not a profile at all but an American saga," Viorst wrote of the article in his book, *Hustlers and Heros*. "It is, as I see it, the story of a man betrayed by the country he had grown to love and the poetry of his revenge. . . . The natural irony in the story was too much to resist. Here was a Chinese who was building missiles for us until he was literally forced, during the McCarthy era, to return home to build missiles for them. This was a story I had to do."

China's nuclear achievements also inspired two Associated Press reporters, William Ryan and Sam Summerlin, to write a long article about Tsien that was published in *Look* magazine on July 25, 1967. It later grew into the book *The China Cloud*, which blamed the success of Chinese atomic development on the witch-hunts of McCarthyism. Published by Little, Brown, in 1967, half of the book is devoted to Tsien's story in the United States. "The shocking truth is that, without the intentional aid of United States authorities, China's nuclear weapons and the rockets to carry them would not have been built until the late 1970s," the book jacket read. The book was the instigator of a *60 Minutes* segment on Tsien entitled "Made in the U.S.A.?" which presented "the story of how the United States, during the Red Scare of the 1950s, handed China much of the know-how to develop nuclear capabilities."

"Had his life developed differently," the *60 Minutes* segment concluded, "he might today be the major figure in our own space program." When Ralph Lapp, a member of the original Manhattan project, was asked by *60 Minutes* about Tsien's role in the development of Chinese delivery systems for the atomic bomb, he said: "I would say it would be fundamental. Actually, with his background in rocketry, I would think he would have been the guiding light in their ballistics program. And the success that they had, and the success that they will have inevitably, I would trace to his basic leadership."

It is impossible to pinpoint all of Tsien's contributions to the missile program because his role in the program was primarily administrative. But Chinese scientists have pointed out specific instances in which Tsien's exceptional foresight and judgment paid off in concrete results.

There was, for instance, Tsien's involvement in the Haiying missile. In April 1965, he headed a meeting held by the Office of National Defense Industry and the Seventh Ministry of Machine Building, during which it was decided that the

Shangyou-1, an improved copy of the Soviet Styx antiship missile, would become the prototype for the Haiying (Sea Eagle) generation of Chinese antiship missiles for coastal defense.

The work on the Haiying was to occupy Tsien's time for the next few years. When the Chinese tested the first Haiying-1 surface-to-ship missile in December 1966 and found problems with its radar system, Tsien and other scientists presided over a meeting in July 1967 to address the problem. After three years of tests and analysis by the Nanchang Aircraft Factory of the Third Ministry of Machine Building and the Third Research Academy of the Seventh Ministry of Machine Building, the Haiying-1 flew successfully in October 1970. Between 1965 and 1970 a longer-range surface-to-ship missile, the Haiying-2, was also being developed and successfully tested. During the 1970s the defense establishment decided to follow Tsien's advice, given in the early 1960s, to replace radar homing with infrared homing for guidance because radar can easily be interrupted, and authorized the design of a new improved missile dubbed the Haiying-2A. The Infrared and Laser Research institute of the Third Research Academy developed for the missile an indium antimonide infrared emission sensor, a small spherical contoured missile head for better structure, and a purified air system to cool the sensor in place of liquid nitrogen. After a number of failures, the Haiying-2A was tested successfully in the early 1980s. The People's Republic of China would later export the Haiying missiles to Middle Eastern countries, where they would become known to the general American public under another designation—the Silkworm missile.

There was also Tsien's proposal in 1966 to design a warhead with the capacity to elude antiballistic missile systems. In the 1960s, when the Chinese were designing an ICBM called the DF-5, they wanted the missile to have the ability to penetrate the antiballistic missile system the Americans were building to protect itself from possible Chinese attack. On January 4, 1966, Tsien proposed the construction of an advanced DF-5 warhead with penetration aids, which resulted in the preliminary design of a missile reentry vehicle equipped with "electronic countermeasures and light exo-atmospheric decoys."

Tsien also had a role in the development of a tracking and control telemetry network, which was instrumental during the 1980 launch of two intercontinental ballistic missiles (ICBMs) over the Pacific Ocean. A telemetry network requires a minimum of three bases to track the location of a missile in the sky. The radar signals emitted by a missile permit the base to calculate its distance from it. A single base can narrow the location of the missile within the space of a sphere, a second base to the intersection between two spheres (a circle), and the third base to the intersection of three spheres (a point). While three major tracking and control bases on the mainland of China would prove adequate for the launching of short-range missiles, those bases would lose sight of an intercontinental missile traveling on the other side of the earth, making it necessary for telemetry devices to be placed on ships in the Pacific Ocean to help the mainland bases complete the triangulation network.

In 1973, Tsien served as a deputy head for a Navy group in charge of development of missile tracking control devices for ships. During a conference held in September 1973, Tsien proposed the establishment of a telemetry network on the mainland of China, using the towns of Xian in Shanxi province, Jiuquan in Gansu province, and Xichang in Sichuan province as the locations for the three bases. He authorized Shangguan Shipan, a missile scientist stationed in Jiuquan, to draw up a proposal for the network. It took Shangguan two years to work out a preliminary plan, which he presented to the government in October 1975. The plan was approved and scheduled for development that year.

In May 1980, two Chinese intercontinental ballistic missiles were launched from the Jiuquan base over the Pacific Ocean. Hundreds of communication devices at more than ten stations across China and on a fleet in the South Pacific reported the velocity, attitude, and altitude of the missile by the second. On May 18, the first missile flew over the Pacific south of the Gilbert Islands, reentered the atmosphere and discarded two missile sections, which disappeared in two bright spots in the clouds. The remaining capsule drifted toward the sea under a parachute, hit the water and spilled a flower of emerald-green dye into the waves so that it could be easily recognized and retrieved by divers in a helicopter. On May 21, 1980, a second ICBM was launched into the sea. The success of these two flights was, according to his protégé Shangguan Shipan, "among Tsien's greatest achievements."

"Tsien had foresight," Shangguan said of his mentor more than a decade later. "If we hadn't started building our own radar control ships ten years earlier this wouldn't have been possible. He was ten years ahead of his time, only we didn't realize it then. He was brilliant."

Aside from the missile program, Tsien's most significant contribution to the China space development was his role in the creation of the first Chinese satellite. He had stressed the importance of building satellites almost as soon as he arrived in China. He set the project in motion in January 1958, when he and other scientists finished the draft of a program to develop an artificial satellite and designated a group to work on the project under the code name "581." By spring and summer of 1958, Project 581 had become a top national priority, especially after the Soviets launched Sputnik III, a cone-shaped satellite weighing more than two thousand pounds.

"We must launch a man-made satellite, too," Mao told his colleagues on May 17, 1958. "We must launch a big one if we do. Or, maybe a smaller one to begin with. However, we must not launch one as small as a chicken egg, something like the American one." During the Great Leap Forward, the entire country planned symbolically to "launch a satellite" in every field. "Fang weixing!" ("Launch a satellite!") became a national slogan, and children throughout the country were beginning to be named Weixing.

In 1962, Tsien started to train scientists for the task. Four top engineers from the Shanghai Institute of Machine and Electrical Design were sent to Beijing to work under Tsien and gain technical mastery of the subject. Tsien held meetings with them for three and a half hours a week, urging them to study English reference materials since there was almost nothing on the subject in the Russian space literature except for one journal entitled *Tehnika Raketa* (Rocket Technology). (This posed difficulties because three of the four engineers had studied Russian, not English, during their middle school and college years.) He also arranged for them to teach his book *Interplanetary Flight*, to host symposiums on satellite development, and to tour rocket assembly and engine test sites at the Fifth Academy. After a year of study, the four engineers went back to Shanghai and headed their own research divisons, passing along Tsien's knowledge to younger technicians and recent university graduates.

His protégés needed state approval to work with factories in Shanghai to build the satellite, so the plan was shelved for a few years. In January 1965 Tsien began to push his satellite proposal before the Party Central Committee, claiming that China had made enough progress in Dongfeng missiles to warrant planning for a comprehensive space program and that the further development of an intermediate long-range missile and intercontinental ballistic missile would permit China to launch its first satellites. Tsien warned Chinese officials that the work involved would be arduous, so it was better to begin as soon as possible.

That spring, Tsien's satellite proposal found favor with the leadership, and on April 29, 1965, the Defense Science and Technical Commission submitted a plan to launch the first Chinese satellite in 1970 or 1971. The Chinese government wanted the Chinese satellite to be visible from the ground and broadcast messages that would be heard by the whole world. On August 10, 1965,

Zhou Enlai formally approved the plan and urged that it be incorporated in state planning.

In May 1966, Tsien and other scientific leaders decided that the first artificial satellite of China would be called the Dongfanghong-1 (The East Is Red–1) and launched by 1970 with the Changzheng-1, or CZ-1: a two-stage liquid-propellent rocket with a third-stage solid-propellent engine. Other research organs in China joined the Seventh Academy in the satellite effort, such as the Chinese Academy of Sciences, which established laboratories and a design institute called "651" to carry out conceptual research for the Dongfanghong-1.

The project was risky. A number of things could have gone wrong, destroying not only Tsien's reputation but that of China. The Chinese were gambling on a partially tested launch vehicle, which had failed only a few months earlier and which still had technical problems. Other things could go wrong: the satellite could get lost, the Americans might detect it before the Chinese could confirm its orbit, the music broadcast from the satellite—the national anthem, "The East Is Red"—could come out garbled and personally offend Mao Tse-tung. The last thing the Chinese wanted was for the satellite to turn into a big international joke, as had the American Project Vanguard satellite in 1957.

In the spring of 1970, Tsien went to the Jiuquan missile base to supervise preparations for the launch. That April, Zhou summoned Tsien back to Beijing in order to discuss with him improvements made on the launch vehicle. Tsien and four others flew to the capital and on April 14 they delivered a report before Zhou at a special meeting. Little more than a week later, on April 24, Mao authorized the launch.

At 9:35 P.M. that evening, the first artificial earth satellite lifted off from the main Chinese rocket center at Shuangchengzhi near Jiuquan in Gansu province. The launch had gone off flawlessly, and Dongfanghong-1 circled the earth about once every 144 minutes, sending a clear and powerful broadcast of "The East Is Red." China became the fifth country to loft a satellite into space, after the Soviet Union, the United States, France, and Japan.

In the PRC, Tsien was hailed as a hero. On May 1, Tsien stood in Tienanmen Square, which was brightly lit and decked with red flags. The band was playing "The East Is Red." A huge portrait of Mao was on the Gate of Heavenly Peace, and Mao himself, along with other leaders, personally commended Tsien for his contributions to the design and launch of China's first satellite.

Tsien's triumph resounded not only in China but internationally. Once again, his name and picture appeared in stories around the world, this time under giant headlines announcing the launch of the first Chinese satellite. "Peking's first satellite was believed planned by a U.S.-trained scientist," the *Wall Street Journal* reported on April 27. "The master builder . . . that did the job is Chien," the *Philadelphia Inquirer* announced. The *Boston Herald Traveler* called Tsien "the man believed responsible for Communist China's rocket development culminating in yesterday's orbit of a satellite around the earth." "Some scientific circles in the United States speculated that Chinese scientist Chien Hsueh-shen may have played a role in this technological and propaganda achievement of China," the *Christian Science Monitor* announced. And the *Washington Evening Star* testified that "Scientist Chien [is] the man believed to be the driving force in the space program."

—᠗

Tsien could reflect with grim satisfaction that it was [his] return to China, along with scores of other talented Chinese, which had been in large measure responsible for the speed and sophistication of the Chinese nuclear program.
—*The China Cloud*, by William Ryan and Sam
Summerlin, 1967

Did Tsien play a key role in the development of the Chinese atomic bomb? Over the past few decades, different press accounts and American scientists have repeatedly—and wrongly—credited Tsien for helping build the Chinese bomb rather than those instruments that would carry such a bomb to its destination. Tsien did, however, play a minor role in the history of the Chinese atomic bomb itself.

For instance, he spoke to the government on atomic issues right from the beginning. During the 1956 meetings for the twelve-year state plan for science, Tsien talked about the importance of neutron reactions and thermonuclear reactions and even the possibility of using atomic energy to propel airplanes and submarines.

Because of his high position in China, Tsien was kept informed about the latest developments concerning the atomic bomb. In 1960, Tsien recommended his protégé Guo Yonghuai for secret atomic work, in the process creating a personal liaison between the Fifth Academy and the Ninth Academy (the government organ for atomic research) to better achieve the goal of mating a Chinese-built missile with a nuclear warhead. Guo, who was serving under Tsien as deputy director at the Institute of Mechanics, joined the Ninth Academy, where within eight years he made significant contributions in the areas of pressure, vibration, structural strength, and environmental and flight testing.

On at least one occasion, news about the Chinese nuclear program came from Mao Zedong himself. Told by a central special committee that the first Chinese atomic bomb would be successfully tested by October, Mao ecstatically announced to Tsien and other scientists in 1964 that the Chinese had made great strides in atomic bomb research: "Women gao yuanzidan hen you chengji a!" Within months, on October 16, 1964, China exploded its first atomic bomb with a yield of roughly twenty kilotons at Lop Nor, a base in the Taklimakan desert of the northwestern Xinjiang province, thus becoming the fifth country in the world to detonate an atomic weapon and to join the international nuclear club, after the United States, Soviet Union, United Kingdom, and France.

Tsien also spent some time consulting on different nuclear projects. On December 28, 1966, the Chinese detonated an atomic device five hundred miles west of Lop Nor in the Tarim desert of Xinjiang province. A few days later, on December 30 and 31, Tsien attended a seminar held at the test base in which he and eight other specialists acknowledged that the test had been successful and urged that the Chinese proceed with a hydrogen bomb test. In addition, it appears that Tsien worked with the nuclear physicist Qian Sanqiang as consultants on a nuclear submarine project before 1970, although details of Tsien's involvement and of the project itself remain vague.

"Atomic development was not Tsien's field," He Zuoxiu wrote of Tsien. "But he favored its development."

Judging from assessments by Chinese and American scientists, Tsien did better than even he could have expected. He inspired and trained those below him—emphasizing the importance of the basics such as fundamental theory and reference materials—while securing support from those above him. He was capable of organizing, decades in advance, gargantuan projects involving thousands of scientists, as well as introducing the kind of engineering systems that could track the tiniest of details within an organization. He played a key role in taking a primitive military establishment and transforming it into one that could deliver nuclear bombs intercontinentally; he initiated and guided numerous projects that brought China into the space age.

One has only to look at the scale of Chinese space development to grasp the full impact of Tsien's leadership since the 1950s. Had Tsien not returned when he did, the central government of China might have hesitated in funding a missile program, which would have caused considerable delays in its development. China's space program is now a vast enterprise, with at least three missile launch sites, a full range of satellites, numerous supersonic and hypersonic wind tunnels, a sophisticated series of Long March launch vehicles, a medical training program for astronauts, the ability to manufacture both liquid and solid propellent fuels, and government-sponsored companies such as the China Great Wall Industry Corporation and the China Precision Machinery Import/Export Corporation to sell its rocket technology abroad.

But behind the triumphs, there was a darker side to Tsien's life, which most people never had the opportunity to see. He had to pay a political price in order to achieve his scientific accomplishments—a price exacted during one of the most violent eras of modern Chinese history.

Becoming a Communist

There is no question that in order for Tsien to keep his position within the burgeoning missile infrastructure he had to make numerous political compromises, each of which brought subtle but definite changes in the nature of his conscience, outlook, and character.

Herewith is the story of how Tsien slowly evolved from scientist to politician: how the loner and individualist became irreversibly tangled in a web of bureaucracy and conformity. I tell the story in full knowledge that much of it remains in shadow: that some sources can never be revealed in order to protect careers and families in China, that some stories can never be confirmed, and that some questions will forever remain unanswered—except, perhaps, by Tsien himself.

In 1955, when Tsien and his family returned to China, they moved into a gray Soviet-style apartment complex in a government housing compound off the busy Zhongguan Cun Boulevard in Beijing. Every building in the compound was identical: rectangular, three stories tall, constructed of earthen brick and concrete. The only way to tell them apart was by number. The lanes between the apartments were dusty and packed with dirt, interspersed with a few trees but no grass. It was a bleak landscape of grays and browns—devoid of any sense of color or individuality.

By Beijing standards, however, Tsien was in an enviable position. His salary as a government scientist was *te yiji*—meaning, literally, "super first class." He owned one of the few Soviet cars in China. His apartment was large—six rooms for a family of four, with material luxuries far beyond the reach of the average Chinese citizen in Beijing. Limited space and salary reduced the possessions of most workers in the city to a few items of furniture, clothing, and small objects like fountain pens, watches, and cameras. In contrast, Tsien had a living room stocked with a large collection of classical music, a phonograph, and a piano; a study with a desk, two tape recorders, three bookshelves, and a glass cupboard with scroll paintings, music scores, and records; and the kitchen with running water. There were two or three big windows for every room, and some had cement balconies.

Although the new government of China had sworn to destroy all the prerogatives of wealth and class, those who occupied the top rungs of the new regime enjoyed live-in chauffeurs, servants, and nannies as well as many other trappings of ruling-class power. Tsien was no exception. He wore simple clothes and gave lip service to the philosophy that everyone was equal while enjoying household help to cook his meals for him and to take care of his children.

His power forever changed the direction of life for his family. Ying was one of the greatest beneficiaries of Tsien's new celebrity status in China. No doubt Tsien's reputation, coupled with her talent, helped her secure a faculty position first at the Center for Experimental Opera and later at the Central Conservatory in Beijing. Free of any cooking or cleaning responsibilities, she had time to teach music at her home two days a week, give public recitals, and tape her singing and piano performances so that Tsien could critique them.

But Ying must have realized that her status in Communist China was unusual, maybe even precarious. Her father had been a Nationalist official— not just any official, but a famous one who had served as a trusted military advisor to Chiang Kai-shek himself. Given her aristocratic lifestyle before the Communist revolution, she was technically a member of the enemy camp. But her marriage to Tsien protected her and, ironically, permitted her to continue living like an aristocrat under the new Communist regime. This is not to say that Ying did not work hard to create a new image for herself to fit the demands of the new government. When the remaining Chinese troops returned from

Korea, Ying went to the train station to welcome them. She toured the country giving performances for coal miners, factory workers, and other members of the working class. She was attentive to the opinions of Beijing's elite. When Premier Zhou Enlai made an offhand comment in 1956 about failing to understand the lyrics to her Mozart and Schubert selections at a concert, Ying resolved to master Chinese opera so that she could sing in her native tongue.

Tsien's children struggled to adjust to their new country. On the surface, it seemed as if life couldn't be better for them. They attended one of the best elementary schools in the city: Beida Fuxiao, a program administered by Beijing University similar to the prestigious experimental school run by Beijing Normal University, where Tsien had received his education as a boy. Neighbors remember them stepping aboard the yellow school bus that took them to school, wearing special red buttons that would permit their entrance to the compound when they returned. But neither Yucon nor Yung-jen could speak Chinese when they first arrived in Beijing. During the first few months they had a difficult time in China, but rapidly they picked up the language and gradually forgot their English.

Tsien's social life during his first few years in China was restricted primarily to his family and a small group of friends. At the end of a long day, Tsien would return home for dinner and spend some time with his children before retreating to his study or going back to the office to do more work. On Saturdays, Tsien usually met with a few other high-ranking scientists to study Communist Party thought. On Sundays the entire family went downtown to visit Tsien's father and Ying's mother, who now both lived in Beijing. Some evenings, Tsien and his family occasionally dined or took walks with other American-trained scientists in the compound. The families of three scientists Tsien had known at Caltech—Guo Yonghuai, C. Y. Fu, and Luo Peilin—lived in apartments nearby; Tsien's wife chatted with their wives during parent-teacher conferences and the children went to school together. For Tsien, they formed a comfortable inner circle: linking the past with the present, and memories of Caltech with China.

Tsien, however, remained an exceedingly private person. The most difficult thing to gauge about him during those early years in China was his personal feelings toward the United States. It is obvious, however, that the details of Tsien's deportation remained a sensitive issue for him. Most of his colleagues in China were not sure whether Tsien had actually been deported, and many thought he had chosen to return to his homeland of his own free will. During interviews with the Chinese press, Tsien emphasized his detention rather than his fight against the INS to lift his deportation order. Tsien, as one of his friends recalled, did not even like to hear the word *deportation* mentioned in his presence. It isn't clear what his motives were in being tight-lipped about the deportation: whether he wanted to forget a painful period in his life or he reasoned that it couldn't hurt to have others believe that he had intended to return to China all along.

On the surface, Tsien bore no ill will toward the country in which he had lived for twenty years. "I will never forget those fair-minded, decent Americans who helped and supported me during my five years of detention in the United States," he said of his Caltech friends to Chinese news reporters when he first returned to China. "They, like all ordinary peace-loving Americans, are quite different from the U.S. government. The actions of the U.S. government are not their actions. The Chinese people have no ill feelings towards the American people. We desire to be friends and coexist peacefully." In retelling the story, Tsien astonished Americans when he told the Chinese press that he had never been a Communist and that he did not smuggle secret technical papers out of the United States.

But he mystified his former colleagues at Caltech with a strange letter, written in response to a request to provide a statement for Kármán's scrapbook on the occasion of his seventy-fifth birthday on May 11, 1956. The letter read:

On this occasion of your seventy-fifth birthday, Dr. von Kármán, what would be the proper words for a greeting? Shall I speak about our happy days together in Pasadena, in your house in Pasadena? No, that would not be proper, for I am not just your friend but, more important, your student. Shall I speak about your great contributions to science and engineering, and wish you will do more in the forthcoming years? No, that would be only a restatement of a world-known fact and a repetition of a very common birthday greeting. I wish to say more, to say something which may have a deeper meaning, because you are my respected teacher.

I presume that at the heart of every sincere scientist the thing that counts is an everlasting contribution to the human society. On this point, Dr. von Kármán, you may not feel as proud as you might feel about your contributions to science and technology. Is it not true that so many of the fruits of your work were used and are being used to manufacture the weapons of destruction, and so seldom were they used for the good of the people? But you really need not think so. For, since I have returned to my homeland, I have discovered that there is an entirely different world away from that world of USA, where now lives 900 million people, more than a third of the world population, and where science and technology are actually being used to help for the construction of a happy life. Here everyone works for the common dedicated aim, for they know only by working together can they reach their goal in the shortest possible time. In this world, your work(s), Dr. von Kármán, are treasured, and you are respected as one who through his contributions to science and technology is helping us to achieve a life of comfort, leisure and beauty. May this statement then be my greeting to you on the occasion of your seventy-fifth birthday.

Kármán was crushed when he received the letter. He was already very sensitive about the entire issue of Tsien's deportation (one of Tsien's friends, William Sangster, recalled how Kármán had tears in his eyes when he told Sangster the full story of Tsien's expulsion). Decades later, a number of scientists remember Kármán standing in a daze and shaking his head, as he read the letter over and over. "That *really* hurt von Kármán!" said Joseph Charyk, one of Tsien's former students. "I saw him talking about the letter, he even showed me the letter, and he was really shook. He said he couldn't believe—here's a guy who he had worked with so closely for years and years and he didn't understand how Tsien could say some of the things he had said in that letter."

Kármán's friends tried to comfort him by pointing out the more positive phrases in the letter and suggested that it had been written with a political slant in order to pass a censored mail system. Some even speculated that Tsien, living as he did under a totalitarian government, may have been forced to write it. Others, however, were furious that Tsien had distressed Kármán with a letter that hinted the elderly scientist had prostituted himself by working for the military. "I had prepared Kármán for a period of joy and celebration," remembered William Zisch, "and would have preferred that he never got any acknowledgment of his birthday from Tsien at all!"

Decades passed, each bringing little more than a few scattered letters or cards from Tsien to his former colleagues. Lacking any substantial contact with Tsien, many wondered how he was faring in Communist China, especially during the numerous political crises of recent history. It was only when the author read news reports from that time and interviewed a number of Tsien's associates in China that she discovered a little more information about him than that commonly known in the United States. The rest of this chapter will discuss Tsien's personal experience in China within the context of four major political upheavals in the People's Republic of China: the anti-rightist movement, the Great Leap Forward, the Cultural Revolution, and the Tienanmen Square massacre.

The Anti-Rightist Movement

In 1956 the government began enacting new policies that seemed to guarantee greater freedom of expression in China. That spring, Chairman Mao announced his "Let a Hundred Flowers Bloom" policy, which was based on the slogan "Let one hundred flowers bloom; let one hundred schools contend." The movement called for the flourishing of art, literature, and science. The following year the government urged intellectuals to openly criticize the Communist party. Mao insisted that they "say whatever they want to say, and in full."

What many intellectuals failed to anticipate was that the Hundred Flowers movement would bring a backlash from the government in which hundreds of thousands of critics were punished, silenced, or killed. Exactly why the government punished the critics so soon after encouraging them to speak out is subject to debate. Some believe that the government was shocked by the onslaught of criticism and decided to clamp down before it got out of hand, while others were later convinced that the Hundred Flowers policy was nothing more than a ploy to root out possible dissidents in the party.

Tsien, the scientist who had fallen victim to the purges of McCarthyism, now witnessed a reprise of the same phenomenon in his home country that was far more severe and far more brutal. He himself never came under serious attack, for during the Hundred Flowers campaign he refused to criticize the Chinese government but proclaimed his loyalty instead. Others were not so cautious. Qian Weichang (also known as Chien Wei-zhang or W. Z. Chien), Tsien's former colleague at JPL, had boldly proclaimed, "I will definitely not join the Communist Party" and in May 1957 became part of a circle of six professors in Beijing who drew up a program demanding more academic freedom for scientists. In June, when the purges began, a series of meetings were held in Beijing so that other scientists could denounce Qian Weichang.

Tsien seemed appalled by the reversal. During one of Qian Weichang's denunciation meetings, people observed that Tsien failed to say a single word. (He later told his secretary that he was too confused and bewildered by the entire situation.) Who would have known better than Tsien the feelings of hurt and helplessness that stemmed from being labeled traitor or spy?

It appears, however, that his past experience made him more callous about persecution than compassionate—or at least more determined that what happened to him in the United States would never happen to him in China. On July 17, 1957, Tsien issued a statement to the *People's Daily* to denounce his former friend. Qian Weichang, Tsien declared, was nothing more than a "liar and political maneuverer, without even a little of the scientist's nature" and "an ambitious politician of the most vile and savage type."

The purges left a psychological imprint on Tsien, rendering him more cautious and politically sensitive than ever. When he first settled in Beijing he wore American-style clothes: a neatly pressed shirt or a checked or flowered sweater, sometimes a suit, a hat, and a walking stick. But after the anti-rightist movement his clothes started to change into the blue or gray of the standard Chinese Communist uniform, helping Tsien blend at least visually into the uniform blue of the city's population.

The Great Leap Forward

Tsien soon encountered a widespread national frenzy even more unexpected and bizarre than the anti-rightist movement.

In January 1958, Chairman Mao had announced, with utmost seriousness, his goal of mobilizing the people to destroy the "four pests" of China: flies, rats, sparrows, and mosquitos. Ever impulsive and romantic, the leader of the most populous country in the world believed that anything was possible if the people in China set their minds to do it. And being distrustful of all intellectuals he dismissed the warnings of potential ecological consequences. Mao's movement to make drastic improvements in agriculture, industry, and the economy was later known as the infamous Great Leap Forward in China.

In 1955, Tsien had returned to China with the expectation of building up its nuclear weapons and space program. How was he to know that three years later, he would be crawling on his hands and knees in a Beijing alley, looking for fly larvae?

Near the Institute of Mechanics in Beijing was a *hutong*: a cool, narrow alley between two low structures of dust-gray brick. There, Tsien could be seen digging up the soil with a spade and probing the ground for tiny wormlike bits of fly larvae, white as a grain of rice. Whenever he spotted one, he snatched it up gingerly with a pair of chopsticks and dropped it into a jar. He tried his hand at adult flies at well. "I still remember Tsien in that *hutong*, swatting at flies with a swatter," one employee remembered. "At the end of the day, we'd count to see how many we had killed. Yes, indeed, we counted them! We'd compare numbers and brag about how well we had done."

Then came the days when virtually the entire population of China mobilized to kill sparrows. On a particular date designated by the government, people all over the country would beat on cymbals or pots, wave sheets tied to bamboo poles, or shriek at the top of their lungs from rooftops or trees to keep the frightened sparrows in the air until they dropped dead from exhaustion. In Beijing, Tsien and several other scientists were responsible for keeping the *hutong* free from sparrows. Decades later, his friends remembered vividly the image of Tsien hollering and thrashing his bamboo stick in wild circles in the air while running up and down the length of the *hutong*.

"He would scream—oh, how he would scream!" remembered his secretary Zhang Kewen, doubling over with laughter at the memory. "We'd all go out— the whole research group—during a specified time. In this respect China was very democratic. For whether you were a high official or a small child, we were all out there, striking at birds. I'd be there, Guo [Yonghuai] would be out there, Tsien would be out there. Of course Tsien wasn't at each one of these activities but he would sometimes take time out of his busy schedule to participate."

Not everybody approved of what Tsien was doing. "It was quite embarrassing," recalled Luo Jin, son of Tsien's close friend Luo Peilin. Luo Jin's brother was on a train when he heard a stranger badmouthing Tsien. The stranger told Jin's brother how he could understand why ordinary citizens would participate in this madness, for they were ignorant, but Tsien, he emphasized, was a scientist and therefore should know better than to willfully promote the destruction of an entire population of birds and ruin the ecosystem of China. Even Tsien's personal secretary was skeptical of his actions. "Now, I felt a great scientist like Tsien shouldn't be engaged in such work!" Zhang said years later. "I told him it was a big waste of time. But he said he wanted to live this new lifestyle, to understand this new society. He was quite willing to do this, and determined to understand what was going on and to try things out."

Perhaps Tsien's behavior would be more in character had he been a slogan shouter and political trend-sniffer in the United States. But he had been no such thing. Rather, Tsien had been remembered as a sarcastic, acid-tongued critic of theories much less idiotic than this. But where was his sarcasm now? Could his experience with McCarthyism in the United States and the anti-rightist movement in China shaken him to such an extent that he was now determined to play it safe? Or did he, being human, simply get carried away by the euphoria of the times?

More surprises lay ahead. During the spring of 1958, Tsien began aggressively to preach the party line to the scientific community. Excerpts of Tsien's comments at the fifth meeting of the Scientific Planning Commission of the State Council appeared in the March 7 issue of the *People's Daily:*

[For] our old scientists—the leaders of the scientific ranks—their responsibility is great. They must be able to mobilize the masses and rely on the masses. But if they are to be able to do this, they must not only resolve to be red, they have to *really* be red, red all the way through. [They] must burn away all vestiges of bourgeois thought, and all sorts of arrogant and self-important, selfish and self-serving ideas. . . .

Of course, as it is clear that this is a burning, burning in the fire of mass and self-criticism, and perhaps [the scientist] will endure many sleepless nights.

But Tsien warned his audience that if the scientist did not change, he would fall behind "objective reality" and end up as garbage, dead though the body would be alive. "So it is better to burn," Tsien concluded. "Let us throw our-selves into the fire, and from it obtain new life. Only with a new life can we coordinate our steps, and then we will truly realize our potential."

A self-critical article by Tsien followed in the April 28 *People's Daily*. In the article, which was entitled "The Utilization of Collective Wisdom Is the Only Truly Good Method," Tsien chastised himself for losing confidence in the Chinese space program, claiming that he had been too bourgeois and too elitist to think that the uneducated working-class masses would actually be able to build a missile. "Mobilizing everyone may look easy, but to an individualist, that is, to someone with bourgeois thoughts like me, it is not easy to accomplish," Tsien confessed. He concluded that anything in China was possible. "Our power will be boundless and there will be no problem we can not overcome," he wrote. "I have now recovered the great optimism I had two years ago and believe that an ambitious scientific leap forward is completely possible."

It is not clear what Tsien's motives were in writing this article. As proud as he had been in the United States, it seems unthinkable that he would have humbled himself willingly in public now. Did Tsien write the article of his own volition? Or did he write it under pressure from the Chinese government? While we will probably never know the truth, observers of the Chinese science community doubt that the government coerced Tsien to write the article. Yao Shuping, formerly a physicist and official historian of the Chinese Academy of Sciences, claims that in 1958 there were so many people eager to debase them-selves to prove their loyalty to the government that publication of such a letter was an honor, not a punishment. Tsien's status as one of the country's leading scientists guaranteed that his letter would be published.

The article seems to have provided a tremendous boost to his career. In

1958, Tsien's stature began to rise considerably. He was seen as a symbol for the entire Chinese space effort: something akin to being the nation's Wernher von Braun. That year, Tsien was asked to write popular science articles and deliver lectures on the Soviet space program. He conferred with Mao during a Chinese Academy of Sciences exhibition. He received a multitude of honorific offices, such as the chairmanship of the compilation, translation, and publication committee of the Chinese Academy of Sciences, the directorship of the Department of Mechanics of the newly founded University of Science and Technology, and membership in the Scientific and Technical Association. He even obtained political office in September 1958, when he was elected deputy for Guangdong province to the legislative National People's Congress.

Then in December came the crowning honor: an invitation to join the Communist Party.

Only a few years before in the United States, Tsien had denounced the philosophy of Communism. Now he wholeheartedly embraced it. "I was so excited at becoming a party member that I could not fall asleep that night," he gleefully declared years later. Other scientists considered the invitation Tsien's reward for his loyalty to the government during the anti-rightist movement. Tsien himself considered it the high point of his life.

The year 1958 seemed to mark the true beginning of Tsien's transformation into a hard-line politician. That year, Tsien found himself doing something that many of his colleagues considered undignified, irresponsible, and downright shocking. It came in the form of a controversial article Tsien wrote for the June 1958 issue of a magazine entitled *Kexue Dazhong* [Science for the Masses]. (A shorter version of the article also appeared that month in *China Youth* magazine.) In the article, Tsien claimed that the "final limit of yield" for agriculture depended on solar light and energy of unit area. It was theoretically possible, Tsien wrote, to increase the yield of a crop by a factor of twenty. "We only need necessary water conservation, manure and labor for the yield of the fields to rise ceaselessly."

The article stunned the scientific community of China. What, they wondered, was Tsien trying to achieve with such an article? Surely he did not really believe that China would increase agricultural production twentyfold—or did he? Was he, like so many others, buying into the propaganda that anything was possible—or was he shrewdly trying to curry favor with Mao? Whatever his motives, other scientists were offended by the article. One of his closest friends in China confided to the author years later that Tsien's article was not only ridiculous but "intolerable."

The article, however, was well received by Mao, who had long dreamed of a utopian society of peasants working together to produce surplus food. In the summer of 1958, Mao ordered that thousands of small peasant collectives be organized into gigantic centralized bureaucracies. Each province established hundreds of large communes, each encompassing some two thousand to twenty thousand households.

Mao also believed that China would overtake Britain in industrial output within fifteen years. In August 1958, Mao announced at the Politburo conference in Peitaiho his goal of increasing steel production in China to 10.7 million tons—a 100 percent increase over 1957 production.

But where was this steel to come from? Mao had a plan. Why not put a steel furnace in *every* Chinese backyard, so that every person in the country could contribute to the steel effort?

Just as the Four Pests campaign had captured the imagination of millions, so did the drive for steel production. In the autumn of 1958, people throughout China began to sacrifice all of their metallic possessions to wood-burning furnaces, convinced that a mighty steel empire would rise from the molten remains of pots, pans, and bedsprings. Tsien was soon confronted with a spectacle the city of Beijing had never seen before or since. On virtually every corner, signs, posters, and loudspeakers hawked the success of the steel drive. The streets were filled with fire and smoke from primitive blast furnaces: crude pyramidal structures of brick, or oil drums lined with clay.

All serious work came to a halt at the Fifth Academy as engineers and researchers searched their homes for metal to feed the furnaces. They smashed woks and pans into pieces, ripped the knobs from the doors, pried loose railings and gates from the earth, broke the metal frames from the windows. "We went to pick up coins from the streets, the courtyards, junkyards—picked up whatever we could get," remembered one rocket scientist from the Fifth Academy. "We stopped working and went out and carried them back. All of the scientists were involved—everyone. Then, like stirfry, we tried to melt the metal. We put the fire on, and stirred the pot. I don't know if Tsien was involved, but other senior scientists were."

As the nation devoted itself to steel production, Mao's ambition grew. He had actually hoped that the amount of steel produced that year would turn out to be 11 or 12 million tons, and that China would produce 120 million tons by

1962 and 700 million tons by the 1970s (twice the per capita steel output of the United States). The fact that the steel that was "produced" in China ended up as cracked, worthless lumps of metal meant nothing to Mao at the time.

The Great Leap Forward was a catastrophe. Before Mao decided to pursue his radical policies, the economy had been prospering, with the summer harvest up 69 percent from the previous year. Now it took a turn for the worse. Peasants nationwide slaughtered their livestock rather than turn them in to the government, parents in the communes beat their children to force them to eat more, and food lay rotting in the fields or piled up in train terminals as locomotive resources were devoted to transporting scrap metal to the furnaces. Ten billion workdays were wasted to pull out almost 100 million peasants from agricultural work and into steel production. The disappearance of sparrows caused crops to be ravaged by insects, and the fuel needed for steel furnaces stripped entire mountains and forests bare of wood.

The consequences of the Leap came the following year. In 1959, flood and drought compounded the problems of food wastage and rotting harvests, and famine swept through the nation. It began in the countryside, where the peasants suffered the worst. The famine moved into the cities and soon grisly stories circulated through China of people killing and eating their own babies, or people abducting and killing children and selling the flesh for food.

Surely Tsien could not help but notice the hunger that permeated Beijing. Many people were stricken with edema, a disease caused by malnutrition that left skin yellow and swollen with abnormal accumulations of fluid. The government, however, saw to it that Tsien and the scientists who worked under him were fed better than the average Chinese citizen. During the famine, engineers at the Fifth Academy received 38 *jin*, or 19 kilograms, of rice and corn a month, in contrast to university professors who received only 28 *jin*. Nie Rongzhen authorized the delivery of special supplies of food to the Fifth Academy, which included the most priceless of commodities: soybeans and fish. But even these extra provisions were not enough to satisfy their appetites. Death may have eluded the engineers, but hunger did not. "We were starving," one scientist remembered. "The Fifth Academy received special treatment because Marshall Nie was in charge and we were considered the precious property of the government. But even so, we were starving."

There remained in their stomachs a constant gnaw, a fugitive craving for fat which the scientists tried to stave off with a few mouthfuls of rice and pickled vegetables. In the Fifth Academy dining hall, built to accommodate a thousand people, there stood wide bins of corn porridge. When food had been plentiful, no one had given the bins much thought. But now that food was rationed, the scientists and workers arrived early in the morning and stood in long lines to grab a share of the porridge, which was free. Around the bins they pushed and jostled and grabbed for the ladles, leaving military caps soaking in the gruel or large white splotches on their green uniforms.

Decades later, Tsien's former American colleagues were concerned that he may have suffered during the time of hunger. In fact, many people in China today believe that Tsien may have been partially *responsible* for the famine.

Critics point to the articles that Tsien wrote during the summer of 1958, in which he claimed that it was possible to increase the yield of land by a factor of twenty. Never, they claim, had Tsien so thoroughly and willfully poisoned the well of truth. It was these articles, they claim, that caused Mao and other top officials to push ahead with unrealistic agricultural policies during the Great Leap Forward. Xu Liangying, a former physicist and historian of science who had followed Tsien's career for years, stated his opinion of him in unequivocal terms: "Tsien, who knew nothing about agriculture, wrote this article to give Mao's programs scientific justification. After this hit the newspapers, Mao followed through with his policies. This article had a *terrible* influence on Mao. And yes, it had an impact on the famine that followed, *in which 30 to 70 million people died.*"

Forty years later, when the author traveled to China to conduct a round of interviews for this book, numerous sources vehemently cursed Tsien for writing the article and possibly causing the famine. The chorus of indignant voices included Chinese scientists, students, journalists, and even friends of Tsien in the United States and China. No one understood how Tsien, a distinguished scientist, could have intentionally written an essay that contained such blatant distortion of fact.

In recent years, two prominent citizens denounced Tsien's actions in private correspondence and books while skillfully avoiding any mention of him by name. One was Li Rui, Mao's former secretary and author of a set of famous memoirs about Mao. Another was physicist Fang Lizhi, later a pro-democracy dissident who defected to the United States.

In his book *Mao Zedong's Merits and Mistakes*, Li Rui pointed out that a certain article written by a famous scientist in China had given Mao confidence to tamper with agricultural policy during the Great Leap Forward. When the Ministry of Agriculture announced in July 1958 that the food crop output for the summer harvest had reached 101 billion catties, a 69 percent increase over the previous year, a "famous scientist [had] added proof in an essay in the June issue of *China Youth*, saying that if the human race's use of solar energy could reach a few percentage points, then there could be outputs even higher than these mythic figures. Mao believed the scientist's words, and for a time he even had a new worry. What to do if there is too much food?"

Mao, Li Rui claims, then thought a way of solving the problem of overproduction: grow crops on one third of land, gardens on another third, and nothing on the last third. This plan made its way into the Sixth Plenum of the Eighth CCP Central Committee in Wuhan on December 1958, but when it was tried out in various parts of the country, it turned out to be a disaster. The *People's Daily* hastily published an editorial stressing that while this kind of planting system was to be implemented in the future, under the present circumstances the country shouldn't be in a rush to promote it.

Fang Lizhi also alluded to Tsien's article, suggesting that its influence on the Chinese leadership was indirectly responsible for the havoc they caused in China. "The result of this calculation was that the Great Leap Forward obtained its scientific basis. However, physics cannot be cheated. The calculation of this gentleman is totally wrong. It was such a pity that no one had the authority and freedom to criticize him, even from the aspect of physics. It was still more tragic that such a big party and great leader were foolish enough to believe such 'science' and to decide on a policy that would involve more than a billion people."

Could Tsien really be held responsible for the famine? This is not, and will never be, an easy question to answer. While blaming him for the deaths of millions of his countrymen may be going too far, it is safe to say that the article destroyed the respect that many intellectuals and scientists had for Tsien up to that point. To them the article symbolized Tsien's willingness to distort scientific fact to gain popularity with those in power. It is one thing for a scientist to comply silently with lunatic policies under a dictatorship; it is quite another to actively and vociferously promote them. At the very least, Tsien's critics say, he should have kept his mouth shut.

The problem with passing judgment on Tsien is that no one really knows the true motive for his writing the article. No one but Tsien knows whether he did it because he was forced to, or wasn't forced to but didn't care about anyone or anything except his own survival, or over time had truly bought into the ant-colony mentality that had swept through China and deluded himself into believing that the article, with its distortion of fact, would ultimately bring good to the people.

Today, Tsien's enemies insist that if he had written the article for political status alone, it was done at a cost of millions of lives. "A lot of people can't forgive Tsien for writing that article in 1958 and ultimately causing the deaths of so many of his countrymen," said one Chinese journalist who asked not to be named. "Tsien should accept some responsibility for the famine that followed the Great Leap Forward. Any man with compassion for others would feel terribly guilty for what he has done. But Tsien himself feels no remorse. To this day, Tsien feels that article is scientifically sound."

The Cultural Revolution

The 1960s ushered in the most secretive decade of Tsien's life. He moved into a heavily guarded compound off Fucheng Road in the Haidian district of Beijing.

Little was known about Tsien's daily routine during the early 1960s. Virtually no one was invited to his home. The few visitors who went to Tsien's home at that time remember little about it except that most of the windows gave a depressing view of the gray concrete of other buildings. Scientists who worked closely under Tsien at the academy caught only glimpses of him after work. On occasion, they saw him walking with his wife near the Summer Palace with their bodyguards, or read in the newspapers how Tsien had been received by Deng Xiaoping and Liu Shaoqi at some party the night before, but that was all. Few people knew the full details of his personal life. Socially, Tsien was simply out of their league.

Tsien lost contact with most of his friends for the next ten to twenty years. Indeed, he had already become something of a stranger to his own family. His life was Spartan. He rose at 6:00 A.M. and went to sleep at 10:00 P.M. He preferred to eat his meals in silence, interrupting the chatter of his children with an abrupt "Just eat." Years later, his son, Yucon, recalled that the only time he really talked with Tsien was when he had school grade reports for him to sign. Without warning, his father would leave town and disappear for weeks at a time—especially during the late 1950s, when China was developing its first short-range missile. Visits to launch facilities could drag on for months or half a year, with neither a phone call nor a letter from Tsien indicating that he was still alive. One prolonged absence left his wife in hysterics. "He doesn't want me anymore?" she said when questioning tightlipped officials about his whereabouts. "Doesn't want the children? And doesn't want this home? Well, then, I'll light a fire and burn the home down!" Tsien finally resurfaced but refused to give any explanation of where he had been or what he had done, and life went on in the mysterious manner it had before.

During his sequestration, Tsien received an unusual opportunity to serve as Chairman Mao's tutor. On February 6, 1964, in the privacy of Mao's quarters, Tsien tutored him on science, while two other distinguished scientists gave Mao lessons in geology, agriculture, meteorology, and ore prospecting. Tsien was the youngest of the three. Tsien never became close friends with Mao; in fact he only met Mao six times in his life. But his relationship to the leader of China was analogous to junior engineers' relationship to Tsien—only on a few special occasions could one be ushered into the presence of the master. Despite his diminished authority, Mao's status remained godlike, and it was considered an extraordinary privilege for Tsien to speak to him in person.

Exactly what it was that Tsien taught Mao is not known, but it did not appear to whet the latter's appetite for more education. Mao was convinced that only the working class held the key to truth, and years later Tsien recalled that Mao's "intention was to urge me to learn from the working people, to take them as my teachers and to make a serious effort to remould my own world outlook." On February 13, 1964, a few days after Tsien's tutoring session, Mao proclaimed: "The present method of education ruins talent and ruins youth. I do not approve of reading so many books. The method of examination is a method of dealing with the enemy. It is most harmful and should be stopped." His words would haunt thousands of scientists two years later.

In 1966, Mao's contempt for education and fear of the Liu-Deng clique crystallized in the Cultural Revolution—a reign of terror launched by Mao to subvert the entire social structure of China and reclaim his own supremacy. In March 1966, Mao urged the youth of China to rise up and partake in a new revolution: "We need determined people who are young, have little education, a firm attitude and the political experience to take over the work. . . . When we started to make revolution, we were mere twenty-three-year-old boys, while the rulers of that time . . . were old and experienced. They had more learning—but we had more truth."

"Then came the ten-year catastrophe," Tsien would write of the Cultural Revolution years later. It plunged China into one of the darkest and most violent eras of modern history. It appeared to have emerged spontaneously, but in reality the movement was carefully organized by Mao and his allies. They had quietly cultivated a network of underground agents in high schools and universities until the movement became official by the end of summer. The first

Red Guards appeared in Beijing, mostly junior high school adolescents aged twelve to fourteen, wearing red cotton armbands with the characters *Hung Wei Ping* ("Red Guards") printed on them in yellow. Big-character posters appeared denouncing university authorities, and within a week some ten thousand students had put up one hundred thousand posters all over the city, some with characters four feet high, most of them carrying accusations, insults, and threats of future violence.

It was Sunday, June 6, when the first poster went up in the Seventh Ministry of Machine Building (the new name of the Fifth Academy after the organization fell under state control). It criticized the leadership of the ministry and included some nonsense about Tsien spreading Nazi propaganda throughout China after his visit to Germany at the end of World War II. Before long, hundreds of posters had sprung up in the ministry carrying equally sensational accusations. Work halted at the Seventh Ministry as everyone avidly read the posters to learn gossip about the others. Many wrote posters in response, signing them with pseudonyms or dictating them to friends so authorship could not be traced back to them by handwriting.

Within a week, the Liu-Deng network struck back with its own campaign. Using a standard party technique to keep disturbances under control, it sent work teams into schools and bureaus nationwide to investigate the situation. The work teams began creating files on people they considered troublemakers, and putting staff members into four categories: good, fair, poor, or counterrevolutionary.

The streets of Beijing were fast becoming dangerous places. Anyone could be accosted, insulted, and slapped by an overeager teenager waving a Mao quotation book. Clothes considered too bourgeois could be slashed off in public, hair deemed too long might be shaved in the streets (the Red Guards were particularly fond of shaving half the head of a victim and calling it a "Yin-Yang"-style haircut). Restaurants, theatres, concert halls, teashops, and coffeehouses were shut down. The Red Guards broke into temples, national museums, and libraries to destroy priceless historical and cultural articles. That summer, Tsien could see bonfires in the streets as the Red Guards consigned books, artwork, fur coats, slit *qipao* dresses, gowns of silk and brocade, musical records, furniture, high-heeled shoes—anything considered western or capitalist—to the flames.

As the months progressed, the situation only grew worse. Children banded together into gangs and publicly denounced their parents, neighbors, and teachers. Houses were broken into and searched, victims forced to walk the streets in dunce caps and placards, beating sessions were held in sports stadiums—victims, gagged and bound hand and foot, were kicked and mauled to death before thousands of spectators. There was no recourse or protection for victims under the law, for the government sanctioned the violence. No one was safe: over the course of the revolution two-thirds of the leading cadres were attacked, humiliated, and sent to labor camps. An estimated four hundred thousand people were murdered.

The missile engineers at the Seventh Ministry were horrified by the speed and intensity of Chinese politics. All serious scientific work at the ministry came to a stop as everyone attended the endless parade of meetings, rallies, and political study sessions. The engineers devoted a substantial part of each day to scrutinizing the political climate in newspaper editorials so they could craft their posters and speeches accordingly. One scientist at the ministry commented years later that life wouldn't have been so bad if the ruling-class elite of the Chinese goverment was toppled once a decade rather than once every few weeks. "The power balance shifted so fast, that even Tsien—who is very fast and adaptive—couldn't adapt fast enough," he said.

The Seventh Ministry soon broke into two major factions: the "915," which supported Liu and Deng, and the "916," which supported Mao. (The names of the factions were derived from the dates on which they were founded: September 15 and September 16, 1966, respectively.) By and large, the technical people joined Faction 916 while the bureaucrats and administrators gravitated to Faction 915. The lowest-echelon workers were divided evenly on the issue and joined both.

At first the staff of the Seventh Ministry clustered into cliques in the halls, whispering among themselves and ignoring others. But the accusations in the big-character posters and the insults graffitied on the walls served as catalysts to open conflict. Before long, people were confronting others directly and screaming at each other. "First they fought with words, and then with fists," one scientist remembered. Seventh Ministry soon lurched into gang warfare. Battles broke out in which factory workers armed themselves with iron sticks. "Luckily, in Beijing, you couldn't get guns," one engineer recalled with a shudder. "In other provinces, with the opposing factions backed by the military and the state, they were fighting with machine guns."

On January 23, 1967, a coup d'état overthrew Tsien and other leaders of the Seventh Ministry. The ringleader of the coup was Ye Zhengguang, a young missile engineer who was the son of the famous Communist general Ye Ting. He had planned the coup for days after reading about Mao's endorsement of similar riots and takeovers in Shanghai. "This was a top-down coup," he was to recall years later. "It was sanctioned by the central government. In fact, we reported to Zhou Enlai and Marshall Nie to ask for permission to take over. When the ministries in Beijing started to rebel against the ministers, we asked permission to do the same from the office of vice premier Li Fuchun."

It was 10:00 P.M. on January 23 when Ye received a phone call from the office of the vice premier, which permitted the seizure of the ministry so long as the rebels did not go "overboard in their enthusiasm." Ye was advised to follow the example of the coup that had toppled the Second Ministry only a week before. When Ye put down the phone, he told his assistants to summon Tsien, Minister Wang Bingzhang, and the four other vice ministers out of bed for a midnight meeting in Wang Bingzhang's office.

They arrived, looking tired and wearing civilian clothes. Ye, backed by eight or nine of his own men, told them they were taking over the ministry.

The blood drained from Tsien's face and he staggered backwards in a near faint. Wang Bingzhang, a physically strong man, had to grab Tsien from behind to keep him from falling to the floor. Ye invited Tsien to sit down. Gently, he told Tsien not to worry because he would be protected and would keep his position as vice minister. Then Ye asked Tsien and his colleagues how they felt about the takeover. Tsien and two vice ministers immediately said they supported it, while Wang Bingzhang and the other two vice ministers opposed it. Wang declared that his power came directly from the central government of China. When Ye asked him to hand over the chops of the ministry, Wang indignantly refused.

The chops were the traditional symbols of authority in China. In the People's Republic, they were wooden disks the size of a fist, imprinted with the constitutional symbol of the country: the image of the Gate of Heavenly Peace in Tienanmen Square within a red circle. Every ministry had its own set of chops, and in the Seventh Ministry of Machine Building they were locked in a metal security box on the third floor of the main building. When Wang refused to relinquish the chops, the rebels simply cut open the security box with a blowtorch. Henceforth, January 23 became known throughout the Seventh Ministry as "The Day 916 Seized Power." Tsien was permitted to stay, Wang was ousted, and Ye acquired the title "General Servant of the People" within the ministry.

Though deeply shaken by the coup, Tsien quickly regained his composure. The next day, he looked serene and even happy. Tsien praised Ye and the rebels in a speech he gave in the Seventh Ministry's second-floor conference room, filled with hundreds of people. "This is the first time I had a breath of fresh air," he said contentedly.

His status, however, plummeted after the coup. With Ye in power, Tsien was now treated like any other employee. For the first time, he had to eat in the cafeteria like everyone else. The Seventh Ministry missile scientists remember Tsien standing in the mess hall looking perplexed, for he had never stood in line there before. He genuinely did not know what to do. One of the bureau directors saw Tsien's dilemma and fetched a bowl of noodles for him. Confused and undoubtedly embarrassed, Tsien ate alone. His position remained lowly until Zhou Enlai announced months later that the seizure of the Seventh Ministry was not appropriate, and that the power and chops Ye held were to be transferred back to Wang Bingzhang.

On June 8, 1968, the violence came to a head when Yao Tongbin, a distinguished metallurgical expert at the Seventh Ministry, was beaten to death with a steel pipe by two men. "After the metal expert died, Zhang sent out an order that no such thing would ever happen again," recalled Zhuang Fenggan. In 1968, Zhou Enlai instructed the Military Control Committee of the Seventh Ministry of Machine Building to draw up a list of experts to be put under state protection. Zhou particularly wanted to protect scientists working on the missile and satellite project and dispatched bodyguards to protect them from physical harm. While other scientific agencies were temporarily disbanded, research on missiles continued as a top national priority. Tsien was one of only fifty top scientists who received protection from the state in Beijing.

During the last years of the 1960s Tsien was clearly one of the most powerful scientists in Beijing. He was given a distinguished position—probably a rank equivalent to general—in the Science and Technology Commission under the State Council, which was later incorporated into the Defense Science and Technology Commission of the military. As a leading comrade of the People's Liberation Army, he attended receptions in the Great Hall, served as an alternative member of the congress of the Party Central Committee, and visited embassies as a distinguished official. The western press reported that Tsien had survived unscathed the ravages of the Cultural Revolution.

Even so, the Cultural Revolution had a profound effect on his entire family. Yucon had just graduated from high school and Yung-jen was in her final year, but college was out of the question for both of them. Virtually all universities in China had been shut down; professors were dispersed across the country to work with the peasants. Fortunately for Yucon and Yung-jen, their father's status exempted them from forced labor in the countryside. Instead, they joined the People's Liberation Army and would not resume their education until more than a decade later.

The Central Conservatory of Music, where Ying served as a professor, shut down during the Cultural Revolution. In 1969, Ying almost went to the countryside to join her faculty colleagues, but on the day of her intended departure the college received a phone call from a top government official forbidding Ying to leave. The official said she had to remain in Beijing for her protection and Tsien's. Ying was assigned to the nursery instead, to take care of the children of professors who had already been dispatched to rural areas. She was remembered as being especially kind: spending her own money to buy the children treats, mending their socks and clothes, taking them to the hospital when they were sick. She even brought a sofa from her own home and donated it to the nursery. She served, in essence, as a substitute mother during the Cultural Revolution.

One difficult moment came for Tsien and his family after the death of Lin Biao, formerly Mao's right-hand man. On September 12, 1971, Lin and his wife were killed during an apparent attempt to flee the country and defect to the Soviet Union. To most citizens in China, the event only served to illustrate the fickleness of power within Mao's inner circle. Only two years earlier, Lin had been lauded for his bravery for his role in the Sino-Soviet border disputes. Now Mao denounced him as a traitor. The Lin Biao incident reverberated throughout Beijing in a purge that resulted in hundreds of arrests and interrogations. No one was immune to close scrutiny—not even lower echelon engineers at the Seventh Ministry, who were appalled to learn that a vibration bed they designed for future cosmonaut training was in reality a part of a secret plan to cure Lin Biao's insomnia. Those with such remote connections to Lin were merely investigated; those with closer relationships could expect worse. Wang Bingzhang, the head of the ministry, had been close enough to Lin to be locked away in isolation for several years.

Only a year earlier, Tsien had unwittingly made a speech supporting Lin Biao, the man he assumed was Mao's heir apparent. "Tsien thought Lin Biao was loyal to Mao," one engineer remembered. "It was all too complicated for him to judge what was what. Then he got into big political trouble. Tsien had to criticize himself and say that he had made a big political mistake in 1970." Information pertaining to Tsien's self-criticism is not available to the public, for the minutes of party meetings are confidential. But rumors circulated that Tsien was forced to write self-criticisms and to submit them to the party.

A hardness seemed to come over Tsien and his wife after the incident. One of the few people who had the opportunity to observe this change was the daughter of one of Tsien's colleagues in China. (For her protection, her name has been omitted in this story.) During the Cultural Revolution, her mother had the misfortune of having previously befriended one of Madame Mao's enemies and was isolated in a *niupeng*, or "cowshed": a loose term to describe a makeshift prison where victims were watched, interrogated by peers, and forced to write self-criticisms. She was under constant interrogation by two women who accused her of being an American spy. The daughter, who had volunteered to work in the countryside, returned to Beijing in 1971. She paid a visit to Tsien's home, expecting empathy.

It was a rude awakening for her. When she was a child, Tsien's wife had usually spoken to her with sugary words while wearing the calm smiling mask of "a western politician." Now Mrs. Tsien looked at her with a scornful, almost indignant expression. It was then that this daughter of Tsien's colleague noticed that Ying had taken to wearing blue military-style uniforms to give the impression that she was a high-ranking cadre. The Tsiens then proceeded to criticize her family. "Getting information out of your mother is like trying to squeeze the last bit of toothpaste out of the tube," Tsien said acidly, in reference to rumors that her mother had proved to be uncooperative in prison. Ying, meanwhile, stared at her without smiling, looking her up and down and uttering "Hum!" or "Oh?" at her every statement. "I found this hard to understand," the protégé's daughter said years later, looking hurt. "I thought they were close friends."

Wrongly assuming that she had been sent to the countryside as punishment, the Tsiens addressed her in cold, contemptuous tones. "If I were you I'd go back to the countryside," they told her. Tsien's colleague's daughter longed to tell them that she had chosen to go to the countryside of her own free will, but instead decided to cut her call short. "I didn't want to play their political games," she recalled later. After that episode, she never had any desire to socialize with the Tsiens again.

One person who had the opportunity to see Tsien about this time was Joseph Charyk, his former student at Caltech who had served as the assistant secretary and undersecretary of the U.S. Air Force and was now the president and director of COMSAT, the Communications Satellite Corporation. In early 1972, Charyk went to China to set up satellite communication facilities so that President Nixon's visit could be televised to the United States. Upon his arrival, Charyk asked Chinese officials if he could see Tsien in person. Shortly afterward, his Chinese hosts ushered Charyk into an elegant restaurant, where Tsien soon arrived with several escorts.

Tsien first asked Charyk if he wouldn't mind if he expressed some formal welcoming remarks in Chinese. He then gave a speech denouncing certain Chinese officials who were out of favor, which the interpreter translated word for word for the guests. When he finished, Tsien said, "Now we can sit down and eat," and proceeded to spend the rest of the evening speaking in perfect English. "There wasn't a doubt in my mind," Charyk said later, "that he was required to make this little speech."

The mood of the dinner was cordial, and Charyk and Tsien talked about old acquaintances and friends. When asked about his activities, Tsien grew vague and changed the subject. After dinner, the two men took a walk around the courtyard of the restaurant. During their stroll, Charyk told Tsien he had heard a rumor that he had been ill. Tsien then made a cryptic remark about how there existed a significant relationship between the mind and the body, and now that things had changed in China he felt fine. "That was a reference, I think," Charyk said, "about not very pleasant experiences he may have had during the Cultural Revolution."

In retrospect, Tsien must have realized how fortunate he was during these times, for he could have easily been sent out to the countryside, thrown in prison, or killed. No one was truly safe. Tsien had sterling scientific credentials, but that was no guarantee from attack: Qian Sanchiang, the nation's top nuclear physicist, had been denounced in 1968 for being a "capitalist roader" and a "secret agent." Tsien had some political power and good relations with the leadership, but that was also no guarantee against attack: he had come from a wealthy family, spent twenty years in an enemy nation, and married a woman who had been the daughter of a top Nationalist official.

Even if Tsien had possessed an unblemished family background of humblest peasant origins, this was still no guarantee against danger. His colleague Guo Yonghuai came from a safe peasant background, yet he perished in an airplane accident one rainy day in December 1968 when an inexperienced pilot misinterpreted the signal from the air traffic controller and tilted the plane as it hit the runway, causing it to burst into flames and kill everyone aboard. The accident, Guo's daughter said later, was chiefly the result of putting incompetent people in positions of authority at the airport through policies enacted during the Cultural Revolution.

Looking back across the ruins of time, Tsien saw the fate that could have been his. Many of his friends had been shut up in *niupengs*. Some of Tsien's friends were tortured. Luo Shijin, who had received his Ph.D. in aeronautics from Caltech in 1951 with Tsien as his thesis advisor, not only watched his wife's co-workers goad her to suicide but endured nearly a year of severe sleep deprivation when imprisoned in his office by the Red Guards. The guards interrogated Luo day and night, accusing him of being an American spy and slapping him awake when he tried to sleep until the dazed Luo began to hallucinate. Miraculously, Luo survived, returning to a ransacked home where his children had lived like hunted animals.

"If Premier Zhou Enlai had not made painstaking efforts to guarantee my safety during the ten-year turmoil," Tsien said years later, "I might have been dead long ago."

The Reign of Deng Xiaoping

In January 1975, Mao appointed Deng Xiaoping vice chairman of the military committee of the chief of staff of the PLA. That year, Zhang Aiping, one of Deng's chief supporters, also rose in power and was appointed minister of the Commission of Science and Technology for National Defense—a position that made him Tsien Hsue-shen's superior in power.

But later that year, Mao began to believe that Deng and Zhou Enlai were conspiring to wipe away all the gains he had made during the Cultural Revolution. By the end of November, Mao had turned against Deng, who was subsequently purged and publicly criticized.

While Deng's power was at its low ebb, Tsien made speeches at a high-level party committee meeting to denounce Zhang Aiping, accusing Zhang of aiding Deng to subvert Mao's plans. He also wrote a critical poster on Zhang and displayed it in the State Commission on Defense, Science, and Technology, near Beihai Park. In the poster Tsien relayed something that had occurred during the 1960s when he accompanied Zhang Aiping to a launch site. Apparently, Zhang had pointed to a map and said, "This is Mongolia. It used to be Chinese territory." Tsien used this statement to accuse Zhang Aiping for being an ambitious national chauvinist who wanted to take Mongolia back from the Soviet Union. The poster, copied word for word, proliferated all over Beijing.

The situation for Deng worsened when Zhou Enlai died on January 8, 1976. When authorities attempted to remove wreaths placed in Zhou's memory in Beijing's main square, one hundred thousand people rioted, hundreds were arrested, and Deng Xiaoping was immediately blamed for the disturbance. That month, Mao officially fired Deng, relieving him of all posts except for party membership. Once again, supporters of the Gang of Four started a movement to criticize Deng and Zhang in the Seventh Ministry of Machine Building. Cadres and staff were persecuted, the ranks were split apart in factions, and the ministry plunged into chaos.

Mao's death came on September 9, 1976. During the fifth day of mourning, Tsien served as a guard of honor at Mao's funeral. On September 16, 1976, the *People's Daily* published an article that Tsien had written, entitled "I Shall Remember Chairman Mao's Kind Teachings All My Life." The article raved about Mao and his influence on Tsien: "Mao not only rescued me from my plight abroad, he personally led me onto the path of revolution and freed me from the shackles of old conventional ideas," Tsien wrote. He also harshly criticized Liu Shaoqi, Lin Biao, and Deng Xiaoping, calling them the "sworn enemies of all scientific workers taking the revolutionary road."

Tsien would soon realize that his article was a terrible mistake. On October 6, the military launched a coup d'état against the Gang of Four, arresting Mao's widow and her allies and accusing them of almost every political crime imaginable. Although Mao's chosen successor was Hua Guofeng, a party secretary from Hunan province, Deng Xiaoping made a second comeback with a rapidity that must have stunned Tsien and would become chairman of the entire Chinese Communist Party in March 1978.

Tsien scrambled to repair the damage to his political standing. He proclaimed his loyalty to the new regime under Deng and denounced the Gang of Four. In July 1977, the journal *Red Flag* published his article "Science and Technology," in which Tsien wrote: "The 'Gang of Four,' of course, opposed our catching up with and surpassing the world's advanced levels." In a Chinese magazine entitled *Economic Management*, he blamed Lin Biao and the Gang of Four for causing China to remain one of the poorest countries of the world.

Unfortunately for Tsien, Deng's rise caused Zhang to regain his power and once again to preside over Tsien. "It really didn't make any sense for Tsien to write that poster about Zhang Aiping," one engineer said. "If he wanted to criticize Zhang he should have simply repeated all the criticism in the newspapers. He should have followed the crowd and repeated what everyone else was saying. Then people would listen to it and forget it. But the problem with Tsien was that he enjoyed thinking up new points and making new comments, and then people would remember those statements forever."

Tsien began to lose his power in the Chinese missile hierarchy, primarily for his criticism of Deng and Zhang. "I believe things went sour for Tsien after 1976," speculated Xu Liangying, a distinguished historian of science and pro-democracy activist. Tsien was slowly pushed into figurehead status, revered but generally ignored during important meetings. In the latter half of the 1970s, as Zhang supervised the new program for missile development, missile scientists noticed Tsien being slowly ostracized at meetings, snubbed by younger, more powerful colleagues. He would sit at meetings, his eyes looking dead, trying to ignore the sarcastic comments aimed his way. "If you had seen him at that time, he was not in a very cheerful mood," recalled Hua Di, a missile scientist who later defected to the United States. "His mood was not good at all. When he attended meetings he became very silent."

Under Deng's direction, China gradually recovered from the ravages of the Cultural Revolution. Scientific and industrial activity, dormant for so long, resumed in the late 1970s. Tsien's family also returned to normalcy. Yucon had become an officer in the army. Yung-jen resumed her education by enrolling in a Shanghai military medical school. Ying went back to the Central Conservatory, where she continued to teach and translate foreign music into Chinese.

Tsien became something of a dilettante. During the late 1970s and 1980s he served as advisor to scientific journals, headed numerous scientific organizations, wrote prefaces for aeronautics magazines, and responded to much of his fan mail by hand. There seemed to be no subject on which Tsien did not have an opinion. He spoke at length on acupuncture and the use of methane and marsh gas energy in the countryside. He urged that UFO studies be included in the teaching of geoscience. He proposed the establishment of a special state committee to work out waste collection systems. He praised Chinese women as being as talented as men.

Tsien also become a mouthpiece for the Chinese government, extolling the virtues of Socialism. At a meeting in the People's Great Hall on July 25, 1981, Tsien announced: "We cannot help saying aloud that the CCP is great, glorious, and correct." He credited Socialism for the advance of Chinese science. He spoke of Socialism as fervently and zealously as a new convert would of religion. Next to Marxism and Maoism, his own ideas were nothing—as insignificant and dismissable as "small bubbles in the sea." At the same time, Tsien stressed that the Chinese were not inferior to the western powers. At the Twelfth National Congress on September 4, 1982, Tsien boasted that the development of Chinese nuclear weapons proved that the Chinese were "definitely not stupid." Three months later, he gave a speech at a panel meeting at the National People's Congress claiming that the Chinese were no worse than their western counterparts.

Nothing about Tsien's propaganda speeches surprised anyone except when they touched on the realm of the paranormal. In 1979, Tsien took a stand on the controversial subject of ESP, which was then being hotly debated throughout China. The controversy began when a twelve-year-old boy named Tang Yu from Sichuan province in southwest China claimed he could read words on a piece of paper placed behind his ears. Newspapers reported that the boy had extrasensory perception that permitted him to "hear" written words on paper and that Yang Fang, the head of the province, believed the boy's story. Suddenly, people all over China discovered that they, too, possessed ESP. It was a trend reminiscent of the UFO craze that had swept through the United States during the same decade.

In an article for the journal *Science and Technology*, Tsien pointed out that almost a thousand other teenagers with ESP had surfaced throughout China. Perhaps ESP was linked with *qigong*, an ancient and mystical Chinese art of healing, he said. Tsien speculated that the most famous doctors of traditional Chinese medicine were probably *qigong* masters. He urged the government to devote more resources to the study of the brain so more could be learned about ESP, *qigong*, and acupuncture. This study, Tsien proclaimed, could be the next big breakthrough in science.

One scientist who witnessed firsthand Tsien's zeal for the topic was Milton van Dyke, Tsien's former student at Caltech and now a professor of aeronautics at Stanford University. Sometime during the late 1970s or early 1980s, van Dyke had the opportunity to meet Tsien in person at his hotel. Tsien, van Dyke recalled, took a piece of paper, jotted down a name or number on it, and then held it to his head. He told van Dyke that a young boy was able to "hear" those words. "I was frankly astonished," van Dyke said. "I nearly said to Tsien, don't you have any magicians in Beijing? But I didn't."

What had become of the Tsien Hsue-shen that Caltech had known, the man strictly intolerant of anything but mathematical proof or scientific fact? "That is so *antithetical*—so *different* from what I remember of him," exclaimed his former student Thomas Adamson when he heard the story of Tsien's obsession with ESP. "With him, it was always 'This is nothing unless you can write it' or 'Prove it!'" Tsien's behavior baffled his former associates. Was he insane? they wondered. Senile? Possibly broken by the strain of Chinese politics? Only one thing seemed certain to them: the Tsien whom people at Caltech had loved, hated, or feared had disappeared for good.

During the spring of 1989 Beijing stirred with a new wave of unrest. Despite economic prosperity, large segments of the Chinese people were voicing their dissatisfaction with the system. In the mid- to late 1980s, there had been numerous economic scandals among the high-ranking cadres, as well as nepotism, tax evasion, and other corruption. In an emerging capitalist society, many Chinese felt the government officials enjoyed an unfair competitive advantage. By the end of the decade, students and professors posed the greatest challenge to Deng's leadership by openly demanding more democracy, freedom, and participation in government.

One of the first scientists to speak out against the government was Fang Lizhi, an astrophysicist and vice president of the University of Science and Technology. In October and November 1986 the *People's Daily* began printing his articles about the importance of free speech and the distribution of power to prevent corruption. In January 1986 student demonstrations began at his university in Hefei and later spread to nearby Beijing. During the latter part of the 1980s students took to waving banners in the streets of Beijing and Shanghai to announce their dream of democracy, despite police opposition and threats from the government.

By May thousands of students decided to go on a hunger strike. If Tsien had walked near Tienanmen Square during this time, he would have seen it filled with makeshift tents and exhausted students. It had become a bustling camp of journalists, onlookers, concerned residents bearing baskets of food, and ambulances rushing starving students to the hospital. The Chinese media defied the government by openly and honestly covering the news, while the students' command of modern technology—such as facsimile machines and computer word processors—ensured that the latest developments were transmitted worldwide.

By the middle of the month, the number of people in the square exceeded one million, and the students were openly demanding that Deng Xiaoping and Li Peng resign. The Chinese leadership held brief and unsuccessful talks with the students, after which they declared martial law and ordered the students to clear the square. By the end of two weeks, the movement was faltering, but it gained renewed energy when Beijing art students used Styrofoam and plaster to craft the symbol of the movement: a woman holding a torch aloft with two hands, whom the Chinese named the Goddess of Democracy and Freedom.

Late at night on June 3, the residents of Beijing awoke to the sounds of machine gun fire. Tanks converged on the square from east and west, moving in columns down Changan Avenue. After shutting off the lights and blocking the exits, armored troops broke into the square, firing at random on the crowd. Other troops burst out of the Forbidden City, also firing. The tanks overran the encampments, crushed the Goddess of Democracy, and smashed through barricades of cars and buses. What ensued was a chaos of shellfire, helicopters, and ambulances in which more than seven hundred people were killed and thousands wounded. The square filled with smoke from torched buses, tanks destroyed by homemade bombs, and what appeared to be large piles of corpses. In the midst of the confusion, Fang Lizhi fled to the U.S. embassy and sought asylum.

The massacre sent shock waves through the international community. On June 10, 1989, Frank Press of the National Academy of Sciences suspended its cooperative exchanges with China. In a telex message to Tsien and other prominent members of the scientific establishment, Press wrote: "We are shocked and dismayed by the action of Chinese government troops against peaceful demonstrations in Tienanmen Square. We must suspend all activities for the time being. We do so in outrage and sadness."

On June 14, Tsien was seen on Beijing television at the Huairen Hall listening to Deng Xiaoping's speech about the "counterrevolutionary rebellion." Two days later, Tsien chaired and addressed a meeting at which leading scientists met to study and discuss the speech Deng had made to generals of martial law troops. Tsien proclaimed Fang Lizhi the "scum of the nation" and condemned his "treasonable act." The students involved in the demonstrations, Tsien said, were nothing more than "evil elements from the Gang of Four" and common "ruffians."

These statements helped Tsien regain some power with the central government. During the late 1980s Tsien was already ascendant when his protégé Song Jian was appointed the state councillor of China. In 1989, Song's influence, coupled with Tsien's support of Deng during the Tienanmen crisis, restored much of the favor that Tsien had lost years earlier with the ruling Chinese elite.

The government rewarded Tsien handsomely for his loyalty. In August 1989, Tsien had won the Willard F. Rockwell Jr. medal—an annual award for technological achievement given by the International Technology Institute in Pittsburgh, Pennsylvania—giving the central leadership an opportunity to shower him with compliments. On August 28, 1989, the *Guangming Daily* carried a lengthy story about Tsien, calling him "The Pride of the Chinese People" and describing his book *Engineering Cybernetics* as a "book from Heaven." On September 15, 1989, the *New York Times* reported that Li Peng, the Chinese prime minister, was publicly praising senior scientists for their support of the Tienanmen crackdown, and that television and newspapers in China were heaping Tsien with praise.

Two years later, in October 1991, the Chinese government bestowed on Tsien the highest honor a scientist could receive in the nation: the title of State Scientist of Outstanding Contribution. That month, the leadership fairly outdid themselves by gushing over Tsien in the press. The Beijing *People's Daily* devoted most of its front page to Tsien, and Li Peng sent Tsien a letter of congratulations. On October 16, an awards ceremony was held for Tsien in the Great Hall of the People, during which Jiang Zemin, then the general secretary of the CP, said that Tsien embodied the true spirit of Chinese patriotism.

He was touted as a role model for other scientists to emulate. Both Li and Jiang urged other intellectuals to learn from Tsien's behavior, and the China Association for Science and Technology issued a circular that proclaimed: "We should follow the example of Comrade Tsien Hsue-shen and learn from his noble national spirit, serious scientific attitude and down-to-earth work style. Like Comrade Tsien, we should be loyal to the party, loyal to the socialist motherland and loyal to the people."

The sad truth, though, was that Tsien no longer served as an idol for the young and idealistic. Precisely those scientists and intellectuals who had held Tsien in high esteem for his directness and honesty in the past now despised him for his support of the government in crushing the pro-democracy movement. While the Chinese government promoted Tsien as a hero who braved the Americans to return to China, many among the younger generation of scientists perceived him as nothing more than a member of the old guard—a man more politician than scientist.

Epilogue

Today, Tsien is known to millions of people in China. News photographs show him as an elderly and revered figure, typically surrounded by distinguished military officials at important ceremonies. He is a plump man in his eighties with a rounded face, bald pate, wrinkled and age-spotted skin; a thick pair of glasses sometimes rests on the tip of his nose; excess weight bulges through his green military uniforms. He is as much of a household name in China as Wernher von Braun was in the United States.

His personal life, however, remains a mystery. Only a handful of his most intimate friends have the opportunity to see him in the privacy of his living quarters, and even for them the meetings rarely occur more than once a year. Usually, they see Tsien only during the annual New Year's parties held in his home. No one dares to call on him casually. A visit to the Tsien home entails talking to the guards who protect the exclusive residential compound and filling out numerous forms. Even Zhuang Fenggan, one of his closest friends, who lives in one of the apartments in Tsien's compound, hesitates to knock on his door: if he has something to discuss, he waits until Tsien comes out for his daily stroll.

Because of the intense privacy with which he conducts his life, there are certain questions about Tsien that we will never be able to answer. We will never learn where Tsien's true loyalties lay during his twenty years in the United States, or the real reason for Tsien's sudden attempt to leave for China in 1950, or whether there is a remote possibility that he could have been a spy.

We will probably never find out the full extent of his contribution to the Chinese missile program, although more details are likely to emerge in the next few years. They will come from individual memoirs of scientists who worked with him as well as from an official biography of Tsien that will probably be written by his former secretary Wang Shouyun after Tsien's death.

We will never know his true feelings toward the United States, whether he is still angry at the government, or nostalgic for a visit. Publicly, Tsien has denounced the country that deported him for its capitalist system, but privately, quietly, unknown to most people in China, Tsien has permitted both his children (who are American citizens by birth) to return to the United States for further education—a sign that he may be far less hostile toward the United States than he makes out to be.

It is likely Tsien will never return to the United States, whether or not he gets his long-awaited apology. His years are numbered. According to those who know him, Tsien suffers from ill health and practices *qigong*.

When Tsien is gone, the question will remain: Did the United States sabotage itself when it deported Tsien? How much of a threat do Tsien's missiles pose to the United States?

The People's Republic of China has never used any of Tsien's missiles to attack another country and is unlikely to do so in the near future. The PRC has sold many of its primitive missiles to other countries, however, some of them hostile to the United States. One can detect traces of Tsien's influence by studying the transfer of missile technology to countries such as North Korea, Iran, Iraq, and Saudi Arabia. So far, of all the missiles developed under Tsien's guidance, the Chinese has sold Silkworm antiship missiles to Iran and Iraq and DF-3 medium-range missiles to Saudi Arabia.

The sales of these missiles raise frightening possibilities. The DF-3, for instance, puts Israel within striking distance of Saudi Arabia. Worse, the DF-3 and Silkworm missiles can be used as seed technology for the Middle East to develop intercontinental ballistic missiles, just as the Soviet R-2 missiles were used in China. Such capability is alarming when viewed in the context of threats Middle Eastern leaders have made in recent years. Iran, which has called the United States the Great Satan, hopes to achieve indigenous nuclear capacity by 1997. Muammar al-Qaddafi, ruler of Libya, once said that if he had the ability to send a nuclear missile flying toward New York he would. The full consequences of Tsien's deportation have yet to be played out—and the thread of the silkworm is still unraveling.

Perhaps it was inevitable that nuclear missile technology would proliferate worldwide, with or without Tsien's help. Who knows how many potential Tsiens there are in universities in the United States today, trying to decide between staying in the country to continue their research, or returning to their home nations to contribute to their defense effort, their fates determined by the whim of an immigration officer or a budget cut. Who is to say that it couldn't happen again, with someone of a different ethnicity, at a different school, over a different international conflict?

The greatest tragedy of the Tsien story is not his deportation from the United States and the subsequent loss and increased threat to U.S. defense, or even the years of quiet suffering he must have endured at the hands of the INS and in China during its various political upheavals. Rather, the real tragedy is the extent to which Tsien himself has apparently betrayed his own principles and bought into the system once he returned to China. There Tsien may have gradually become his own worst enemy—the very kind of rigid, unquestioning bureaucrat that he had once so despised within the INS and the U.S. government during the McCarthy era.

How stark the contrast between the young Tsien and the old. The young Tsien had devoted his early life to scientific fact. The older Tsien, whether voluntarily or involuntarily, helped not just one dictator but two spread lies and chaos. The young Tsien dreamed of a world of peace and equality. The older Tsien lived in a world governed by regimented hierarchy and helped manufacture the weapons of world destruction. The young Tsien was both Chinese and American, at heart a citizen of two countries. The older Tsien felt alienated by both. Perhaps it is no wonder that in the last days of his life the elderly scientist who once scorned superstition is now embracing it, seeking solace in *qigong*.

Notes

Introduction

xii *things began to happen in China at an incredible speed:* William Ryan and Sam Summerlin, *The China Cloud: American's Tragic Blunder and China's Rise to Nuclear Power* (Boston: Little, Brown, 1967), p. 170; John Wilson Lewis and Xue Litai, *China Builds the Bomb* (Stanford: Stanford University Press, 1988), foreword and p. 50; P. S. Clark, "The Chinese Space Programme," *Journal of the British Interplanetary Society,* 37, no. 5 (May 1984): 195; telephone interview with Ernest Kuh; interview with Zhuang Fenggan.

xiv *"deathbed":* Raymond Chuan, letter to author, May 29, 1993.

xvii *an acknowledgment from someone in government:* Personal interview with Yucon Tsien, July 13, 1991.

Chapter 1: Hangzhou (1911–1914)

1 *For the first three years of his life:* Interview with Hu Guoshu, professor of history, Zhejiang Academy of Social Sciences, Hangzhou, June 1, 1993. Hu, who is knowledgeable of Tsien's family history, is perhaps the world's greatest authority on the history of Hangzhou.

1 *In the tenth century:* "Yidai mingzhu gongji houren—Qian Liu yu Hangzhou" [An enlightened ruler's achievements for posterity—Qian Liu and Hangzhou], placard in the Hangzhou museum.

1 *By the time Marco Polo arrived:* Jacques Garnet, *Daily Life in China on the Eve of a Mongol Invasion 1250–1275* (Stanford: Stanford University Press, 1970).

2 *On the east side of the lake:* Author's visit to Hangzhou, 1993; telephone interview with Tsien Hsue-chu, cousin of Tsien Hsue-shen, December 10, 1992.

2 *From the Phoenix mountain:* Author's visit to Hangzhou, 1993.

2 *The original nine-level structure:* Zhou Feng (chief ed.), *Yuan Ming Qing mingcheng Hangzhou* [Famous Hangzhou in the Yuan, Ming, and Qing dynasties], Hangzhou History Series, no. 5 (Hangzhou: Zhejiang renmin chuban she, 1990), p. 220.

2 *The father, Tsien Chia-chih:* Personal interview with Qian Xuemin, associate professor, Renmin University of China, and distant cousin of Tsien's, and with her mother, July 3, 1993. Tsien refers to his father as Tsien Chia-chih in the questionnaires that he filled out in the United States, although the name in pinyin is Qian Jiazhi.

3 *Born in Hangzhou in 1882:* Author's visit to Hangzhou; interview, Hu Guoshu, 1993; Qian Jiazhi, "Quishi shuyuan zhi chuangshe yu qi xuefeng" [Establishment of the Qiushi Academy and its academic spirit], (n.p., n.d.) originally in *Guali Zhejiang daxue xiaokan fukan* [National Zhejiang University Journal], supplement no. 151 (May 1947).

3 *There, in his native city of Hangzhou:* Personal interview, Hu Guoshu, June 1, 1993.

3 *On December 11, 1911:* It is not clear if these are Tsien's actual place and date of birth. Different records give both Shanghai and Hangzhou as Tsien's birthplace and between 1909 and 1912 as his birthday. For instance, the *China Pictorial* (November 1989) claims Tsien was born in Shanghai, while the *Beijing Review* 34, no. 48 (December 2–8, 1991) claims it was Hangzhou. Tsien himself has given inconsistent information to news reporters and on U.S. federal forms. The December 11, 1911, birthday is given by Wang Shouyun, *People's Daily,* November 1, 1991; Wang worked as Tsien's secretary and is now Deputy Secretary General of the Science and Technology Committee for the Commission of Science, Technology, and Industry for National Defense.

4 *East of the family temple:* Personal interview, Qian Xuemin's mother, July 3, 1993; interview, Tsien Hsue-chu, December 10, 1992.

4 *A day in the Tsien household began:* Personal interview, Hu Guoshu, June 1993.

4 *The food was stored:* Personal interview, Qian Xuemin's mother, July 3, 1993.

4 *The feet of Chang Langdran:* Personal interview, Qian Xuemin's mother, July 3, 1993.

5 *Hangzhou was a market city:* Zhejiang sheng Xinhai geming shi yanjiu hui and Zhejiang Sheng tushuguan, *Xinhai geming Zhejiang shiliao xuanji* [Selected historical materials on the 1911 revolution in Zhejiang], (Hangzhou: Zhejiang Renmin chuban she, 1981), pp. 3, 17–18.

6 *The Tongyi Gong Cotton Mill opened in 1897:* Hu Guoshu, "Chongxin renshi xinhai xinhai geming shiqi de Zhejiang" [A reappraisal of Zhejiang in the (1911) revolutionary period], *Zhejiang fangzhi,* 5 (1991), pp. 9–10.

6 *Although Tsien would have been too young:* Photographs from the public library of Hangzhou.

6 *It was Sun Yat-sen's dream:* Jonathan Spence, *The Search for Modern China* (New York: W. W. Norton, 1990), pp. 280, 284, 285.

7 *appointment in the Ministry of Education:* Personal interview, Hu Guoshu, June 1, 1993; telephone interview, Tsien Hsue-chu, December 10, 1992, and April 1993.

Chapter 2: Beijing (1914–1929)

8 Population statistics from History Department of Beijing University, *History of Beijing* compilation group, *Beijing shi* [History of Beijing] (Beijing: Beijing chuban she, 1985).

8 Description of clothing from Ding Bingsui, *Beijing, Tianjin, ji qita* [Beiping, Tianjin, and others] (Taipei: n.p., 1977).

9 Description of the water and facilities situation from Bai Tiezheng, *Lao Beiping de gugu dianer*(Sanchang, Taiwan: Huilong chuban she, 1977).

9 Description of ancient architecture from L. C. Arlington and W. M. Lewisohn, *In Search of Old Peking* (Beijing: Henri Vetch, 1935); Osvald Siren, *The Walls and Gates of Peking* (London: John Lane, The Bodley Head, 1926); and *Encyclopedia Brittanica.*

9 Description of housing, living conditions, and street life in 1920s Beijing from Ting Bingsui, *Beijing*; David Strand, *Rickshaw Beijing: City People and Politics in the 1920s* (Berkeley: University of Califoria Press, 1989); *Beijing Jiedao De Gushi* [The stories of Beijing's streets] (Beijing: Beijing Press, 1960); Hedda Morrison, *A Photographer in Old Peking* (New York: Oxford University Press, 1985).

10 Information on high literacy in Beijing from *Beijing Fengwu Zhi* [On the scenery of Beijing] (Beijing: Beijing Travel Press, 1981).

11 The architectural description of Beijing Experimental Primary School No. 2 from the author's observations during a visit to Beijing in summer 1993; old photographs from the school's private collection; personal interview with Huo Mianzheng, a former teacher at the school, June 26, 1993.

11 The official history of Beijing No. 2 is sketched out in a handwritten document kept in the school archives.

11 The dates of Tsien's attendance of elementary school are from the memories of his surviving classmates and may not be accurate.

11 Information about the entrance examinations and daily school life in Tsien's time from personal interview, Huo Mianzheng, June 26, 1993.

13 Description of Tsien's precocity from Zhang Wei, "Tongchuang, Tonghong, Tongzhi—Canjian Qian Xuesen Xuezhang Shoujiang yishi you gan" [Classmate, colleague, and comrade—Thoughts on attending elder classmate Qian Xueshen's award ceremony], *Qinghua xiaoyou tongxun congshu* [Qinghua Alumni Bulletin series] 24 (1991): 136–39; Ji Tao, "Zhang Wei jiaoshou tan tongchuang Qian Xuesen: Cong fei zhibiao de haizi dao hangtian xianqu" [Professor Zhang Wei talks about his classmate Qian Xuesen: From a boy flying paper airplanes to an aeronautical pioneer," clipping from unknown newspaper.

13 *His hobbies flourished under his father's direction:* Interviews with Yucon Tsien (Tsien's son) and other family members.

13 *"My father was my first teacher":* Zhong Jihe, "Qian Xuesen's Artistic Destiny," *People's Daily,* January 9, 1992.

14 Information about the entrance exam to Beijing Experimental Primary School No. 1 from interview, Huo Mianzheng, June 26, 1993.

14 Description of Beijing No. 1 from author's observations during visit to Beijing in June 1993; interviews with alumni.

15 Description of China's test system from Y. S. Han, "Civil Service Examination System in China," *China Quarterly,* 3, no. 2 (spring 1938): 167–77; John Cleverley, *The Schooling of China: Tradition and Modernity in Chinese Education* (Boston: George Allen & Unwin, 1985).

18 Physical description of Beijing Normal University High School from interviews with alumni at the school during their sixtieth high school reunion, June 27, 1993.

18 *"fell in love:"* Tan Yinheng, former high school classmate of Tsien's, letter to author, October 5, 1993.

18 *taboo to cram the night before:* Personal interview, Yucon Tsien, July 13, 1991.

18 *appeared to study all the time:* Bai Yongxue, letter to author, August 7, 1993; Tan Yinheng, letter to author, October 5, 1993.

18 *empty classroom:* Li Xiyan, letter to author, August 23, 1993.

19 *"I heard classmates say":* Tan Yinheng, letter to author, October 5, 1993.

19 *He painted:* Hu Qianshan, professor of Dong Nan University and former high school classmate of Tsien's, letter to author, August 27, 1993; author's visit to Tsien's high school, 1993.

19 *He discussed literature:* Wang Di, professor, Tangshan Institute of Technology, and former classmate of Tsien's, letter to author, September 1993.

19 *took up debating:* Interviews with alumni during a sixtieth high school reunion, June 27, 1993.

19 *Beijing was a city in decay:* Strand, *Rickshaw Beijing.*

20 *"It was a miracle!":* Qian Xuesen, "Recollections and Hopes," *Beijing Review,* 34, no. 48 (December 2–8, 1991): 19.

20 *studying the laboratories very carefully:* Wang Di, letter to author, September 1993.

20 History of China's railroads from Spence, *The Search for Modern China;* interview with Luo Wen Zhun, railway historian, Jiaotong University, Beijing, 1993.

21 The date when Tsien entered college and his rank in the entrance examination from Zhang Xu, college classmate, personal interview, June 12, 1993; Qian Xuesen, "Huigu yu zhanwang" [Looking back and looking to the future], *Shanghai Jiaotong daxue tangxun* [Bulletin of Shanghai Jiaotong University] (January 1992): 4.

Chapter 3: Shanghai (1929–1934)

22 Physical description of Jiaotong circa 1929 from photographs, Jiaotong University Archives.

22 History of Jiaotong University from "History of Jiaotong University" compilation group, *Jiaotong Daxue Xiaoshi (1896–1949)* [History of Jiaotong University (1896–1949)], (Shanghai: Shanghai jiaoyu chuban she, 1986), pp. 3–11.

23 Information on Tsien's dormitory from telephone interview, Tsien Hsue-chu, December 10, 1992; Zhang Xu, letter to author, April 25, 1994; author's visit to Jiaotong University, summer 1993.

23 *routine at the university:* Personal interview, Luo Peilin, June 30, 1993; personal interview, Zhang Xu, June 12, 1993.

24 *"At that time, Tsien was already well-known as the best student":* Personal interview, Luo Peilin, June 30, 1993.

24 *He did not talk much:* Personal interview, Zhang Xu, June 12, 1993.

24 Description of library from author's visit to campus in June 1993; photographs, Jiaotong University Archives.

24 *He would read for hours:* Personal interview, Zhang Xu, June 12, 1993; telephone interview, Tsien Hsue-chu, December 10, 1992.

24 *"That was his peculiarity":* Personal interview, Zhang Xu, June 12, 1993.

24 *most brilliant instructor on campus:* Personal interview, Luo Peilin, June 30, 1993.

24 *Beautiful tiny print on the blackboards:* Personal interview, Zhang Xu, June 12, 1993.

25 *weekly walk through Shanghai:* Personal interview, Luo Peilin, June 30, 1993.

25 Population statistics from Hu Huanyong (ed.), *China Population: Shanghai* (Beijing: Chinese Financial Economy Press, 1987).

25 Details of Shanghai street life from telephone interview, Dr. Frederic E. Wakeman, Jr., Institute of East Asian Studies, University of California at Berkeley; Baruch Boxer, "Shanghai," *Encyclopedia Britannica*, 15th ed. (Chicago: 1992), vol. 27, p. 271; Ernest Hauser, *Shanghai City for Sale,* (Shanghai: Chinese-American, 1940); Gail Hershatter, "The Hierarchy of Shanghai Prostitution, 1870–1949," *Modern China,* 15, no. 4 (October 1989).

26 *Mario Pacci:* Tsien's INS hearings, April 16, 1951, Tsien's IRR file (Investigative Repository Records, case no. X1461732, box 234, RG 319), military reference branch, National Archives, Washington, D.C., p. 275.

26 *not only auditory but visual:* Personal interview, Luo Peilin, June 30, 1993.

27 *came down with typhus:* Qian Xuesen, "Huigu yu zhanwang" [Looking back and looking to the future], *Shanghai Jiaotong daxue tangxun* [Bulletin of Shanghai Jiaotong University] (January 1992): 4.

27 *My outlook on life changed:* Ibid; Luo Peilin, "Puhui, jishi, yancheng—yi Qian Xuesen xiang de ersan shi" [Recollections of Qian Xuesen], *Haiwai Xueren [Overseas Scholar],* (February 1991): 15–16.

27 Description of student activism from Jeffrey Wasserstrom, *Student Protest in Twentieth-Century China: The View from Shanghai* (Stanford: Stanford University Press, 1991); John Israel, *Student Nationalism in China, 1927–37* (Stanford: Stanford University Press, 1966).

29 *He loathed the Monday morning 8 a.m. gatherings:* Personal interview, Zhang Xu, June 12, 1993; Qian Xuesen, "Looking Back," p. 4.

30 *"beat you up with iron bars":* Telephone interview, Tsien Hsue-chu, December 10, 1992.

30 *"We'd see dark vans":* Personal interview, Luo Peilin, June 30, 1993.

30 *"I was awakened shortly after four o'clock":* A. T. Steele, *Shanghai and Manchuria, 1932: Recollections of a War Correspondent* (Tempe, AZ: Center for Asian Studies, Arizona State University, 1977), p. 5.

30 Information on air wars between China and Japan from Julian Bloom, *Weapons of War, Catalyst for Change: The Development of Military Aviation in China, 1908–41* (Ph.D. diss., University of Maryland, n.d.), San Diego Aerospace Museum, doc. no. 18-146; Rene Francillon, *Japanese Aircraft of the Pacific War* (London: Putnam, 1970); Eiichiro Sekigawa, *Pictorial History of Japanese Military Aviation,* ed. David Mondey (London: Ian Allan, 1974); Robert Mikesh and Shorzoe Abe, *Japanese Aircraft, 1910–1941* (Annapolis: Naval Institute Press, 1990).

33 *Tsien confided his worries:* Personal interviews, Luo Peilin, November 1991, June 30, 1993. Information about Tsien and Luo's relationship comes from these interviews, which is also discussed in some detail in Luo Peilin, "Recollections of Qian Xuesen," p. 15.

Chapter 4: Boxer Rebellion Scholar (1934–1935)

35 Information about the history of the scholarship from Michael Hunt, "The American Remission of the Boxer Indemnity: A Reappraisal," *Journal of Asian Studies* (May 1972); *Qinghua Daxue Shiliao Xuanbian* [Selected works of historical materials of Qinghua University] (Beijing: Qinghua University Press, 1991). The second book contains detailed information about the test itself.

38 Some details of the tour come from telephone interviews with former Boxer Rebellion Scholarship recipients Kai-loo Huang (March 17, 1995), Tunghua Lin (July 15, 1992), and C. C. Chang (July 22, August 2, August 18, 1992) as well as from interviews and correspondence with Ku Yu-Hsiu (April 8, 1993), the dean of engineering at Qinghua University when Tsien was a Boxer Scholarship recipient. Tsien's visit to airplane factories in Hangzhou and Nanchang is mentioned briefly in Qian Xuesen, "Huigu yu zhanwang" [Looking back and looking to the future], *Shanghai Jiaotong daxue tangxun* [Bulletin of Shanghai Jiaotong University] (January 1992): 4.

38 One person who recalled meeting Tsien at Qinghua University was C. C. Chang, the assistant to Tsien's mentor, S. C. Wang, working on what was then China's largest wind tunnel for Generalissimo Chiang Kai-shek. Tsien, Chang recalled, was a "very shy, very polite young man" who wore a traditional light blue or gray *zhangpao*.

Chapter 5: MIT (1935–1936)

40 Historical information about the MIT aeronautics program comes from Shatswell Ober, "The Story of Aeronautics at MIT 1895 to 1960"(unpublished ms., MIT Museum, 1965); War Records Committee of the Alumni Association of the Massachusetts Institute of Technology, *Technology's War Record* (Camridge, MA: Murray, 1920); Francis E. Wylie, *MIT in Perspective: A Pictorial* (Boston: Little, Brown, 1975).

41 *We were all told by our advisors:* Telephone interview with Tunghua Lin.

41 *"When they arrived at the professor's office":* Webster Roberts, letter to author, May 31, 1993.

42 Information about Tsien's friendship with William Sangster from William Sangster, letter to author, May 4, 1992; telephone interview with William Sangster, July 7, 1992.

42 *"There is a general disdain for handiwork":* Dr. James Mueller, narrative report, Section of Neuropathology, Department of Pathology, Indiana University.

43 Information about Tsien's work with W. H. Peters from telephone interview with W. H. Peters, November 2, 1993; Tsien Tsue-shen, "Study of the Turbulent

Boundary Layer" (master's thesis, MIT Department of Aeronautical Engineering, 1936).

44 *"kind of lost"*: Eric Mollo-Christensen, letter to author, June 9, 1993; Mollo-Christensen discusses his conversation with an older MIT faculty member who remembered Tsien's experience at the university in 1935.

44 *"It was my only form of entertainment"*: William Ryan and Sam Summerlin, *The China Cloud: American's Tragic Blunder and China's Rise to Nuclear Power* (Boston: Little, Brown, 1967), p. 43; Tsien's INS hearings, April 16, 1951, Tsien's IRR file, military reference branch, National Archives, Washington, D.C., p. 275.

44 *"Look, if you don't like it here"*: Personal interview with Andrew Fejer, December 7, 1991.

44 Information about reasons for Tsien's leaving MIT from telephone interview with Judson Baron, November 20, 1992.

45 *"In those pre–World War II days"*: Hisayuki Kurihara, letter to author, July 24, 1993.

45 *"We didn't want to go home"*: Telephone interview with Tunghua Lin.

Chapter 6: Theodore von Kármán

47 *"Horse-drawn droshkies"*: Theodore von Kármán and Lee Edson, *Wind and Beyond: Theodore von Kármán, Pioneer in Aviation and Pathfinder in Space* (Boston: Little, Brown, 1967), p. 14.

48 *The city was divided in two*: Richard Rhodes, *The Making of the Atomic Bomb* (New York: Simon & Schuster, 1986), p. 105; Steve Heims, *John von Neumann and Norbert Wiener: From Mathematics to the Technologies of Life and Death* (Cambridge: MIT Press, 1980), p. 27; Laura Fermi, *Illustrious Immigrants* (Chicago: University of Chicago Press, 1971), pp. 38–39.

48 *Contributing to this renaissance*: Heims, *John von Neumann and Norbert Wiener,* p. 29.

48 Information regarding Kármán's childhood from Kármán and Edson, *Wind and Beyond*; Michael Gorn, *The Universal Man: Theodore von Kármán's Life in Aeronautics* (Washington, D.C.: Smithsonian Institution Press, 1992).

50 *Kármán arrived in Göttingen in October*: Gorn, *The Universal Man,* p. 15; Daniel Fallon, *The German University: A Heroic Ideal in Conflict with the Modern World* (Boulder: Colorado Associated University Press, 1980), p. 7.

51 *"It was not uncommon"*: Fallon, *The German University,* p. 42.

51 *Kármán's earliest years at Göttingen*: Kármán and Edson, *Wind and Beyond,* pp. 35–41.

52 *Twice he almost quit*: Gorn, *The Universal Man,* pp. 18–25.

52 *Then came a chance encounter*: Ibid., pp. 18–19; Will Johnson, *Saturday Review,* February 2, 1963; Hugh Dryden, "Theodore von Kármán, May 11, 1881–May

7, 1963," *Biographical Memoirs,* 38 (New York: National Academy of Sciences, Columbia University Press, 1965).

53 *For the next four years:* Gorn, *The Universal Man,* pp. 20–25.

53 *began to see the school as part of a flawed system:* Kármán and Edson, *Wind and Beyond,* p. 71; Fallon, *The German University,* pp. 43–44.

53 *Fuming about his situation:* Ibid., p. 71.

54 *the eminent Felix Klein:* Ibid., pp. 71–73.

54 *called to active duty:* Ibid., p. 80.

54 *political career lasted two months:* Ibid., pp. 82, 89–95.

55 *"saved me for all my life":* Ibid.

55 *his life returned to normal:* Ibid., pp. 96–97, 103–4, 107.

55 *"enough to tell jokes":* Shirley Thomas, *Men of Space: Profiles of Leaders in Space Research, Development and Exploration* (Philadelphia: Chilton, 1960), vol. 1, p. 158.

55 *Hyper-inflation drained research funds:* Rhodes, *The Making of the Atomic Bomb,* p. 18.

56 *"of the anti-Semitic composition of the natural science faculty":* Paul Hanle, *Bringing Aerodynamics to America* (Cambridge: MIT Press, 1982), p. 109.

56 *"What is the first boat":* Gorn, *The Universal Man,* p. 38.

56 Information on Robert Millikan from ibid., pp. 39–43; Robert Kargon, *The Rise of Robert Millikan: A Portrait of a Life in Science* (Ithaca, NY: Cornell University Press, 1987).

56 *When no reply was received:* Gorn, *The Universal Man,* p. 44.

56 *he and his sister Pipö journeyed to New York:* Ibid., p. 45.

57 *Kármán agonized over the offer:* Ibid., p. 51.

57 For details of Kármán's encounters with anti-Semitism and concerns for the future of Germany, see Kármán and Edson, *Wind and Beyond.*

57 *a suitable home:* Author's visit to 1501 South Marengo Avenue, Pasadena, CA; Gorn, *The Universal Man,* p. 62; telephone interview with Liljan Malina Wunderman, March 6, 1992.

58 *local aircraft engineers:* Gorn, *The Universal Man,* p. 57.

58 *significant new contributions to aeronautics:* Ibid., pp. 58–59, 63.

58 *one of Caltech's most popular professors:* Ibid., pp. 65, 69.

59 *somewhat paradoxical life:* Interview with Liljan Malina Wunderman, March 6, 1992; interviews with Caltech alumni.

59 *"Young lady":* Dryden, "Theodore von Kármán."

59 *"like his children":* Telephone interview with C. C. Chang.

60 *"I think this is what you want!":* Gorn, *The Universal Man,* p. 69.

60 *Kármán insisted that they call him "Grandpa":* Correspondence between C. C. Chang's family and Kármán, von Kármán collection, Caltech.

60 *took the oath of U.S. citizenship:* Gorn, *The Universal Man,* p. 64.

60 *followed a familiar schedule:* Ibid., pp. 64–65; Shirley Thomas, "Theodore von Kármán's Caltech Students" (paper presented at the World Space Congress, Washington, D.C., 1992).

Chapter 7: Caltech (1936)

61 *"One day in 1936":* Theodore von Kármán and Lee Edson, *Wind and Beyond: Theodore von Kármán, Pioneer in Aviation and Pathfinder in Space* (Boston: Little, Brown, 1967), p. 309.

61 *wrote a letter to Luo Peilin:* Personal interview with Luo Peilin, November 1991.

62 *"Pasadena is ten miles from Los Angeles":* Morrow Mayo, "Croesus at Home," *American Mercury,* 27, no. 106 (October 1932), p. 230.

62 Information about Tsien's housing situation from interview with James Bonner, December 15, 1992; interview with Shao-wen Yuan, December 10, 1992; Pasadena Historical Society.

62 *school on a tight budget:* William Dodge Jr., Reuban Moulton, Harrison Sigworth, Adrian Smith, *Legends of Caltech* (Caltech Alumni Association, 1982), p. 9.

62 *began as Throop University:* Judith Goodstein, *Millikan's School: A History of the California Institute of Technology* (New York: W. W. Norton, 1991); Harold Carew, *History of Pasadena and the San Gabriel Valley,* 1 (S. J. Clarke, 1930), ch. 15.

63 *"every luncheon, every dinner":* James Gleick, *Genius* (New York: Pantheon, 1992), p. 281.

63 *conducted his own study of aeronautics:* Gao Yijin, "Qian Xuesen's Family," *Wen Hui Bao,* March 27, 1957.

63 *rather mysterious figure on campus:* F. H. Allardt, letter to author, July 1, 1993; Hisayuki Kurihara, letter to author, June 25, 1993.

63 *astute, complex questions:* Hisayuki Kurihara, letter to author, June 25, 1993.

63 *"Jewish blood?":* Kármán and Edson, *Wind and Beyond,* p. 309. According to William Sears, Epstein was convinced that Tsien was a descendant of a group of Jews who had migrated to China long ago; see William Rees Sears, *Stories from a 20th Century Life* (Stanford, CA: Parabolic Press, 1993), p. 67.

64 *"my revered teacher":* Kármán and Edson, *Wind and Beyond,* p. 310.

64 *"He worked with me on many mathematical problems":* Ibid., p. 309.

65 *"He was Kármán's right-hand man":* Telephone interview with Martin Summerfield, March 12, 1992.

65 *"Kármán could look at a problem":* Personal interview with Frank Marble, July 16, 1991.

65 *During his first year as a graduate student:* Interview with Shao-wen Yuan, December 10, 1992; interviews with other Caltech alumni.

66 *"Tsien definitely made von Kármán more productive"*: Personal interview with Hans Liepmann, July 17, 1991.

66 *long and tedious process*. Ibid.

67 *"at least once a week"*: Interview with Liljan Malina Wunderman, December 21, 1993.

67 *"Kármán was a lucky man"*: Personal interview with Homer Joseph Stewart, July 17, 1991.

67 *"Tsien was the greatest collaborator"*: Personal interview with Frank Marble, July 16, 1991.

Chapter 8: The Suicide Squad (1937–1943)

68 *"not talkative"*: Personal interview with Apollo Milton Olin Smith, July 17, 1991.

68 *group was started by Frank Malina*: Judith Goodstein and Carol Buge, *The Frank J. Malina Collection at the California Institute of Technology: Guide to a Microfiche Edition* (Institute Archives, Robert A. Millikan Memorial Library, California Institute of Technology, 1986); Frank Malina, "The Beginning of Rocketry and JPL," in *Guggenheim Aeronautical Laboratory at the California Institute of Technology: The First Fifty Years,* ed. F. E. C. Culick (San Francisco: San Francisco Press, 1983), p. 68.

68 *labor of love*: "Excerpts on Rocket Research and Development from Letters Written Home by Frank Malina Between 1936–1946," Jet Propulsion Laboratory Archives; personal interview with A. M. O. Smith, 1991.

69 *in late October 1936*: "Excerpts on Rocket Research," November 1, 1936, p. 4.

69 *"He was curious"*: Personal interview with A. M. O. Smith, 1991.

69 *"If you're interested in rockets"*: William Ryan and Sam Summerlin, *The China Cloud: America's Tragic Blunder and China's Rise to Nuclear Power* (Boston: Little, Brown, 1967), p. 47.

69 *"A Chinese graduate student"*: "Excerpts on Rocket Research," May 22, 1937, p. 10.

70 *Konstantin Tsiolkovsky*: The New Encyclopedia Brittanica, 15th ed. (Chicago: Encyclopedia Brittanica, 1990), vol. 12, p. 16.

70 *Robert Goddard*: Milton Lehman, *This High Man: The Life of Robert Goddard* (New York: Farrar Straus, 1963).

70 *Hermann Oberth and Max Valier*: Michael Neufeld, "Weimar Culture and Futuristic Technology: The Rocketry and Spaceflight Fad in Germany, 1923–1933," *Technology and Culture* 31, no. 1 (January 1990); Wernher von Braun and Frederick Ordway II, *History of Rocketry and Space Travel* (New York: Crowell, 1975).

71 *William Bollay*: Frank Malina, "The GALCIT Rocket Research Project

1936–1938" in F. C. Durant and G. S. James, *First Steps Towards Space: Proceedings of the First and Second History Symposia of the International Academy of Astronautics* (Washington, D.C., 1974).

71 *two other young men:* Ibid.

72 *In early 1936:* Oral history interview of Frank Malina by James Wilson, June 8, 1973 (JPL Archives), p. 23; Malina, "GALCIT Rocket Research Project."

72 *raise the funds themselves:* "Excerpts on Rocket Research," March 16, 1936, June 29, 1936, April 24, 1937; personal interview with A. M. O. Smith, 1991; Frank Malina, "The Rocket Pioneers: Memoirs of the Infant Days of Rocketry at Caltech," *Engineering and Science,* 31, no. 5 (February 1968): 30.

72 *Weld Arnold:* "Excerpts on Rocket Research," May 22, 1937, May 29, 1937.

73 *"How do we open up a fund":* Malina, "The Beginning of Rocketry and JPL," p. 68.

73 *warehouse for accumulating tetranitromethane:* Malina, "The Rocket Pioneers," p. 30; oral history interview of Frank Malina by James Wilson, p. 7.

73 *On May 29, 1937:* "On the GALCIT Rocket Research Project, 1936–1938," *Smithsonian Annals of Flight,* no. 10 (Washington, D.C.: Smithsonian Institution, n.d.).

73 *In June 1937:* Ibid.

73 *disaster struck:* Personal interview with Smith, 1991.

74 *"If there were any rats in the building":* "Excerpts on Rocket Research," August 7, 1937.

74 *"event was considerably horrifying":* Telephone interview with Martin Summerfield, October 6, 1992.

74 *"Out!":* Malina, "The Beginning of Rocketry and JPL," p. 68.

74 *atop a concrete loading platform:* Personal interview with Smith, 1991.

74 *"I remember looking out of my window":* Personal interview with Hans Liepmann, July 17, 1991; "The Jet Propulsion Laboratory: Its Origins and First Decade of Work," *Spaceflight,* 6, no. 5: 160–65.

74 *In late 1937:* Frank Malina and A. M. O. Smith, "Flight Analysis of the Sounding Rocket," *Journal of Aeronautical Sciences,* 5, no. 5 (March 1938): 199–202.

74 *Kármán surprised Malina:* "Excerpts on Rocket Research," January 21, 1938.

75 *extremely well received in New York:* Time, 31 (February 7, 1938): 30.

75 *"This analysis":* Associated Press, January 26, 1938 (doc. no. 3-94D, JPL archives).

75 *"Scientists Plan to Shoot Rocket":* Houston Chronicle, January 26, 1938 (doc. no. 3-94E, JPL archives).

75 *"Needless to say":* "Excerpts on Rocket Research," February 6, 1938.

75 *"The rocket has emerged from the realm of fiction":* Herb Strong, "Graduates Conduct Research on Rocket Efficiencies," *California Tech,* February 3, 1938, p. 1.

76 *"We were even more surprised":* "Excerpts on Rocket Research," February 6, 1938, p. 16.

76 *time of intense teamwork:* Ibid., February 27, 1938, March 6, 1938, March 13, 1938, April 25, 1938, p. 17.

76 *work began to pay off:* Ibid., April 4, 1938, April 10, 1938, May 8, 1938, May 30, 1938, June 7, 1938, June 25, 1938, July 5, 1938; Scholer Bangs, "Rocket Altitude Record Sought," *Los Angeles Examiner,* July 15, 1938.

76 *immediate and sensational:* E. H. Tipton, Associated Press, April 1938, JPL HF 3-94a, JPL Archives; "Excerpts on Rocket Research," May 22, 1938, June 25, 1938, July 5, 1938.

77 *"blasted away":* "Excerpts on Rocket Research," May 7, 1938.

77 *rewards of his efforts:* Wang Shouyun (ed.), *Collected Works of H. S. Tsien: 1938–1956* (Beijing: Science Press, 1991); Theodore von Kármán and Hsue-shen Tsien, "Boundary Layer in Compressible Fluids," *Journal of the Aeronautical Sciences,* 5 (1938): 227–32; Hsue-shen Tsien, "Supersonic Flow over an Inclined Body of Revolution," *Journal of the Aeronautical Sciences,* 5 (1938): 480–83; Hsue-shen Tsien and Frank J. Malina, "Flight Analysis of a Sounding Rocket with Special Reference to Propulsion by Successive Impulses," *Journal of the Aeronautical Sciences,* 6 (1939): 50–58.

77 *"Tsien really should have had his name on it":* "Excerpts on Rocket Research," December 10, 1938, p. 24.

78 *"brilliant at art":* Telephone interview with Liljan Malina Wunderman, March 12, 1993.

78 *"Son of heaven":* Personal interview with Hans Liepmann, 1991.

78 *"The Caltech fellows":* William Rees Sears, *Stories from a 20th Century Life* (Stanford, CA: Parabolic, 1993).

78 *patron in a movie theatre:* Gu Xiong, *Women De Qian Xuesen* [Our Qian Xuesen] (Hong Kong: Modern, 1970), p. 34.

78 *"I can't help the fact that China is poor":* Personal interview with Yucon Tsien, July 13, 1991.

78 *"He was very stubborn":* Telephone interview with Shao-wen Yuan, December 10, 1992.

79 *threw Tsien a surprise party:* "Excerpts on Rocket Research," May 30, 1938, p. 19.

79 *old, dirty-gray Chevy:* Telephone interview with Liljan Malina Wunderman, December 21, 1993.

79 *their excursions:* Interview with Liljan Malina Wunderman, March 12, 1993, and March 6, 1992; interview with Andrew and Edith Fejer, December 7, 1991; Tsien's INS hearings, April 16, 1951, Tsien's IRR file, military reference branch, National Archives, Washington, D.C.

79 *intellectual discussion group:* Ibid.

80 *sharply aquiline features:* Telephone interview with John Wiggins, May 27, 1993.

80 *something of a Renaissance man:* "Sidney Weinbaum: Politics at Mid-Century." *Engineering and Science,* 55, no. 1 (Fall 1991): 30.

80 *after supper about 8:00:* Telephone interview with John Wiggins, May 27, 1993. Tsien recalled that the parties took place once a week on Wednesday nights; see "Interview of Hsue-shen Tsien at United States Immigration Bureau," U.S. Customs file, National Archives, p. 5.

80 *neatly dressed in vest:* Telephone interview with Liljan Malina Wunderman, March 6, 1992.

80 *tall lean figure:* Telephone interview with John Wiggins, May 27, 1993.

80 *tattered sleeves, cigar ashes:* Ibid.

80 *predictions of Karl Marx:* Telephone interview with Liljan Malina Wunderman, March 6, 1992.

80 *group sympathetic to the plight of China:* Tsien's INS hearings, IRR file, National Archives; "Interview of Dr. Hsue-shen Tsien at United States Immigration Bureau," U.S. Customs file, National Archives, September 20, 1950, p. 39..

81 *scholarship stipends to all Boxer Scholars:* Telephone interview with Kai-loo Huang, April 8, 1993.

81 *somewhat more formal than a party:* Telephone interview with Martin Summerfield, March 12, 1992.

81 *a coffee table:* Telephone interview with John Wiggins, May 27, 1993.

81 *prepare book reports:* Telephone interview with Liljan Malina Wunderman, March 6, 1992.

81 *books included works:* Interviews with former members; Tsien's INS hearings, IRR file, National Archives, pp. 182, 202; John Whiteside Parson's FBI file.

81 *end with music and refreshments:* Telephone interview with Martin Summerfield, March 12, 1992; Tsien's INS hearing, IRR file, National Archives, February 16, 1951, p. 183.

81 *large community of musicians:* Personal interview with Andrew Fejer, December 7, 1991.

81 *join Weinbaum at the piano.* Telephone interview with Liljan Malina Wunderman, December 21, 1993.

81 *play the piccolo:* Telephone interview with John Wiggins, May 27, 1993.

81 *alto recorder:* Telephone interview with Liljan Malina Wunderman, March 6, 1992; Sears, *Stories from a 20th Century Life,* p. 69.

81 *game of krigspiel:* Telephone interview with Liljan Malina Wunderman, December 21, 1993.

81 *"We used to meet to listen to music":* oral history interview of Frank Malina by James Wilson, p. 32.

82 *nothing he loved more than a good argument:* Telephone interview with Liljan Malina Wunderman, March 12, 1993.

82 *"Malina would actually":* Personal interview with A. M. O. Smith, 1991.

82 *slid into the doldrums:* Malina, "GALCIT Rocket Research Project," p. 124.

83 *unbearably cold and arrogant:* Personal interview with Jeanne Forman, November 18, 1992.

83 *"Jack and Forman":* oral history interview of Frank Malina by James Wilson, p. 29.

83 *Gen. Arnold paid a surprise visit:* Clayton Koppes, *JPL and the American Space Program: A History of the Jet Propulsion Laboratory* (New Haven, CT: Yale University Press, 1982), p. 8.

83 *"Buck Rogers job":* Kármán and Edson, *Wind and Beyond,* p. 243.

83 *presentation for the National Academy of Sciences:* Frank Malina, "Report on Jet Propulsion for the National Academy of Science Committee on Air Corps Research" (doc. no. 3-86, JPL Archives, December 21, 1938).

83 *NAS awarded Caltech one thousand dollars:* "The Jet Propulsion Laboratory."

83 *"My enthusiasm vanishes":* "Excerpts on Rocket Research," October 24, 1938.

83 *wanted to see the first report by June:* Michael Gorn, *The Universal Man: Theodore von Kármán's Life in Aeronautics* (Washington, D.C.: Smithsonian Institution Press, 1992), p. 85.

84 *near-fatal accident:* Ibid.

84 *banished all rocket experiments from campus:* Ibid.

84 *ten-thousand-dollar grant:* Frank Malina, "The Beginning of Rocketry and JPL," in *Guggenheim Aeronautical Laboratory at the California Institute of Technology: The First Fifty Years,* ed. F. E. C. Culick (San Francisco: San Francisco Press, 1983), p. 70.

84 *graduated on June 9, 1939:* 1939 commencement program, Institute Archives, Caltech.

85 *Two-Dimensional Subsonic Flow of Compressible Fluids:* See Theodore von Kármán and Hsue-shen Tsien, "Boundary Layer in Compressible Fluids," *Journal of the Aeronautical Sciences,* 5 (1938): 227–32.

85 *"The Kármán-Tsien formula":* Interview with A. M. O. Smith, 1995.

85 *a Lockheed test pilot died:* Ibid.; James Hansen, *Engineer in Charge: A History of the Langley Aeronautical Laboratory, 1917–1958* (Washington, D.C.: National Aeronautics and Space Administration, 1987), p. 249.

85 *Curtis SB2C Helldivers:* Hansen, *Engineer in Charge,* p. 260.

86 *"There seemed to be some confusion":* Sears, *Stories from a 20th Century Life,* p. 67.

86 *member of the aeronautics staff:* Biographical resume of H. S. Tsien, NASA Historical Archives.

86 *euphemism jet propulsion:* Malina, "The Beginning of Rocketry and JPL," p. 70.

86 *"bloody fool"*: Ibid., p. 69.

86 *"Do you honestly believe"*: Kármán and Edson, *Wind and Beyond*, p. 244.

87 *On some nights, Tsien worked at Malina's home*: Telephone interviews with Lil-jan Malina Wunderman, March 12, 1993, and December 21, 1993.

87 *"Early one day"*: Ellis Lapin, letter to author, June 30, 1993.

87 *concrete conclusions about the rocket fuel problem*: Kármán and Edson, *Wind and Beyond*, p. 245.

88 *fifty thousand combat planes a year*: George Brown Tindall, *America: A Narrative History*, vol. 2 (New York: W. W. Norton, 1984).

88 *$17 billion in defense funds*: Ibid.

88 *$22,000 for the fiscal year 1941*: Koppes, *JPL and the American Space Program*, p. 10.

88 *pen-and-ink cartoon*: Pen-and-ink cartoon of Theodore van Kármán and colleagues seated at table, birthday 1940 (TVK 184), Caltech archives.

89 *"I would like to emphasize"*: Kármán, letter to Lt. Col. Tsoo Wong, April 20, 1940, folder 30.37, Kármán collection, Caltech archives.

89 *asked the Department of Immigration*: Edward Barrett, letter to Commissioner of Immigration and Naturalization Service, December 4, 1940, folder 30.37, Kármán collection, Caltech archives.

89 *A Method for Predicting the Compressibility Burble*: Technical report no. 2, Aeronautical Research Institute, Chengtu, China, 1941. See Wang Shouyun (ed.), *Collected Works of H. S. Tsien: 1938–1956* (Beijing: Science Press, 1991).

89 *Chinese Natural Science Association*: Hsue-shen Tsien, "Heat Conduction Across a Partially Insulated Wall," *Bulletin of the Chinese Natural Science Association* [U.S. West Coast chapter], 1 (1942): 7–11; interview with Jack C. N. Tsu, May 6, 1992.

89 *"listened to them talk"*: Interview with Andrew Fejer, December 7, 1991.

89 *especially close to Chieh-Chien Chang*: Interview with Chieh-Chien Chang.

90 *shared an office with Tsien*: Interview with C. C. Chang, July 22, 1992.

90 *"Tsien and I were very close friends"*: Ibid.

90 *December 2, 1940*: Tsien's alien registration form, INS files, INS Regional Office, Los Angeles.

90 *subject of structural buckling*: Shouyun (ed.), *Collected Works of H. S. Tsien*.

91 *metal shell about ten feet long*: Personal interview with Bernard Rasof, December 7, 1991.

91 *"Sure enough"*: Ibid.

91 *design of a small wind tunnel*: Personal interview with Allen Puckett, February 25, 1992; Allen Puckett, "Early Supersonics and Beyond," in *Guggenheim Aeronautical Laboratory at the California Institute of Technology: The First Fifty Years*, ed. F. E. C. Culick (San Francisco: San Francisco Press, 1983), p. 74.

91 *first continuously operating supersonic wind tunnel*: The Guggenheim Aeronautical

Laboratory of the California Institute of Technology: The First Twenty-Five Years (Caltech, 1954), p. 31.

91 *larger wind tunnel that Puckett designed:* Kármán and Edson, *Wind and Beyond,* p. 230; interview with Allen Puckett.

91 *"On the Design of a Contraction Cone": Journal of the Aeronautical Sciences,* 10 (1943): 68–70.

92 *local aircraft companies moved valuables inland:* James Richard Wilburn, *Social and Economic Aspects of the Aircraft Industry in Metropolitan Los Angeles During World War II* (Ph.D. diss., UCLA, 1971).

92 *from student to visiting scientist:* Memo, August 25, 1941, folder 30.37, Kármán collection, Caltech archives.

92 *"I haven't the slightest doubt":* Kármán, letter to Chief of Ordnance Department, War Department, November 16, 1942, folder 30.37, Kármán collection, Caltech archives.

92 *Tsien's clearance was approved:* Clark Millikan, letter to Western Procurement District, December 4, 1992, folder 30.37, Kármán collection, Caltech archives.

Chapter 9: The Jet Propulsion Laboratory (1943–1945)

93 *assistant professor of aeronautics:* Biographical resume, NASA Historical Archives, security questionnaires, IRR file, National Archives.

93 *statistics were simply astounding:* Richard Wilburn, *Social and Economic Aspects of the Aircraft Industry in Metropolitan Los Angeles During World War II* (Ph.D. diss., UCLA, 1971).

94 *"It's fun to work in an aircraft factory":* Ibid.

94 *changes brought on by the war:* Interviews with Caltech alumni.

94 *"no-nonsense undergraduate life":* William Dodge Jr., Reuban Moulton, Harrison Sigworth, Adrian Smith, *Legends of Caltech* (Caltech Alumni Association, 1982), p. 18.

94 *Arroyo Seco changed:* Photographs from JPL Archives; interviews with former employees; Clayton Koppes, *JPL and the American Space Program: A History of the Jet Propulsion Laboratory* (New Haven, CT: Yale University Press, 1982), p. 15.

94 *"you were liable to lose your teeth":* Koppes, *JPL and the American Space Program,* p. 15.

94 *forced to use his car as an office:* Ibid., p. 11.

94 *string of liquid- and solid-propellent test shacks:* Interview with Chester Hasert, February 7, 1993.

94 *shatterproof glass:* Interview with Homer Joseph Stewart.

95 *reminded the engineers of Niagara Falls:* Ibid.

95 *"Half the time they blew up":* Ibid.

95 *primitive little rockets:* Koppes, *JPL and the American Space Program,* p. 11.

95 *rushed to March Field:* Theodore von Kármán and Lee Edson, *Wind and Beyond: Theodore von Kármán, Pioneer in Aviation and Pathfinder in Space* (Boston: Little, Brown, 1967), pp. 249–51.

95 *"the plane shot off the ground":* Ibid.

95 *increased it to $125,000:* Koppes, *JPL and the American Space Program,* p. 15.

95 *Navy gave Caltech a contract:* Ibid., p. 12.

95 *Parsons invented one in 1942:* Ibid., pp. 12–13.

96 *red fuming nitric acid:* Ibid., p. 14.

96 *tried aniline:* Ibid., p. 14.

96 *tested two JATOs supplied with liquid fuel:* Ibid.

96 *"as though scooped upward by a sudden draft":* Ibid.

96 *more than eighty employees worked for Malina:* John Bluth, JPL historian.

96 *some labored over a hydrobomb:* Leonard Edelman, letter to author, June 18, 1993.

96 *grow to $650,000:* John Bluth.

96 *free Haley up from a complicated legal case:* Information about founding of Aerojet from Kármán and Edson, *Wind and Beyond,* pp. 257–58.

97 *company expanded to 120 employees:* John Bluth.

97 *first big break:* Public relations division, Aerojet.

97 *largest manufacturer of rockets:* Kármán and Edson, *Wind and Beyond,* p. 258.

97 *part of the corporate conglomerate GenCorp:* Public relations division, Aerojet.

97 *"an excellent chemist and delightful screwball":* Kármán and Edson, *Wind and Beyond,* p. 257.

97 *In June 1942:* Information about Parsons moving into the Arthur Fleming mansion on 1003 S. Orange Grove from Brad Branson and Susan Pile (unpublished ms.).

97 *front porch:* Personal interview with Jeanne Forman, November 1993.

97 *"It was always open":* Ibid.

98 *FBI recorded Parsons' wild parties:* Doc. no. 100-189320-2, Oklahoma City letter to Bureau, March 31, 1944, FBI file of John Whiteside Parsons, p. 58.

98 *"All I could think at the time":* Telephone interview with Liljan Malina Wunderman, 1992.

98 *naked pregnant women:* Mike Davis, *City of Quartz* (New York: Vintage, 1992), p. 59.

98 *Ode to Pan:* Interview with Jeanne Forman, November 1993.

98 *well-known writers such as Lou Goldstone:* Ibid.

98 *"He was a mechanic working with Jack":* Oral history interview of Frank Malina by James Wilson, June 8, 1973 (JPL Archives), p. 17.

99 *untimely death at the age of 37:* William Ryan and Sam Summerlin, *The China*

Cloud: American's Tragic Blunder and China's Rise to Nuclear Power (Boston: Little, Brown, 1967), p. 149.

99 *In April 1943:* Information about the reports from GALCIT Project No. 1, progress reports no. 7 and 17, JPL Archives.

99 *began to teach a special group:* Interviews with Caltech alumni.

99 *Tsien taught two courses:* Ibid.

100 *"Why! On Sunday, I expect":* Ellis Lapin, letter to author, June 30, 1993.

100 *"If someone asked him a stupid question":* Interview with Homer Joseph Stewart, July 17, 1991.

100 *one unfortunate student:* Interview with A. M. O. Smith, July 17, 1991.

100 *"Most students lived in dread of it":* Interview with Chester Hasert, February 7, 1993.

100 *"He'd always be a couple minutes late":* Ibid.

100 *"There was no erasing of errors":* Webster Roberts, letter to author, May 31, 1993.

100 *"I don't understand the third equation":* Interview with Joseph Charyk, March 9, 1992.

100 *"Only fools need foolproof methods":* Interview with Bernard Rasof, December 9, 1991.

100 *"Sometimes he came out with a sentence":* Interview with Chester Hasert, February 7, 1993.

101 *"His notes were bare bones mathematics":* Ibid.

101 *"got a flat zero":* Ibid.

101 *"There were blackboards":* Interview with Robert Bogart.

101 *"One day I happened to walk into his office":* Interview with anonymous alumnus.

101 *"I'm not teaching kindergarten!":* Interview with Chester Hasert, February 7, 1993.

102 *"Many of these students had been shot down":* Interview with Bernard Rasof.

102 *"Students were scared stiff of him":* Interview with Hans Liepmann, 1991. Half a century later, when the author mailed a questionnaire to Tsien's former students, some of the replies were so riddled with expletives they were unprintable.

102 *"Out of the goodness of his heart":* Leonard Edelman, letter to author, June 18, 1993.

102 *"The biggest crime you could commit":* Interview with Andrew Fejer, December 7, 1991.

103 *"Von Kármán would start developing":* Interview with Bill Davis, October 13, 1992.

103 *"Tsien attempted to do to his students":* Interview with Joseph Charyk.

103 *"I guess that his goal":* Leonard Edelman, letter to author, June 18, 1993.

103 *"Tsien thought Kármán was such a compassionate":* Interview with R. B. Pearce, August 2, 1993.

103 *top secret aerial photographs:* Kármán and Edson, *Wind and Beyond,* pp. 263–64.

104 *In September 1943, Army Ordnance:* Koppes, *JPL and the American Space Program,* p. 19.

104 *memorandum that Kármán . . . submitted:* "The Possibilities of Long-Range Rocket Projectiles by Kármán and a Review and Preliminary Analysis of Long-Range Rocket Projectiles by H.S. Tsien and F.J. Malina" (memorandum, JPL Archives, November 20, 1943).

104 *Kármán suggested a four-stage research program:* Ibid.

104 *"first official memo in the U.S. missile program":* Kármán and Edson, *Wind and Beyond,* p. 265.

104 *Army Air Force turned the proposal down:* Koppes, *JPL and the American Space Program,* p. 19.

105 *Col. Sam B. Ritchie flew to Pasadena:* Ibid.

105 *"The Ordnance Department is very anxious":* G. W. Trichel, letter to Kármán, January 15, 1944, doc. no. 3-978, JPL Archives.

105 *"The scale of operations threw us into a proper dither!":* Frank Malina, "America's First Long Range Missile and Space Exploration Programme: The ORDCIT Project of the Jet Propulsion Laboratory, 1943–1946" (doc. no. 3-714, JPL Archives).

105 *rough blueprints:* Doc. no. 3-979, JPL Archives.

105 *comparative study of various jet propulsion systems:* Theodore von Kármán, Frank Malina, Martin Summerfield, H. S. Tsien, "Comparative Study of Jet Propulsion Systems as Applied to Missiles and Transonic Aircraft" (memorandum JPL-2, JPL Archives, March 28, 1944).

106 *obtaining $1.6 million:* John Bluth.

106 *guided missile work officially began:* Koppes, *JPL and the American Space Program,* p. 20.

106 *mass construction:* Interview with Homer Joseph Stewart.

106 *manpower doubled:* John Bluth.

106 *chief of the first research analysis section:* Frank Malina, memo, August 18, 1944, doc. no. 3-57, JPL archives.

106 *the private A:* Interview with Homer Joseph Stewart.

106 *About a dozen scientists:* Ibid.

106 *On Wednesday afternoons:* Ibid.

107 *nine sections in all:* Malina, "America's First Long Range Missile."

107 *Private A was ready for testing:* Photographic negatives 293-12, 293-13, 293-70, 293-207C, JPL Archives; Koppes, *JPL and the American Space Program,* p. 22; Kármán and Edson, *Wind and Beyond,* p. 265.

107 Information about the Private F from Koppes, *JPL and the American Space Program,* p. 22.

107 Information about the WAC Corporal from ibid., pp. 23–24.

107 *first man-made object to escape the earth's atmosphere:* Kármán and Edson, *Wind and Beyond*, p. 265.

107 *Corporal E:* For description of missile, see Koppes, *JPL and the American Space Program.*

107 *stomach surgery:* See Gorn, *The Universal Man,* ch. 6; John Bluth, letter to author.

108 *Malina had long disliked Millikan:* Telephone interview with Liljan Malina Wunderman, 1992; Millikan diary notes of incident, Caltech Archives.

108 *"Tsien did not like Clark Millikan":* Interview with Hans Liepmann, 1991.

108 *secret meeting with Kármán:* Kármán and Edson, *Wind and Beyond*, p. 267–68.

108 *letter signed by Tsien:* H. S. Tsien, C. C. Lin, W. Z. Chien, and Y. H. Kuo, letter to Kármán, November 7, 1944, folder 30.37, Kármán collection, Caltech Archives.

109 *asked him to join him:* Gorn, *The Universal Man,* p. 101.

109 *Tsien resigned his position:* Interview with Homer Joseph Stewart.

Chapter 10: Washington and Germany (1945)

110 *reeling from four years of wartime chaos:* David Brinkley, *Washington Goes to War* (New York: Ballantine, 1988).

110 *issued a gold badge and a top secret clearance:* Memo, January 22, 1945, MID 200.2 China, file 201, National Archives.

110 *"For twenty years, the Air Force was built":* Theodore von Kármán and Lee Edson, *Wind and Beyond: Theodore von Kármán, Pioneer in Aviation and Pathfinder in Space* (Boston: Little, Brown, 1967), p. 271.

111 *"regard the equipment":* Gorn, *The Universal Man,* p. 102.

111 *toured RCA Laboratories:* "Req. visit to the RCA laboratory, Princeton, New Jersey. By H.S. Tsien," memorandum, MID 335.11 China, file 201, National Archives.

111 *"It was not unusual to say":* Telephone interview with Chester Hasert, February 7, 1993.

111 *"All that we were writing":* Ibid.

111 *Future Trends of Development of Military Aircraft:* Telephone interview with Chester Hasert, February 7, 1993; Gorn, *The Universal Man,* p. 101.

111 *"I really got to know Tsien":* Telephone interview with Chester Hasert, February 7, 1993.

111 *"Why not go to Germany":* Kármán and Edson, *Wind and Beyond*, p. 272.

111 *"All I need":* Tsien, telegram to "Miss Fey Martin, Room 4D 1070 Pentagon," April 13, 1945, folder 90.2, Kármán collection, Caltech Archives.

112 *Fredrick Glantzberg wrote to the INS:* Fredrick Glantzberg, letter to Commissioner of Immigration and Naturalization Service, April 17, 1945, folder 90.2, Kármán collection, Caltech Archives.

112 *INS granted Tsien a special waiver:* T. B. Shoemaker, Acting Commissioner of the Immigration and Naturalization Service, letter to Fredrick Glantzberg, April 23, 1945, folder 90.2, Kármán collection, Caltech Archives.

112 *assimilated rank of Colonel:* Fredrick E. Glantzberg, "Issuance of AGO Cards," memo to Adjunctant General, folder 90.2, Kármán collection, Caltech Archives; Fredrick Glantzberg, memo, April 19, 1945, Library of Congress.

112 *memo from a top Pentagon official:* G. T. McHugh, memo (n.d.), folder 90.2, Kármán collection, Caltech archives.

112 *"unlikely but pleasant":* Kármán and Edson, *Wind and Beyond,* pp. 272–75.

112 *boarded a C-54 transport:* Gorn, *The Universal Man,* p. 104.

112 *Tsien met with von Braun in person:* R. Cargill Hall, *Earth Satellites, A First Look by the United States Navy,* Fourth History Symposium of the International Academy of Astronautics, German Federal Republic, October 1970.

112 *"Survey of Development of Liquid Rockets":* See *Peenemunde East: Through the Eyes of 500 Detained at Garmish* (U.S.A.F. Academy Library).

113 *"I remember one of them, Dr. Tsien":* Rudolph Hermann, unpublished memoirs (Archives of the U.S. Space and Rocket Center, Huntsville, AL). For more information, contact Mitch Sharpe.

113 *"The whole thing was incredible":* Kármán and Edson, *Wind and Beyond,* pp. 272–75.

113 *used metal detectors:* Personal interview with Abe Hyatt.

114 Information concerning Robert Jones's swept-back theory from James Hansen, *Engineer in Charge: A History of the Langley Aeronautical Laboratory, 1917–1958* (Washington, D.C.: National Aeronautics and Space Administration, 1987), pp. 279–85; personal interview with Robert Jones, January 14, 1992.

114 *Jones complained of the situation:* Telephone interview with George Schairer, January 11, 1992.

114 *in a dry well:* Kármán and Edson, *Wind and Beyond,* p. 273.

115 *the papers, slightly damp:* Interview with A. M. O. Smith; Smith served as one of translators of the papers.

115 *George Schairer as the courier:* Telephone interview with George Schairer, January 11, 1992; Kármán and Edson, *Wind and Beyond,* p. 277.

115 Description of Dora from interviews with George Schairer, Paul Dane, and A. M. O. Smith.

115 *"something fantastic":* Ibid., p. 94.

115 *"Hell of all concentration camps":* Jean Michel, *Dora* (London: Weidenfeld and Nicolson, 1979), p. 2.

115 *"Compared to Dora, Auschwitz was easy":* Ibid., p. 64.

116 *"This was not a handclasping, welcoming get-together":* Telephone interview with Paul Dane, October 24, 1992.

116 *"Nordhausen was fresh on my mind"*: Kármán and Edson, *Wind and Beyond,* pp. 277–83.

116 *"If we had to be conquered"*: Ibid.

116 *"I couldn't tell whether"*: Ibid.

116 *"I'll learn from the Germans"*: Telephone interview with Judson Baron, November 20, 1992.

116 *inspected wind tunnel facilities:* H. S. Tsien, "Reports on the Recent Aeronautical Developments of Several Selected Fields in Germany and Switzerland," vol. 3, *Toward New Horizons* (Dayton, OH: Headquarters Air Materiel Command, Publications Branch, Intelligence T-2, Wright Field).

116 *preparations to return to Washington:* Allan Juhl, INS investigator, report filed September 7, 1950, U.S. Customs file, National Archives.

116 *assistant professor to associate professor:* Tsien's biographical resume, NASA Historical Archives.

117 *Jet Propulsion:* Hsue-shen Tsien, ed., "Jet Propulsion: A Reference Text Prepared by the Staff of the Guggenheim Aeronautical Laboratory and the Jet Propulsion Laboratory, GALCIT. California Institute of Technology for the Air Technical Service Command" (JPCHF no. 5-392), JPL Archives; personal interview with Allen Puckett.

117 *Toward New Horizons:* Tsien's contributions to *TNH* can be found in vols. 4, 6, 7, and 8 and in the technical intelligence supplement. See Pentagon Library.

118 *"It was a comprehensive scope"*: Telephone interview with Joseph Charyk.

118 *Possibility of Atomic Fuels:* H. S. Tsien, "Possibility of Atomic Fuels for Aircraft Propulsion of Power Plants," *Toward New Horizons,* vol. 7, pt. 5.

118 *"undisputed genius"*: Kármán and Edson, *Wind and Beyond,* p. 308.

118 *"excellent and complete"*: Citation in Milton Viorst, personal files.

118 *similarly praised by James Conant:* Ibid.

118 *"outstanding performance"*: Ibid.

119 *"SuperAerodynamics"*: See *Journal of the Aeronautical Sciences,* 13 (1946): 653–64.

119 *"Wake up! Wake up!"*: Interview with Albert de Graffenried, November 2, 1992.

119 *"I believe C.I.T."*: J. C. Hunsaker, letter to James Killian, June 14, 1946, MIT Office of the President, 1930–1958 (AC4), Institute Archives and Special Collections, MIT Libraries, Cambridge, MA.

119 *"It was a crucial transition"*: Interview with Homer Joseph Stewart.

119 *"Tsien had this vision"*: Interview with C. C. Lin, 1991.

120 *On June 17, 1946:* Photographic negative no. 4A13540, NASM Archives.

120 *August 1946:* Roger Wolcott, FBI report, folder LA23-24, "Investigative Case Files Relating to Neutrality Violations," Records of the U.S. Customs Service, record group 36 (National Archives—Pacific Southwest Region, Laguna Niguel, CA), p. 6.

120 *terminated his employment:* Interview with Homer Joseph Stewart.

Chapter 11: Return to MIT (1946–1947)

121 *large, red-brick Georgian Colonial:* Lynne Marcus, current resident; address from Tsien's IRR file, National Archives.

121 *"I have yet":* Tsien, letter to Caltech, October 1, 1946, folder 30.37, Kármán collection, Caltech Archives.

122 *building number 33:* Security questionnaires, Tsien's IRR file, National Archives.

122 *own office on the third floor:* Interviews with alumni.

122 Information about the drive from Newton to MIT and Tsien's view of MIT from his office from current photographs; Phil Bergen of the Bostonian Society; photographs from the MIT Museum.

122 *"Research contract activities":* Kathy Halbreich, *Art and Architecture at MIT: A Walking Tour of the Campus* (Cambridge: MIT Press, 1984), p. 27.

122 Details of MIT during the war from the MIT Museum; Halbreich, *Art and Architecture at MIT,* p. 27; Robert Wattson, letter to author, August 4, 1993, p. 3; John Dugundji, letter to author, August 20, 1993; Claude Brenner, letter to author, August 30, 1993, pp. 9–10.

123 Information about the MIT aeronautics faculty from Shatswell Ober, "The Story of Aeronautics at MIT 1895 to 1960" (unpublished ms., MIT Museum, 1965), p. 26.

123 Enrollment figures from *MIT Report of the President* (MIT Archives, 1947), p. 83

123 Staff information from ibid., p. 84.

123 *"There was considerable excitement":* Interview with Bob Summers, November 15, 1993.

123 *Tsien applied for security clearance:* Tsien's IRR file and U.S. Customs file, National Archives.

123 *applied for top secret clearance:* Ibid.

123 *"Atomic Energy":* Mention of Tsien's involvement in the Manhattan District project in a personnel security questionnaire dated October 17, 1946, Tsien's IRR file, National Archives.

124 *"That paper didn't have a practical influence":* Interview with Yuan-cheng Fung, November 13, 1992.

124 *lecture on nuclear powered rockets:* Interview with Horton Guyford Stever, April 9, 1992.

124 *"The difficulty of constructing a simple nuclear fuel rocket":* Hsue-shen Tsien, "Rockets and Other Thermal Jets Using Nuclear Energy with a General Discussion on the Use of Porous Pile Materials," seminars 54 and 55, May 13 and 15, 1947, in Wang Shouyun (ed.), *Collected Works of H. S. Tsien;* see p. 534.

124 *"Dr. Tsien is certainly one of the leading men":* Kármán, letter to Hunsaker, February 21, 1947, folder 30.38, Kármán collection, Caltech Archives.

124 *taught his first course at MIT*: Interviews with MIT alumni.

125 *"He seemed very young"*: Interview with Jim O'Neill, May 4, 1992.

125 *"Tech is Hell"*: Claude Brenner, letter to author, August 30, 1993, p. 9.

125 *"He had the reputation for being an egotistical loner"*: Jim Marstiller, letter to author, August 20, 1993.

125 *"He appeared ill at ease"*: Daniel Fink, letter to author, August 19, 1993.

125 *"He appeared to be a loner"*: Robert Chilton, letter to author, August 1993.

125 *"He was a very cold and unemotional person"*: Frederick Smith, letter to author, August 10, 1993.

125 *"He stood out as the only aloof"*: Leonard Sullivan, letter to author, August 10, 1993.

125 *"As a teacher Professor Tsien was a tyrant"*: Claude Brenner, letter to author, August 30, 1993.

125 *"generally disliked and even feared him"*: James van Meter, letter to author, August 27, 1993.

125 *"At least one good man"*: Robert Wattson, letter to author, August 4, 1993.

125 *shut himself in his office*: Telephone interview with Rene Miller, August 5, 1992.

125 *"There were no text or notes available"*: Edgar Keats, "The Blackboard Jumble," *Saturday Review,* January 27, 1968.

127 *"Tsien would rush up to his office"*: Interview with Holt Ashley, August 14, 1992.

127 *"He had an intense look"*: Interview with Bob Summers, November 15, 1993.

127 *"Shortly after the start of the second semester"*: Mrs. Robert Postle, letter to author, August 8, 1993.

127 *"You had to be clever enough"*: Interview with MIT alumnus.

127 *only one doctoral candidate*: Claude Brenner, letter to author, August 30, 1993, p. 5.

127 *how difficult the tests could be:* Jim Marstiller, letter to author, August 20, 1993.

128 *lectures in "masamatics"*: Ibid.

128 *"Where did you get this handbook formwula?"*: E. Eugene Larrabee, letter to author, August 5, 1993.

128 *"I had often wondered if he had learned spoken English"*: Ibid.

128 *"copy down every smidgen"*: Interview with Leo Celniker, May 4, 1992.

128 *"I ended up with a book"*: Edwin Krug, letter to author, August 11, 1993.

128 *Mack was initially so intimidated by Tsien:* Interview with Leslie Mack, November 4, 1992.

128 *"I was writing away"*: Ibid.

128 *"He was an extremely hard worker"*: Ibid.

129 *brush them off with a comment:* Erik Mollo-Christensen, letter to author, June 9, 1993.

129 *"He would sit in his office with the door closed"*: Larry Manoni, letter to author, August 23, 1993.

129 *"Inevitably, the students found small ways"*: Claude Brenner, letter to author, August 30, 1993, p. 5.

129 *"Dare we invite Tsien?"*: Ibid, p. 6.

129 *Judson Baron, a former student of Tsien's*: Interview with Judson Baron, November 20, 1992.

130 *"Tsien's courses were invaluable for the first ten years of my career"*: Interview with Leo Celniker.

130 *promoted from associate professor to full professor*: MIT Technology Review (May 1947): 412.

130 *"For Tsien to get tenure that early"*: Interview with John Bluth, JPL historian. Tsien was the youngest full professor on the faculty of MIT at the time; see Milton Viorst, *Hustlers and Heroes* (New York: Simon & Schuster, 1971), p. 141.

130 *possibility of becoming president of his alma mater*: Viorst, *Hustlers and Heroes*, p. 141.

131 *"One of the things Tsien brought to MIT"*: Interview with Leslie Mack.

131 *"The old-timers like Shatswell Ober"*: Ibid.

131 *"different class of guy altogether"*: Interview with Jim O'Neill, May 4, 1992.

131 *applied for permanent residency status*: Robert Wolcott, FBI report, September 22, 1950, folder LA23-24, "Investigative Case Files Relating to Neutrality Violations," Records of the U.S. Customs Service, record group 36 (National Archives—Pacific Southwest Region, Laguna Niguel, CA).

131 *flew to Montreal*: Ibid, p. 3.

131 *spent a few weeks with friends in Pasadena*: See letters from Sidney Weinbaum and his wife to Frank Malina in the Frank Malina collection, Caltech Archives.

131 *left for China in July*: Allan Juhl, INS report, September 7, 1950, folder LA23-24, "Investigative Case Files."

Chapter 12: Summons from China (1947)

132 According to Tsien's IRR file, Tsien's father was living on Lane 1032 #111 Yu-Yuan Road, Shanghai.

132 *mother had died from typhoid*: Interviews with Yucon Tsien and Qian Xueming. The date of her death is unknown.

132 *visited three major cities*: Hsue-shen Tsien, "Engineering and Engineering Sciences," *Journal of the Chinese Institute of Engineers*, 6 (1948): 1–14. The contents of Tsien's speech are reprinted there.

133 *"The famine in China"*: Lloyd Eastman, *Seeds of Destruction: Nationalist China in War and Revolution, 1937–1949* (Stanford: Stanford University Press, 1984), p. 73.

135 *official reason given later*: Who's Who in Communist China (Hong Kong: Union Research Institute), from Milton Viorst's personal file.

Chapter 13: Jiang Ying

136 Biographical information about Jiang Fangzhen from Howard Boorman and Richard Howard, *The Biographical Dictionary of Republican China* (New York: Columbia University Press, 1967–79), pp. 312–17.

137 *"the five flowers"*: Qian Xuemin, "Riyue bihe, fengyu tongzhou" [A perfect match, sharing a common fate], nine-part series in *People's Daily,* overseas edition, March 31–April 10, 1992, part 1.

137 *begged Chiang for permission to adopt her:* Ibid, part 4.

137 *took on the new name, "Tsien Hsue-yin":* Ibid.

137 *He told her stories:* Ibid.

137 *Ying entered the Zhongxi Women's School:* Gan Jiayu, "Haiwai ershi nian, guiguo sishi chun—ji zhuming nugaoyin gechangjia, shengyue jiaoyujia Jiang Ying jiaoshou" [Overseas for twenty years and home for forty years—A record of the renowned soprano and vocal music educator Jiang Ying] (unpublished manuscript, 1991), p. 2.

137 *became a skilled equestrienne and swimmer:* Ibid., p. 2; personal interview with Gao Yun and Guo Dai Zhao, former colleagues of Jiang Ying, June 28, 1994.

137 *learned to sing and play the piano:* Ibid., pp. 2–3.

137 *sang Wagnerian operas at home:* Ibid, p. 2.

137 *read poetry from the Tang dynasty:* Ibid.

137 *father wrote books about Western history:* Ibid.

137 *edited a literary journal:* Ibid.

137 *invited world-class authors:* Ibid.

137 *three-month grand tour of Europe:* Ibid, p. 3.

137 *study the structure of national mobilization:* Boorman and Howard, *Biographical Dictionary,* p. 316.

137 *dazzled by the architecture and music:* Gan Jiayu, "Overseas for Twenty Years," p. 3.

137 *enrolled Ying in an aristocratic German high school:* Ibid.

137 *University of Berlin:* Ibid, p. 4.

137 *"After I entered college":* Ibid.

137 *she sang at parties:* Ibid.

137 *learned that her father had died of a heart attack:* Ibid.

137 *fled to Switzerland:* Ibid.

138 *debuted with a recital:* Ibid, p. 9.

139 *"How about it?":* Qian Xuemin, "A Perfect Match," part 5.

139 *it would be Ying or nobody:* Ibid.

139 *wedding held on September 17, 1947:* Allan Juhl, INS report, September 7, 1950, folder LA23-24, "Investigative Case Files Relating to Neutrality Violations,"

Records of the U.S. Customs Service, record group 36 (National Archives—Pacific Southwest Region, Laguna Niguel, CA).

139 *a large opera house in Italy:* Qian Xuemin, "A Perfect Match," part 5.

139 *a long, graphic letter to Theodore von Kármán:* Theodore von Kármán and Lee Edson, *Wind and Beyond: Theodore von Kármán, Pioneer in Aviation and Pathfinder in Space* (Boston: Little, Brown, 1967), p. 310.

139 *urged his friends not to remain in China:* Personal interview with Xu Liangying, June 21, 1993. Xu derives this information from his conversations with those scientists.

Chapter 14: Ascent (1947–1948)

140 *lived in an apartment on 9 Chauncy Street:* Theodore Pian, letters to author, September 6, 1993, September 21, 1993.

140 *joined the American Academy of Arts and Sciences:* Allan Juhl, INS report, September 7, 1950, folder LA23-24, "Investigative Case Files Relating to Neutrality Violations," Records of the U.S. Customs Service, record group 36 (National Archives—Pacific Southwest Region, Laguna Niguel, CA), p. 2.

140 The Tsiens' musical activities described in Gan Jiayu, "Overseas for Twenty Years."

140 *spent an evening with Rene Miller and his wife:* Interview with Rene Miller, August 5, 1992.

140 *huge teaching load:* Shatswell Ober, "The Story of Aeronautics at MIT 1895 to 1960" (unpublished ms., MIT Museum, 1965), p. 37.

140 *presented a paper on wind tunnel testing problems:* Hsue-shen Tsien, "Wind Tunnel Testing Problems in Superaerodynamics," *Journal of the Aeronautical Sciences,* 15, no. 10 (October 1948): 573–80; presented at the sixteenth annual meeting of the IAS in New York, January 26–29, 1948.

140 *coauthored journal articles:* With C. C. Lin and Eric Reissner, "On Two-Dimensional Non-Steady Motion of a Slender Body in a Compressible Fluid," *Journal of Mathematics and Physics,* 27 (1948): 220–31; with Judson Baron, "Airfoils in Slightly Supersonic Flow," *Journal of the Aeronautical Sciences,* 16 (1949): 55–61; with M. Finston, "Interaction Between Parallel Streams of Subsonic and Supersonic Velocities," *Journal of the Aeronautical Sciences,* 16 (September 1949): 515–28.

140 *served on the aerospace vehicles panel:* Thomas Sturm, *The USAF Scientific Advisory Board: Its First Twenty Years, 1944–1964* (Office of Air Force History, 1986), p. 146.

141 *Tsien told his friends at MIT:* Ryan and Summerlin, *The China Cloud,* p. 72.

141 *blessed with a son:* Birth certificate no. 088341, registry division, City of

Boston. Tsien's son was named Yucon Tsien on the birth certificate, although his name in pinyin is Qian Yonggang.

141 *at Princeton:* Telephone interview with Joseph Charyk, March 9, 1992.
141 *"All of your many friends here":* DuBridge, letter to Tsien, September 29, 1948, folder 100.4, DuBridge collection, Caltech Archives.
142 *resigned from his position as acting director:* Judith Goodstein and Carol Buge, *The Frank J. Malina Collection at the California Institute of Technology: Guide to a Microfiche Edition* (Institute Archives, Robert A. Millikan Memorial Library, California Institute of Technology, 1986).
142 *"I had long been convinced that":* Frank Malina, "America's First Long Range Missile and Space Exploration Program: The ORDCIT Project of the Jet Propulsion Laboratory, 1943–1946" (prepared for the Fifth International Symposium on the History of Astronautics, International Academy of Astronautics, Brussels, Belgium, September 23, 1971; JPL Archives), p. 1.
142 *"The atmosphere here is too business-like":* Tsien, letter to Malina, December 27, 1948, section 1, file 3.29, Frank Malina microfilm collection, Caltech Archives.
143 *he applied for U.S. citizenship:* FBI report (file number 65-4857), September 22, 1950, folder LA23-24, "Investigative Case Files"; Declaration of intention to become a U.S. citizen (no. 329866 at U.S. District Court in Boston), April 5, 1949, National Archives, New England Regional Branch.

Chapter 15: Caltech (1949)

144 *he wanted to buy a house:* Personal interview with Bill Davis, October 13, 1992.
144 *Tsien decided to rent:* Tsien had previously rented this house from 1941 to 1946. Pasadena Historical Society; personal interview with Luo Peilin, June 30, 1993.
144 *a long hall bisected the house:* Interview with Luo Peilin, June 30, 1993.
144 *the center of a tiny social circle:* Milton Viorst, *Hustlers and Heroes* (New York: Simon & Schuster, 1971), p. 142; telephone interview with Frank Goddard, November 18, 1992.
145 *Perhaps his closest friend.* Interview with Luo Peilin, November 1991.
145 *She was more outgoing:* Viorst, *Hustlers and Heroes,* p. 142.
145 *working on the Sergeant test vehicle:* Clayton Koppes, *JPL and the American Space Program: A History of the Jet Propulsion Laboratory* (New Haven, CT: Yale University Press, 1982), p. 62.
146 *to consult three Mondays a month:* Roger Wolcott, FBI report, September 22, 1950, folder LA23-24, "Investigative Case Files Relating to Neutrality Violations," Records of the U.S. Customs Service, record group 36 (National Archives—Pacific Southwest Region, Laguna Niguel, CA), p. 9.

146 *applied for security clearance:* Security questionnaires in Tsien's U.S. Customs file and IRR file at National Archives.

146 *heavy teaching load at Caltech: Bulletin of Caltech* (October 1949), Institute Archives, California Institute of Technology.

146 *nine out of the ten students who graduated:* H. S. Tsien, "Guggenheim Jet Propulsion Center," *GALCIT: The First Twenty-five Years* (California Institute of Technology, 1954), p. 49.

146 *they were taught according to Kármán's principle:* Ibid., p. 51.

146 *"His advice was that":* William Ward, letter to author, December 26, 1992.

147 *"It made us realize that we didn't understand":* Frederic Hartwig, letter to author, May 27, 1993.

147 *"But it's wrong!":* Personal interview with Thomas Adamson, December 10, 1991.

147 *"Stop! This is a bunch of":* Telephone interview with Robert Evans.

147 *argument so loud:* Interview with Hans Liepmann, 1991.

147 *to promote the peacetime and commercial uses of rockets:* H. S. Tsien, "Guggenheim Jet Propulsion Center," p. 49.

147 *transcontinental rocketliner: New York Times,* December 2, 1949, p. 52; *Time,* December 12, 1949, p. 46.

148 *"not at all beyond the grasp of present-day technology": New York Herald Tribune,* December 2, 1949.

148 *Newspapers described in detail: New York Sun,* December 1, 1949.

148 *Popular Science and Flight: Popular Science,* February 1950; *Flight,* March 9, 1950, p. 307.

148 *"The rocketliner could give virtually": New York Times,* December 2, 1949.

148 *trip to the moon would be possible: Pittsburgh Sun Telegraph,* June 19, 1949.

148 *Popular Mechanics ran a drawing:* See *Popular Mechanics,* May 1950.

148 *drunk or mentally unbalanced:* Murrey Marder, "The Chinese Scientist We Expelled, and His Pupil in the Pentagon," *Washington Post,* June 14, 1981.

Chapter 16: Suspicion (1950)

149 Information about Tsien's initial encounter with the FBI from FBI file, folder LA23-24, "Investigative Case Files Relating to Neutrality Violations," Records of the U.S. Customs Service, record group 36 (National Archives—Pacific Southwest Region, Laguna Niguel, CA), p. 14.

150 *Caltech administrators received a hand-delivered letter:* Jack Young, Headquarters, Sixth Army, Presidio of San Francisco, letter to Caltech, folder LA23-24, "Investigative Case Files."

150 *ninety percent of all research at JPL:* John Bluth, JPL historian, letter to author.

150 *they contacted Tsien again:* FBI file 65-4857, folder LA23-24, "Investigative Case Files," p. 17.

150 *Tsien gave the FBI a prepared statement:* Ibid.

151 *"didn't believe that he should have to prove to the authorities":* Theodore von Kármán and Lee Edson, *Wind and Beyond: Theodore von Kármán, Pioneer in Aviation and Pathfinder in Space* (Boston: Little, Brown, 1967), p. 311.

151 *about to undergo a serious stomach operation:* "Interview of Dr. Hsue-shen Tsien at United States Immigration Bureau, Terminal Island, California," September 20, 1950, folder LA23-24, "Investigative Case Files," p. 7.

151 *also wanted to spend time with Tsien's children.* Personal interview with Frank Marble, 1992.

151 *"He was confused as to what to do":* Telephone interview with Martin Summerfield, October 6, 1992.

152 *half-suspected that his father was being pressured:* Ibid.; DuBridge, letter to Peyton Ford, Deputy Attorney General, September 18, 1950, folder 100.4, DuBridge collection, Caltech Archives. In his letter, DuBridge said that Tsien "stated frankly to many people many months ago that he realized there was a possibility that his father might be under pressure from the Communist regime in China to write these letters." See also Roger Wolcott, FBI report, folder LA23-24, "Investigative Case Files," p. 19.

152 *out of Red China and into Hong Kong:* Telephone interview with William Zisch, 1991.

152 *Kármán was in Paris:* Michael Gorn, *The Universal Man: Theodore von Kármán's Life in Aeronautics* (Washington, D.C.: Smithsonian Institution Press, 1982), p. 125.

152 *arrest of Sidney Weinbaum:* "FBI Arrests Caltech Scientist for Perjury," *Pasadena Star News,* June 16, 1950, p. 1.

152 *application for the position of mathematician:* Oral history of Sidney Weinbaum by Mary Terrall (Caltech Archives, 1986), p. 56.

152 *Tsien had originally recommended Weinbaum:* Roger Wolcott, FBI report, September 22, 1950, folder LA23-24, "Investigative Case Files," p. 16.

152 *"turned the FBI's attention to him":* Kármán and Edson, *Wind and Beyond,* p. 311.

152 *Luo Peilin wanted to return to China:* Personal interview with Luo Peilin, November 1991.

153 *birth of his second child:* Certification of Vital Record, State file no. 50-101246, State of California, Department of Heath Services. The name on the certificate is Yung-jen Tsien, although in pinyin the name is now spelled Qian Yongzhen.

153 *wrote to the State Department:* DuBridge, letter to Kimball, September 18, 1950, folder 100.4, DuBridge collection, Caltech Archives.

153 *visited Washington to secure official permission:* Personal interview with Luo Peilin, November 1991.

153 *tried to make reservations:* Wiliam Ryan and Sam Summerlin, *The China Cloud: America's Tragic Blunder and China's Rise to Nuclear Power* (Boston: Little,

Brown, 1967), p. 85; "Interview of Dr. Hsue-shen Tsien," folder LA23-24, "Investigative Case Files," pp. 11–12.

153 *wrote to the International Trade Services:*Ibid.

153 *latter thought he was "crazy":* Personal interview with Hans Liepmann, July 17, 1991.

153 *"that he couldn't fulfill his responsibilities":* William Rees Sears, *Stories from a 20th Century Life* (Stanford, CA: Parabolic Press, 1993), pp. 68–69.

153 *In July, he wrote to Kármán:* DuBridge, letter to Kármán, July 21, 1950, folder 100.4, DuBridge collection, Caltech Archives.

153 *whole thing was nothing more than a witch-hunt:* "EB" [DuBridge's secretary?], memo to DuBridge about Clark Millikan's phone call, July 14, 1950, folder 100.4, DuBridge collection, Caltech Archives.

153 *"This is a ridiculous situation":* DuBridge, letter to Kármán, July 21, 1950, folder 100.4, DuBridge collection, Caltech Archives.

153 *"in the hands of the Russians":* E. C. Watson, letter to James Page, June 22, 1950, folder 100.4, DuBridge collection, Caltech Archives.

153 *scheduled a hearing for Tsien:* DuBridge, telegram to Col. F. M. Wray, August 14, 1950, folder 100.4, DuBridge collection, Caltech Archives.

154 *denied clearance not only by the Air Force, Navy, and Army:* Tsien's IRR file and U.S. Customs file.

154 *"The authorities at Caltech":* "Interview of Dr. Hsue-shen Tsien," September 20, 1950, folder LA23-24, "Investigative Case Files."

154 *hired a packing company:* "Statement of C. Harold Sexsmith, Made in the Office of the Customs Agent in Charge, Room 308, H.W. Hellman Building, 354 So. Spring St. Los Angeles 13, California, at 9:20 am, September 20, 1950," folder LA23-24, "Investigative Case Files."

154 *time was running out:* Ibid, p. 28.

154 *"The shipment appeared":* Ibid, p. 27.

154 Information about Dan Kimball from Kármán and Edson, *Wind and Beyond,* p. 312.

155 *started to tell him everything:* Milton Viorst, *Hustlers and Heroes* (New York: Simon & Schuster, 1971), p. 146.

155 *burst into tears:* Ibid.

155 *seeing Tsien's anxiety:* Ibid.

155 *determined to see Tsien restore his clearance:* Ibid.

155 *"He was so overwhelmed":* "Interview of Dr. Hsue-shen Tsien," September 20, 1950, folder LA23-24, "Investigative Case Files,"

155 *Tsien met with Porter:* Ibid.

155 *Tsien met with Kimball again:* Ibid.

155 *"You can't leave":* Ryan and Summerlin, *The China Cloud,* p. 87.

155 *warned Tsien to think over the matter very seriously:* "Interview of Dr. Hsue-shen Tsien," September 20, 1950, folder LA23-24, "Investigative Case Files," p. 14.

155 *boarded a plane for Los Angeles:* Ibid.

155 *Kimball called the Justice Department:* Ibid, p. 15; Viorst, *Hustlers and Heroes,* p. 146.

155 *Kimball believed that the Chinese government:* "Interview of Dr. Hsue-shen Tsien,"September 20, 1950, folder LA23-24, "Investigative Case Files," p. 15.

155 *agent handed him a paper:* Ibid, p. 17; H. R. Landon, District Director of INS, letter served to Tsien at Los Angeles International Airport, 9:03 P.M. PST, August 23, 1950 (INS agency files in Los Angeles).

155 *cancelled his reservation:* Ibid.

156 *considered having her and his children:* Ibid.

156 *already inspected and impounded the luggage:* Folder LA23-24, "Investigative Case Files"; DuBridge, letter to Kimball, September 5, 1950, p. 3, Folder 100.4, DuBridge collection, Caltech Archives.

156 *descended on the Bekins warehouse:* Roy Gorin, "Dr. H.I. Tsien, possible espionage, Export Control, and Neutrality Act Violations," memorandum to Collector of Customs, August 24, 1950, folder LA23-24, "Investigative Case Files"; Ernest Glazer and Roy Gorin, U.S. Customs, letter to U.S. Attorney, September 11, 1950.

156 *State department officials recommended seizure:* Roy Gorin, memorandum to Collector of Customs, August 24, 1950, p. 2, folder LA23-24, "Investigative Case Files."

156 *U.S. attorney's office in Los Angeles wanted Tsien put under surveillance:* Ibid. On January 4, 1951, the U.S. Attorney determined that there was no evidence of intent to violate espionage laws of the United States.

156 *Customs officials applied for civil warrant:* Box 64, folder 12180-HW, Civil Case Series, Southern District of California Central Division (Los Angeles), U.S. District Courts, record group 21 (National Archives—Pacific Southwest Region, Laguna Niguel, CA).

156 *a federal judge granted it:* "Caltech Scientist Held on U.S. Count," *Los Angeles Times,* September 8, 1950.

156 *Immigration officials issued a warrant:* Folder LA23-24, "Investigative Case Files."

156 *Mrs. Tsien called Sexsmith:* "Statement of C. Harold Sexsmith, Made in the Office of the Customs Agent in Charge, Room 308, H.W. Hellman Building, 354 So. Spring St. Los Angeles 13, California, at 9:20 am, September 20, 1950," folder LA23-24, "Investigative Case Files," p. 26.

156 *"I told Mrs. Tsien that I was rather surprised":* Ibid.

156 *Apparently, Tsien had been unaware:* DuBridge, letter to Dan Kimball, September 5, 1950, folder 100.4, DuBridge collection, Caltech Archives.

156 *"It was my personal property":* Los Angeles Times, August 25, 1950.

157 *"There are no code books"*: Statement of Dr. Tsien, August 25, 1950, folder 100.4, DuBridge collection, Caltech Archives; *San Francisco Chronicle,* August 26, 1950.

157 *"Secret Data Seized"*: *Los Angeles Times,* August 26, 1950.

Chapter 17: Arrest (1950)

158 *accusation first reached Sidney Weinbaum*: Oral history of Sidney Weinbaum by Mary Terrall (Caltech Archives, 1986), pp. 36–37.

158 *Army Intelligence heard accusations*: Handwritten document signed "aim," Tsien's IRR file, National Archives.

159 *"greatest confidence"*: *Los Angeles Examiner,* June 22, 1950.

159 *several scientists wrote letters*: Personal files of Betty Weinbaum.

159 Information about trial from *United States of America vs. Sidney Weinbaum,* box 1261, folder 21408, Criminal Case Files 21383–21408, Central Division 1929–1938, Records of the District Court, Southern District (RG 21) (National Archives—Pacific Southwest Region, Laguna Niguel, CA).

159 *"biting his nails"*: "FBI Expert Links Weinbaum to Reds on Handwriting," *Los Angeles Mirror,* September 1, 1950, p. 2.

159 *use of code names*: Tsien's IRR file; Weinbaum trial transcript.

159 *"He might know things"*: Weinbaum trial transcript, September 12, 1950, 10:00 A.M.

160 *sheer bulk was intimidating*: Folder LA23-24, "Investigative Case Files Relating to Neutrality Violations," Records of the U.S. Customs Service, record group 36 (National Archives—Pacific Southwest Region, Laguna Niguel, CA).

160 *three men arrived from Wright-Patterson*: William Jennings Bryan, Jr., Collector of Customs, letter to Col. Harold Watson, Air Materiel Command, September 8, 1950, ibid.

160 *outline inventory*: Folder LA23-24, "Investigative Case Files."

160 *four hundred page collection*: Ibid.

161 *"The importance of the entire library"*: R. J. Trauger, letter to Roy Gorin, September 6, 1950, ibid.

161 *warrant had been issued weeks before*: Folder LA23-24, "Investigative Case Files."

161 *apparently eluded FBI surveillance*: State Department, telegram, September 7, 1950, 793.5211/9-750, Civil Reference Division, National Archives.

161 *almost wired a telegram to the Mexican government*: Ibid.

161 *holding her baby daughter*: William Ryan and Sam Summerlin, "The Incredible Story of How China Got the Bomb," *Look,* 31, no. 15 (July 25, 1967).

161 *"cowering in a corner"*: Ibid.

161 *"Well, it's finally over"*: Ibid.

161 *gave him a standard interview*: Allan Juhl, report, September 7, 1950, folder LA23-24, "Investigative Case Files."

161 *"we drove directly to his house"*: Telephone interview with Frank Goddard, October 19, 1992.

162 History of Terminal Island from Mary Zang, "Terminal Island History," *San Pedro Bay Historical Society Shoreline,* 29, no. 1 (June 1991).

162 *three-story beige stucco building*: Interview, San Pedro Bay Historical Society.

162 *not put in these crowded quarters*: Interview with Kenneth John Elwood, Assistant District Director, INS, Los Angeles.

Chapter 18: Investigation (1950)

163 *"For fifteen days"*: Qian Xuesen, "My Hard Lot in the US," *People's Daily,* January 2, 1956.

163 *"perfectly comfortable:"* "Lee DuBridge," interview by Judith Goodstein, 1982, p. 74, Caltech archives.

163 *Theodore von Kármán tried to communicate*: Muellen, telephone communication to Emerick, September 15, 1950, folder LA23-24, "Investigative Case Files Relating to Neutrality Violations," Records of the U.S. Customs Service, record group 36 (National Archives—Pacific Southwest Region, Laguna Niguel, CA).

163 *Tsien could be seen smiling and waving*: Interview with Frank Goddard, October 18, 1992.

164 *"still alive and hadn't committed suicide"*: Telephone interview with H. T. Yung, November 24, 1992.

164 *Tsien wrote a statement*: DuBridge, letter to Kimball, September 18, 1950, folder 100.4, DuBridge collection, Caltech Archives.

164 *interviewed at the INS Bureau*: "Interview of Dr. Hsue-shen Tsien at United States Immigration Bureau, Terminal Island, California," September 20, 1950, folder LA23-24, "Investigative Case Files."

164 *"It had a rather drastic effect"*: Ibid., p. 9.

165 *held a conference about Tsien*: DuBridge, letter to Deputy Attorney General Peyton Ford, September 27, 1950, folder 100.4, DuBridge collection, Caltech Archives.

165 *Tsien was released on bail*: Milton Viorst, *Hustlers and Heroes* (New York: Simon & Schuster), p. 148.

165 *"As compared to the ransoms"*: *People's Daily,* January 2, 1956.

165 Details of Tsien's long talk with Zisch from telephone interview with Zisch, 1991.

166 *"has a very high regard"*: Robert Oppenheimer, letter to DuBridge, September 26, 1950, folder 100.4, DuBridge collection, Caltech Archives.

166 *"the very fact that his father is still in China"*: DuBridge, letter to Peyton Ford, September 27, 1950, folder 100.4, DuBridge collection, Caltech Archives.

166 *wanted to deport Tsien*: Tsien's IRR file; folder LA23-24, "Investigative Case Files."

Chapter 19: Hearings (1950–1951)

167 *explanation given by the INS*: Elizabeth Berrio, Chief of Historical Reference Library and Reading Room Section of the INS, letter to the author, March 11, 1994, p. 3. Letter with conclusions of Marian Smith, INS historian.

167 *"Obviously, Dr. Tsien"*: Ibid.

168 *DuBridge wrote to Norman Chandler*: DuBridge, letter to Chandler, October 18, 1950, folder 100.4, DuBridge collection, Caltech Archives.

168 *Clark Millikan wrote to Harry Guggenheim*: Millikan, letter to Guggenheim, October 23, 1950, folder 100.4, DuBridge collection, Caltech Archives.

168 *Theodore von Kármán and other top aerodynamicists*: Statement from Kármán, November 14, 1950, Milton Viorst file.

168 *detailed investigation*: Tsien's INS hearings, IRR file, Military Reference Division, National Archives.

168 *two retired police officers*: Ibid.

169 *procedure by which they had pilfered the card*: Ibid., pp. 56–58, 103, 121.

169 *saw Tsien's registration card*: Ibid., pp. 74, 98.

169 *they did not recognize the two policemen*: Ibid., p. 134; Frank Marble, letter to Kármán, May 17, 1951, papers of Dan Kimball, Harry S. Truman Library, Independence, MO.

170 *"They always argued"*: Tsien's INS hearings, November 15, 1950, p. 26.

170 *"In the event of conflict"*: Ibid., p. 45.

170 *"My essential allegiance"*: Ibid., p. 47.

170 *emphasized that he was not a Communist*: Tsien's INS hearings, April 15, 1951, p. 281.

170 *"Were you unfavorable to Marxism"*: Ibid.

170 *a photograph of a man named Richard Lewis*: Tsien's INS hearings, November 16, 1950, p. 140.

170 *INS officials began to question Richard Lewis*: Associated Press, January 29 and 30, 1951.

170 *"If I am required to answer questions"*: Richard Lewis, letter to Carl Niemann, January 8, 1951, Milton Viorst file, p. 2.

171 *Caltech biologist Jacob Dubnoff*: Tsien's INS hearings, April 16, 1951, p. 237.

171 *considered recruiting aeronautics professor Clark Millikan*: Ibid., p. 156.

171 *"It was so ludicrous"*: Interview with Thomas Adamson, December 10, 1991.

171 *saw Tsien at a party meeting*: Tsien's INS hearings, April 11, 1951, p. 216.

171 *letter from Frank Marble to von Kármán:* May 17, 1951, papers of Dan Kimball, Harry Truman Library.

171 *"an alien who was a member of the Communist Party":* "Decision of the Hearing Officer," Tsien's INS hearing, April 26, 1951, p. 1.

Chapter 20: Waiting (1951–1954)

172 *convinced of Tsien's innocence:* Cooper, letter to DuBridge, July 24, 1951, folder 100.4, DuBridge collection, Caltech Archives.

172 *complications for some of the engineers at JPL:* Interview with Adamson, December 10, 1991.

172 Hsue-shen Tsien, T. C. Adamson, E. L. Knuth, "Automatic Navigation of a Long Range Rocket Vehicle," *Journal of the American Rocket Society,* 22 (1952): 192–99.

172 *forbidden to travel outside the boundaries of Los Angeles:* Bo Sheng, *People's Daily,* November 3, 1955; DuBridge, letter to Assistant Attorney General Stanley Barnes, November 5, 1953, folder 100.4, DuBridge collection, Caltech Archives.

173 *required to appear before the Immigration and Naturalization Service:* Ibid.

173 *resigned to the fact:* Frank Marble, letter to Kármán, October 29, 1951, folder 19.32, Kármán collection, Caltech Archives.

173 *turned out a scientific paper once a month:* "Physical Mechanics, a New Field in Engineering Science," *Journal of the American Rocket Society,* 23 (1953): 14–16; "The Properties of Pure Liquids," *Journal of the American Rocket Society,* 23 (1954): 17–24; "Similarity Laws for Stressing Heated Wings," *Journal of the Aeronautical Sciences,* 20 (1953): 1–11; "Take-off from Satellite Orbit," *Journal of the American Rocket Society,* 23 (1953): 233–36.

173 *foresaw a day when people would travel in rocket passenger ships:* Tsien, letter to Kármán, May 2, 1952, folder 30.38, Kármán collection, Caltech Archives.

173 *constantly watched: China Today,* February 1992, p. 11. The article claims that Tsien's home and office were constantly broken into, his letters opened, and phone tapped.

173 *friends who called him:* Interviews with Caltech alumni.

173 *phone would ring repeatedly:* Interview with Yucon Tsien, July 13, 1991.

174 *flattened herself on the floor of the car:* Interview with Gao Yun and Guo Dai Zhao, June 28, 1994.

174 *"We were very nervous":* Gao Yijin, "Qian Xuesen's Family," *Wen Hui Pao,* March 27, 1957.

174 *often acted as if nothing was wrong:* Interview with C. C. Chang.

174 *listen to records of Beethoven and Mozart: People's Daily,* March 31 to April 10, 1992.

174 *Immigration authorities rejected Grant Cooper's argument:* INS decision regarding appeal, April 15, 1952, IRR file, National Archives.

174 *"the conclusions are not based":* DuBridge, letter to Cooper, February 25, 1952, folder 100.4, DuBridge collection, Caltech Archives.

174 *Cooper sent word to DuBridge:* Cooper, letter to DuBridge, March 6, 1952, DuBridge collection, Caltech Archives.

175 *last appeal was denied:* Tsien's IRR file, National Archives; "Deportation Plea Denied: Immigration Unit Rejects Appeal of Dr. Tsien, Rocket Expert," *New York Times*, December 3, 1952, p. 15.

175 *could be picked up at any time:* Grant Cooper, letter to Tsien, January 2, 1953, folder 100.4, DuBridge collection, Caltech Archives.

175 *"I am personally convinced":* DuBridge, letter to Barnes, November 5, 1953, folder 100.4, DuBridge collection, Caltech Archives.

175 *Barnes wrote back:* Barnes, letter to DuBridge, November 12, 1953, folder 100.4, DuBridge collection, Caltech Archives.

175 *"Perhaps we should relax":* "Fred" [Fred Lindvall?], memorandum to DuBridge attached to November 22, 1953 letter from DuBridge to Barnes, folder 100.4, DuBridge collection, Caltech Archives.

175 *study of games and economic behavior:* Tsien, letter to Kármán, January 16, 1953, folder 30.38, Kármán collection, Caltech Archives.

175 *published a textbook:* Hsue-shen Tsien, *Engineering Cybernetics* (New York: McGraw-Hill, 1954).

175 *"an extraordinary achievement":* Wallace Vander Velde, electronic mail message to author, April 15, 1995.

176 *"Just keeping that much material in his head was awesome":* Frederic Hartwig, letter to author, May 27, 1993.

176 *"I always enjoyed talking to him":* Interviews with Robert Meghreblian, January 8, 1992, and 1995.

176 *ask him crucial questions:* Interviews with Caltech alumni.

176 *sought permission from the Navy:* Mrs. Carl Holmquist, letter to author, June 2, 1993.

177 *"increasingly impatient and irritable":* Yusuf A. Yoler, letter to author, June 2, 1993.

177 *frequently ignored them:* Interview with Bernard Rasof, December 9, 1991.

177 *"He looked worn out and under stress":* F. W. Diederich, letter to author, June 7, 1993.

177 *"At times, I actually feared for his mind":* Theodore von Kármán and Lee Edson, *Wind and Beyond: Theodore von Kármán, Pioneer in Aviation and Pathfinder in Space* (Boston: Little, Brown, 1967), p. 313.

178 *Atomic Energy Commission withdrew a sixteen hundred dollar fellowship:* Jessica Wang, "Science, Security and the Cold War in the Case of E. U. Condon," *Isis*, 83, no. 2 (June 1992): 258, citation 53.

178 *playing Russian language records:* "Hearing Before the Loyalty Board of the

National Advisory Committee for Aeronautics: In the case of Robert Jones," transcript for hearing held in conference room of Ames Aeronautical Laboratory, Moffett Field, California, July 24, 1950 (Robert Jones personal collection); personal interview, Robert Jones.

178 *James Bonner, a distinguished professor of biology:* Telephone interview with James Bonner, July 28, 1992.

178 *"That's the horrifying part":* Personal interview with Homer Joseph Stewart, April 9, 1993.

179 Information on Frank Oppenheimer from his FBI file.

179 Information on Frank Malina from telephone interview with Liljan Malina Wunderman.

179 Information on Sidney Weinbaum from personal interview with Betty Weinbaum, 1992; oral history of Sidney Weinbaum by Mary Terrall (Caltech Archives, 1986); *United States of America vs. Sidney Weinbaum,* box 1261, folder 21408, Criminal Case Files 21383–21408, Central Division 1929–1938, Records of the District Court, Southern District (RG 21) (National Archives— Pacific Southwest Region, Laguna Niguel, CA).

179 Information on Selina Bendix from telephone interview with Selina Bendix, March 19, 1992.

179 *"Every two years, someone would bring it up":* Interview with Ena Dubnoff, February 26, 1992.

179 *cast a shadow on Frank Malina's first wife:* Telephone interview with Liljan Malina Wunderman, March 6, 1992.

180 *Martin Summerfield . . . lost his security clearance:* Martin Summerfield, letter to William Zisch, December 10, 1952, folder 29.2, Kármán collection, Caltech Archives; telephone interview with Martin Summerfield, March 12, 1992. The loss of clearance for Summerfield almost prompted him to retire to his beach house and live off his Aerojet dividends.

180 *INS and FBI joined forces:* David Kaplan, *Fires of the Dragon: Politics, Murder and the Kuomingtang* (New York: Atheneum, 1992).

180 *Chinese nationals studying:* James Reston, *New York Times,* March 9, 1951.

181 *"a whole series of small interdepartmental wars":* Ibid.

181 *would have three or four hours to put up a $1,000 bond:* DuBridge, letter to INS, June 19, 1952, DuBridge collection, Caltech Archives; telephone interview with Zheng Zheming, July 14, 1993; telephone interviews with H. T. Yung and Eugene Loh, November 23, 1992.

182 *three small suitcases packed:* Gao Yijin, "Qian Xuesen's Family," *Wen Hui Pao,* March 27, 1957.

182 *"Do you expect anyone in the Caltech administration":* Tsien, letter to Malina, section 2, box 6, folder 29, Frank Malina microfiche collection, Caltech Archives.

182 *briefly eluded the FBI: People's Daily,* March 31, 1992.

Chapter 21: The Wang-Johnson Talks (1955)

184 Description of the President's Room of the Palais des Nations from U. Alexis Johnson and Jef Olivarious McAllister, *The Right Hand of Power* (Englewood, NJ: Prentice-Hall, 1984).

185 Information about the Korean War from ibid.; Albert Biderman, *March to Calumny* (New York: Macmillan, 1963); Rosemary Foot, *A Substitute for Victory* (Ithaca, NY: Cornell University Press, 1990); Rosemary Foot, *The Wrong War* (Ithaca, NY: Cornell University Press, 1985); Burton Kaufman, *The Korean War* (Philadelphia: Temple University Press, 1984); Eugene Kinkead, *In Every War But One* (New York: W. W. Norton, 1959); James Sanders, *Soldiers of Misfortune* (Washington, D.C.: National Press Books, 1992); Howe Vetter, *Mutiny on Koje Island* (Vermont: C. C. Tuttle, 1965); documents from the Harry Truman Library.

187 Description of diplomatic maneuvering from Johnson and McAllister, *The Right Hand of Power*, and telephone interviews with Johnson.

187 *more than half were told they could leave:* Johnson and McAllister, *The Right Hand of Power*, p. 237.

187 *again tried to resume talks:* Ibid.

187 *released on May 30, 1955, four American airmen:* Ibid.

187 *On July 11:* Ibid.

187 *eleven more American airmen:* Ibid., p. 241.

187 *"Despite the background":* Ibid., p. 242.

187 *"For sheer endurance":* Kenneth Young, *Negotiating with the Chinese Communists: The United States Experience, 1953–1967* (New York: McGraw-Hill, 1958), p. 7.

187 *strict ritual of negotiation:* Ibid.

187 *Johnson handed Wang a list:* Ibid., p. 243.

188 *On August 8, 1955:* U. Alexis Johnson, telegram to Department of State, Department of State central files, 611.93/8-855, National Archives.

188 *In June 1955:* Secretary of Defense, memorandum to President Eisenhower, June 11, 1955, Eisenhower Library.

188 *Eisenhower had probably never heard of Tsien:* Milton Viorst, *Hustlers and Heroes* (New York: Simon & Schuster, 1971), p. 152.

189 *"give them all back":* "Telephone Conversation with Secretary Anderson," memorandum dated 9:55 P.M., June 12, 1955, General Telephone Conversations, Dulles Papers, Eisenhower Library.

189 *The next day, June 13, 1955:* Handwritten note on Secretary of Defense memo to President Eisenhower, June 11, 1955, Eisenhower Library.

189 *By August 3:* William Godel, Deputy Assistant to the Secretary of Defense, memorandum to Secretary of State, August 3, 1955, Eisenhower Library.

189 *informed Tsien that he was free to leave:* Albert Del Guercio, letter to Tsien, August 4, 1955, Immigration and Naturalization Service files, Los Angeles.

189 *worked out a formal agreement:* Johnson and McAllister, *The Right Hand of Power,* p. 250.

189 *"I note the newspaper suggestion":* William Ryan and Sam Summerlin, *The China Cloud: America's Tragic Blunder and China's Rise to Nuclear Power* (Boston: Little, Brown, 1967), p. 154.

189 *"prime swapping goods":* Telephone interview with U. Alexis Johnson, October 8, 1992; Dulles, memorandum to Eisenhower, April 1, 1955, Eisenhower Library.

189 *"too much 'loss of face' ":* R. B. Pearce, letter to author, July 22, 1993.

189 *"Of seventy-six American prisoners":* Johnson and McAllister, *The Right Hand of Power,* p. 261.

190 *eventually got some ninety-four Chinese scientists:* "Made in the USA?", *60 Minutes,* October 27, 1970, CBS Archives.

190 *"We had won back Tsien Hsue-shen":* Lin Jiyuan, ed., *China Today: Space Industry* (Beijing: Astronautic Publishing House, 1992), p. 452.

Chapter 22: "One of the Tragedies of This Century"

191 *"I remember Tsien talking about his imminent departure":* Telephone interview with S. S. Penner, July 20, 1992.

192 *does not remember seeing Tsien at a single official meeting:* Telephone interview with Liljan Malina Wunderman, March 6, 1992.

192 *insists that Tsien was never a member:* Personal interview with Betty Weinbaum, 1992.

192 *duped into thinking Unit 122 was a music group:* Personal interview with Andrew Fejer, December 7, 1991.

192 *"a dirty trick on him":* Personal interview with Homer Joseph Stewart, July 17, 1991.

192 *"royally railroaded":* Liljan Malina Wunderman, letter to author, April 6, 1992.

193 *"While I don't rule out the possibility":* "Made in the USA?", *60 Minutes,* October 27, 1970, CBS Archives.

194 *"I think . . . Tsien thought":* Theodore von Kármán and Lee Edson, *Wind and Beyond: Theodore von Kármán, Pioneer in Aviation and Pathfinder in Space* (Boston: Little, Brown, 1967), p. 314; Michael Gorn, *The Universal Man: Theodore von Kármán's Life in Aeronautics* (Washington, D.C.: Smithsonian Institution Press, 1982), pp. 170–72.

194 *questioned him about his role as minister:* Maj. Gen. Joseph F. Carroll to Lt. Gen. Donald L. Putt, November 26, 1954; Putt to Carroll, no date; Kármán to Putt, December 22, 1954, enclosure "Statement of Theodore von Kármán Requested in the Letter of the U.S.A.E.C.," U.S. Air Force Academy Archives.

194 *evidence that Kármán tried to help*: Kármán, letter to Marble, October 22, 1951, folder 19.32, Kármán collection, Caltech Archives; Kármán, letter to Marble, November 11, 1951, folder 19.32, Kármán collection, Caltech Archives.

195 *"Tsien had broken communication"*: Telephone interview with Rene Miller, August 5, 1992.

195 *"Tsien had no intention of going back to China"*: Telephone interview with Shao-wen Yuan, December 10, 1992.

195 *"You don't ship 1,700 pounds of books"*: Personal interview with Hans Liepmann, July 17, 1991.

195 *if Tsien was truly determined not to go back*: Oral history interview of Fritz Zwicky by R. Cargill Hall, May 17, 1971.

195 *"It was very clear that he felt a certain duty"*: Interview with Chester Hasert, February 9, 1993.

197 Information about Prevention of Departure authority and the present situation of deportees from Marian Smith, INS historian, report within a letter to the author from Elizabeth A. Berrio, Chief of the INS History Office, May 16, 1995.

197 *"my mother-in-law"*: Ryan and Summerlin, *The China Cloud*, p. 108.

197 *"That the government permitted this genius"*: "Made in the USA?", *60 Minutes*, October 27, 1970, CBS Archives.

Chapter 23: A Hero's Welcome (1955)

199 *"I do not plan to come back"*: "Jet Propulsion Scientist Sailing to Red China: Dr. Hsue-shen Tsien Ends Long, Honorable Career Here to Help People of Own Nation," *Los Angeles Times*, September 18, 1955.

200 *"I'd rather shoot Tsien"*: "Made in the USA?", *60 Minutes*, October 27, 1970, CBS Archives.

200 *"It was the stupidest thing this country ever did"*: Milton Viorst, *Hustlers and Heroes* (New York: Simon & Schuster, 1971), p. 153.

200 *"To say that I was shocked"*: Grant Cooper, letter to INS, September 28, 1955, Milton Viorst personal files. It was reprinted in Viorst, *Hustlers and Heroes*, p. 153, and William Ryan and Sam Summerlin, *The China Cloud: America's Tragic Blunder and China's Rise to Nuclear Power* (Boston: Little, Brown, 1967), p. 153.

200 *would not be responsible for his safety*: Xu Guozhi, Shu Songui, and He Guozhu, "Qian's Family on the Ship Returning to China," *Senzhou xueren*, (n.d.), pp. 20–21.

201 *celebrated the sixth anniversary*: Ibid.

201 *Xu was impressed by Tsien's mind*: Personal interview with Xu Guozhi, June 25, 1993.

201 *"I was looking out so eagerly"*: Tsien Hsue-shen, "A Chinese Scientist Comes Home," *People's China,* February 1, 1956, p. 20.

201 *"flood" of newspapermen:* Ibid.

201 *"What about the confiscation of your papers":* Ryan and Summerlin, *The China Cloud,* p. 156. This interview was reconstructed from the files of Union Research in Hong Kong.

203 *"Same questions, same mentality":* Tsien, "A Chinese Scientist Comes Home," p. 20.

203 *"Yes, it was our flag!":* Ibid., p. 20.

203 *"Outside, on the stations":* Ibid., p. 25.

203 *In Guangzhou, Tsien was treated like a celebrity:* Xinhua, October 9, 1955.

204 Description of Guangzhou from Simone de Beauvoir, *The Long March* (Cleveland: World, 1958), pp. 463–75.

204 *a visit to two museums of Communist history:* Tsien, "A Chinese Scientist Comes Home," p. 25.

204 *father, now 74, met him:* Bo Sheng, "Scientist Qian Xuesen Loves Motherland," *People's Daily,* November 3, 1955.

204 *coincided with Yucon's birthday:* Ibid.

204 *visited his alma mater:* Ibid.

204 *pay his respects to his deceased mother:* Ibid.

204 *"[Shanghai] was no longer familiar":* Tsien, "A Chinese Scientist Comes Home," p. 26.

205 *delegation of twenty distinguished scientists:* Xinhua, October 30, 1955.

205 *lavish banquets:* Xinhua, November 11, 1955.

205 *"Today there are no more prostitutes":* Beauvoir, *The Long March,* pp. 42, 44, 52.

206 *"The vista was simply breathtaking":* Tsien, "A Chinese Scientist Comes Home," p. 26.

Chapter 24: Missiles of the East Wind

208 *no factories in China:* Interviews with former missile scientists; *China Today: Space Industry,* (Beijing: Astronautic Publishing House, 1992), pp. 6, 467.

208 *"We had no research team personnel":* Qian Xuesen, "Fahui jiti zhihui shi weiji hao banfa" [The utilization of collective wisdom is the only good method] *People's Daily,* April 29, 1958, p. 7.

209 *"Tsien's role was symbolic":* Interview with Lin Jin, September 2, 1992.

211 *founded the Institute of Mechanics:* "Institute of Mechanics, Academia Sinica, 1956–1986," pamphlet (official history of institute); Qian Xuesen, "Reading a Letter from Sun Yefang Again," *People's Daily,* July 30, 1984, p. 5.

211 *facilities at the fledging institute:* Personal interview with Xu Guozhi, June 25, 1993.

211 *deputy director was Guo Yonghuai.* Pamplet entitled "Institute of Mechanics, Academia Sinica, 1956–1986"

211 *held weekly workshops:* Personal interview, Xu Guozhi, June 25, 1993.

211 *"played a decisive role":* He Zuoxiu, "Professor Qian Xuesen and the Twelve-year Development Plan for Science and Technology," *Materials and Research on the History of CAS,* no. 3 (1992): 23–31.

212 *arriving at the office at 7:30:* Personal interview with Zhang Kewen, June 27, 1993.

212 *"My philosophy was that":* Ibid.

212 *consulted for Minister of Defense Peng Dehuai:* "Dangdai Zhongguo" congshu bianji bu ["Modern China" series editorial department] *Dangdai Zhonguo de guofang keji shiye* [Modern China: Scientific and technological undertakings of national defense]. 2 vols. (Beijing: Dangdai Zhonguo chuban she, 1992), vol. 1, pp. 28–29; *People's Daily,* March 31, 1992, p. 7.

212 *a secret proposal to establish research facilities:* Wang Shouyun (ed.), *Qian Xue-sen Wenji* [Collected works of H. S. Tsien] (Beijng: Kexue chuban she, 1991), pp. 802–9.

213 *special conference: China Today: Space Industry,* p. 5.

213 Information about the beginnings of the Fifth Academy from ibid., pp. 5–6; Liu Jingzhi, "Qian Xuesen: Zhongguo ren de jiaoao" [Qian Xuesen: Pride of the Chinese], *Guangming Daily,* August 28, 1989, pp. 1–2; personal interview with Zhuang Fenggan (October 2, 1991) and other scientists.

213 *"First we recognized":* Tsien Hsue-shen, "A Chinese Scientist Comes Home," *People's China,* February 1, 1956.

213 *invited section leaders:* Qian Xuesen, "Recollections and Hopes," *Beijing Review,* 34, no. 48 (December 2–8, 1991).

214 *agreed to sell China two R-1 missiles:* Personal interview with Zhuang Fenggan, October 2, 1991; *China Today: Space Industry,* p. 6; Hua Di and John Lewis, "China's Ballistic Missile Programs," *International Security,* 17, no. 2 (Fall 1992): 7.

214 *demanded something more sophisticated:* Personal interview with Zhuang Feng-gan, October 2, 1991; Di and Lewis, "China's Ballistic Missile Programs," p. 8.

214 *went to the Soviet Union:* Interview with Xu Guozhi, Zhang Kewen, and missile scientists, summer 1993. In Russia, Tsien was seen giving a talk on nuclear-powered rockets at the Moscow Aviation Institute. According to Boris Kogan, a former Soviet scientist who had met Tsien in China and introduced the ana-log computer to Chinese aerodynamicists in the 1950s, Tsien commanded tremendous respect among Soviet scientists after his book *Engineering Cyber-netics* was translated into Russian (personal interview with Boris Kogan, November 1991).

214 *Sino-Soviet New Defense Technical Accord: China Today: Space Industry,* p. 7; Di and Lewis, "China's Ballistic Missile Programs," p. 8.

214 *Soviets sold China two R-2 missiles;* Di and Lewis, "China's Ballistic Missile Programs," p. 8.

214 *delivered a total of 10,151 volumes:* Ibid.

214 *transferred more than three thousand technical professionals: China Today: Space Industry,* p. 452.

215 *typical recruit:* Interviews with anonymous Chinese missile scientists.

215 *atmosphere of overconfidence:* Interviews with missile scientists; *China Today: Space Industry,* p. 8.

215 *too short a range to hit American bases:* Di and Lewis, "China's Ballistic Missile Programs," p. 13.

215 *By early 1959, Nie Rongzhen announced:* Interviews with Lin Jin and Sun Jingliang, June 12, 1993.

215 *code name "1059":* Interview with Zhuang Fenggan; Di and Lewis, "China's Ballistic Missile Programs," p. 8.

215 *In January 1959: China Today: Space Industry,* p. 552.

215 *negotiate the delivery of more machines:* Ibid.

215 *fourteen manufacturers: China Today: Space Industry,* pp. 90–91.

216 *lack of tools:* Ibid.

216 *makeshift conditions: China Today: Space Industry,* pp. 92, 473.

216 *former hospital:* Interviews with anonymous missile scientists.

216 *windowless airplane hangar:* Personal interview with Sun Jingliang, June 12, 1993.

216 *completed a two-stage, uncontrolled rocket: China Today: Space Industry,* pp. 72–74.

216 *flat, desolate area: China Today: Space Industry,* p. 73.

217 *arrived at the Shanghai Jiangwan airport:* Ibid., p. 74.

217 *the T-7 was launched successfully:* Ibid., p. 74.

217 *limit the documents they showed the Chinese:* Interview with Zhuang Fenggan, October 2, 1991.

217 *"the mute monks who would read but not speak":* John Lewis and Xue Litai, *China Builds the Bomb* (Stanford, CA: Stanford University Press, 1988), p. 160.

217 *"We may lose more than three hundred million people":* Paul Johnson, *Modern Times: The World from the Twenties to the Eighties* (New York: Harper & Row, 1983), p. 546.

217 *Some 1,390 Soviet experts:* Spence, *The Search for Modern China,* p. 589.

217 *"treachery by the socialist imperialists":* Paul Craig and John Jungerman, *Nuclear Arms Race: Technology and Society* (New York: McGraw-Hill, 1990), p. 37.

217 *On August 12:* Photograph, Lin Jin personal collection, Beijing.

218 *Chinese students majoring in rocketry:* Di and Lewis, "China's Ballistic Missile Programs," pp. 13–14; interviews with missile scientists.

218 *launched a Soviet-made R-2 rocket: China Today: Space Industry,* p. 11.

218 *headed a committee to organize the first flight test:* Ibid.

218 *moved by train to the Shuangchengzhi base:* Interviews with anonymous missile scientists.

218 Information about Jiuquan from *A Guide to All China* (Chicago: Rand McNally, 1981), p. 119; Zhou Shunwu, *China Provincial Geography* (Beijing: Foreign Language Press, 1992), p. 443; *China Today: Space Industry*, p. 364.

218 *dispatched the 20th Corps: China Today: Space Industry*, p. 365.

219 *Tsien journeyed to the base:* "Modern China" series editorial department, *Modern China*, vol. 1, pp. 318–19; *China Today: Space Industry*, p. 11 (for description of the base, see pp. 364–66).

219 *successfully launched the Chinese R-2:* Ibid.

219 *successfully launched two other homemade R-2 missiles: China Today: Space Industry*, p. 11.

219 *a rocket called the DF2 was moved to the Jiuquan base:* "Modern China" series editorial department, *Modern China*, vol. 1, p. 320.

219 *"After the launch":* Ibid.

219 *appointed himself chief designer of the DF3:* Di and Lewis, "China's Ballistic Missile Programs," p. 14.

220 *last time he appointed himself chief designer:* Personal interview with Hua Di, January 13, 1992.

220 *"sha xuelu":* Personal interview with Lin Jin, September 2, 1992.

220 *"What the Americans can do":* Personal interview with Zhuang Fenggan, October 2, 1991.

220 *"We all felt like small pupils":* Personal interview with Lin Jin, September 2, 1992.

220 *valuable guide for engineers:* Interview with Lin Jin and other missile scientists. Liu Jingzhi referred to Tsien's *Engineering Cybernetics* as "a book from heaven"; see Liu Jingzhi, *Guangming Daily*, August 28, 1989.

221 *Tsien created a plan to facilitate communication: China Today: Space Industry*, p. 443; interviews with missile scientists.

221 *active participant in several important meetings:* Interviews with missile scientists.

221 *revised their goals for the Dongfeng program:* Ibid.

221 *clash of philosophy:* Ibid.

222 *"We should not":* Ibid.

222 *Several successful launches were accomplished:* "Modern China" series editorial department, *Modern China*, vol. 1, p. 320.

222 *launching of the DF2A: New York Times,* October 28, 1966; Di and Lewis, "China's Ballistic Missile Programs," p. 15. *China Today* claims that it was Tsien who proposed an improved medium-range missile to carry a nuclear warhead. The article notes that it took Tsien and his colleagues two years of work to build a control system that worked after the March 1962 failure. "It took the

Americans 13 years from firing their first atomic bomb to their first nuclear warhead missile. It took the Chinese only two years to do the same thing." See *China Today* (February 1992): 12.

222 *most dangerous nuclear experiment:* The United States Arms Control and Disarmament Agency Library and Los Alamos National Laboratory Library ran into problems of classified information when trying to determine if China was the only country in the world to have tested an atomic weapon and a missile at the same time. Jack Carter, the librarian at Los Alamos, said that as far as he knew China was the only country that had done it. "China was the only country cavalier enough to try it," Carter said. "Doing this would bring down the governments in other countries that might try it." (Benjamin Sawyer and Olivia Flisher, Black Gold Information Center, Santa Barbara Public Library, letter to author, March 11, 1993).

222 *"If by any chance":* Nie Rongzhen, "How China Develops its Nuclear Weapons," *Beijing Review,* no. 17 (April 29, 1985): 19.

222 *"irony of cold war history":* "Key Chinese Scientist: Tsien Hsue-shen," *New York Times,* October 28, 1966.

223 *"Properly speaking":* Milton Viorst, *Hustlers and Heroes* (New York: Simon & Schuster, 1971), p. 136.

223 *"Had his life developed differently":* "Made in the USA?", *60 Minutes,* October 27, 1970, CBS Archives.

223 *involvement in the Haiying missile:* "Modern China" series editorial department, *Modern China,* vol.2, pp. 67, 73–75, 78–80.

224 *capacity to elude antiballistic missile:* Di and Lewis, "China's Ballistic Missile Programs," p. 21.

224 *tracking and control telemetry network:* Personal interview with Shangguan Shipan, July 2, 1993; "Modern China" series editorial department, *Modern China,* vol. 1, p. 47.

225 *two Chinese intercontinental ballistic missiles: China Today: Space Industry,* p. 104.

225 *"Tsien had foresight":* Personal interview with Shangguan Shipan, July 2, 1993.

225 *code name "581": China Today: Space Industry,* p. 551.

226 The information about the development of the first Chinese satellite from Zhu Yilin (unpublished manuscript); Hua Di and John Lewis, "China's Missile and Space Programs: Technologies, Strategies, Goals" (unpublished draft manuscript given to author by Hua Di), p. 40; Dangdai Zhongguo congshu bianji bu ["Modern China" series editorial department], *Dangdai Zhongguo de Hangtian Shiye* [Modern China's space program] (Beijing: Chinese Academy of Social Sciences, 1986), pp. 27, 562.

227 *In the spring of 1970:* "Modern China" series editorial department, *Modern China's Space Program,* pp. 53–54, 256–60.

227 *hailed as a hero:* Ibid.; *Xinhua,* May 1, 1970 (FBIS).

228 *appeared in stories around the world: Wall Street Journal,* April 27, 1970, p. 3-E; *Philadelphia Inquirer,* April 27, 1970; *Christian Science Monitor,* April 29, 1970; *Boston Herald Traveler,* April 26, 1970.

228 *wrongly—credited Tsien:* See *Xinhua,* May 11, 1989 (item no. 0511042) and October 4, 1986 (item no. 1004179).

228 *talked about the importance of neutron reactions:* He Zuoxiu, "Professor Qian Xuesen and the Twelve-year Development Plan for Science and Technology," *Materials and Research on the History of CAS,* no. 3 (1992): 23–31.

228 *recommended his protégé Guo Yonghuai:* Lewis and Xue, *China Builds the Bomb,* p. 147.

229 *came from Mao Zedong himself:* "Modern China" series editorial department, *Modern China,* p. 67.

229 *exploded its first atomic bomb:* Ibid.

229 *attended a seminar:* Ibid., p. 258.

229 *nuclear submarine project:* Ibid, p. 353.

229 *"Atomic development was not Tsien's field":* He Zuoxiu, "Professor Qian Xuesen," pp. 23–31.

Chapter 25: Becoming a Communist

231 *moved into a gray Soviet-style apartment complex:* Personal interview with Qin Guo, January 11, 1993; personal interview with C. Y. Fu, June 19, 1993; personal interview with Jin Luo, September 5, 1991; author's observation of building during visit to China in summer 1993.

232 *enviable position:* Gao Yijin, "Qian Xuesen de yijia" [Qian Xuesen's family], *Wenhui Bao,* March 23, 1957; personal interview with Qin Guo; Pen Ziqiang, *The Secret Behind Two Atomic Bombs and a Satellite in China* (Beijing: Jueshi Yuwen Press, 1993).

232 *Ying was one of the greatest beneficiaries:* Personal interviews with Tsien's friends and Ying's former colleagues Gao Yun and Guo Daizhao, June 28, 1993.

232 *tape her singing:* Personal interview with Zhang Kewen, June 27, 1993; Gao Yijin, "Qian Xuesen's Family."

233 *went to the train station to welcome them:* Gan Jiayu, p. 13.

233 *Zhou Enlai made an off-hand comment:* Ibid., p. 12.

233 *Beida Fuxiao:* Personal interview with Mr. and Mrs. C. Y. Fu, June 18, 1993.

233 *aboard the yellow school bus:* Ibid.

233 Tsien's family activities are discussed in Gao Yijin, "Qian Xuesen's Family."

233 Details of Tsien's Saturday morning political meetings can be found in Qian Xuesen, "Recollections and Hopes," *Beijing Review,* 34, no. 48 (December 2–8, 1991).

233 Information pertaining to Tsien's social life with families in his compound from interviews with Guo Qin (January 11, 1993), Luo Jin (September 5,

1991, June 29–30, 1993), Luo Peilin (September 5, 1991, June 29–30, 1993), Mr. and Mrs. C. Y. Fu (June 18, 1993).

234 *"I will never forget those fair-minded"*: Milton Viorst, *Hustlers and Heroes* (New York: Simon & Schuster, 1971), p. 154.

234 *"On this occasion"*: Ibid., p. 156.

235 *"That really hurt von Kármán"*: Telephone interview with Joseph Charyk, March 9, 1992.

235 *"I had prepared Kármán"*: Telephone interview with William Zisch, 1991. Years after Kármán's death, Tsien sent an even more cryptic note to Dean Earnest Watson of Caltech. On a Christmas card decorated with a picture of flowers, Tsien had written neatly: "This is a flower that blooms in adversity."

236 For more information on the Hundred Flowers movement, see Roderick Mac-Farquhar (ed.), *The Hundred Flowers Campaign and the Chinese Intellectuals* (New York: Praeger, 1960); Roderick MacFarquhar, *Origins of the Cultural Revolution: Contradictions Among the People, 1956–1957,* vol. 1 (New York: Columbia University Press, 1974); Yao Shuping, "Chinese Intellectuals and Science," *Science in Context* (1989): 443–45.

236 *"I will definitely not join the Communist Party"*: MacFarquhar, *The Hundred Flowers Campaign,* p. 168.

236 *failed to say a single word:* Personal interview with Zhang Kewen, June 27, 1993.

236 Tsien's denunciation of Qian Weichang can be found in "Jianjue weihu dang dui kexue gongzuo de lingdaoquan: kexuejia jiechuan Zeng Zhaolun Qian Weichang zhiding fandong kexue gangling de yin mou" [Resolutely defend the Party's leadership of scientific research: Scientist exposes Zhen Zhaolun and Qian Weichang's plot to establish a counterrevolutionary science program], *People's Daily,* July 17, 1957, p. 1.

237 *a neatly pressed shirt:* Interview with Zhang Kewen, June 27, 1993.

237 *blue or gray:* Interviews with missile scientists.

237 *to destroy the "four pests" of China:* Roderick MacFarquhar, *Origins of the Cultural Revolution: The Great Leap Forward, 1958–1960,* vol. 2 (New York: Columbia University Press, 1983), pp. 21–24.

237 *"I still remember Tsien"*: Personal interview with Zhang Kewen, June 27, 1993.

238 *vividly the image of Tsien hollering:* Ibid.; interviews with Tsien's friends.

238 *"He would scream"*: Personal interview with Zhang Kewen, June 27, 1993.

238 *"It was quite embarrassing"*: Personal interview with Jin Luo, September 5, 1991.

238 *"Now, I felt a great scientist"*: Personal interview with Zhang Kewen, June 27, 1993.

239 *"Our old scientists"*: Qian Xuesen, "Zhende hong, hongde tou" [Really red, thoroughly red], *People's Daily,* March 7, 1958, p. 7.

239 *tremendous boost to his career:* Chu-yuan Cheng, *Scientific and Engineering Man-power in Communist China, 1949–1963* (National Science Foundation booklet, NSF 65-14, 1965); Viorst, *Hustlers and Heroes,* p. 155; "Qian Xuesen," *Who's Who in the People's Republic of China* (Armonk, NY: M. E. Sharpe, 1981), pp. 294–96.

240 *"I was so excited":* Qian, "Recollections and Hopes," p. 20.

240 *"We only need necessary water conservation":* Qian Xuesen, *Kexue Dazhong* (June 1958): 228–30.

240 *streets were filled with fire and smoke:* Interviews with missile scientists; "China's Backyard Steel Boom and the Way It Turned Out," *U.S. News and World Report,* January 30, 1958.

241 *"We went to pick up coins":* Interview with missile scientist.

242 Information about the Great Leap Forward and its catastrophic aftermath from MacFarquhar, *Origins of the Cultural Revolution*; Jung Chang, *Wild Swans* (New York: Simon & Schuster, 1991); Harrison Salisbury, *The New Emperors: Mao and Deng, A Dual Biography* (New York: HarperCollins, 1993).

242 *fed better than the average Chinese citizen:* Qian, "Recollections and Hopes," p. 17; "Huang Weilu—Father of China's Carrier Rockets," *Xinhua,* October 8, 1989, item no. 1008050.

242 Description of the hunger in Beijing and in the Fifth Academy from interviews with former missile scientists; *China Today: Space Industry* (Beijing: Astronautic Publishing House, 1992), p. 472.

243 *"Tsien, who knew nothing about agriculture":* Interview with Xu Liangying, June 21, 1993.

244 *"famous scientist [had] added proof":* Li Rui, *Mao Zedong de gongguo shifei* [Mao Zedong's merits and mistakes] (Hong Kong: Tiandi) p. 278. Tsien's article can be found in *Zhongguo qingnian bao* [China youth], no. 1363 (June 16, 1958): 4. It is similar in content to his article in *Kexue Dazhong* (June 1958): 228–30.

244 *"The result of this calculation":* Fang Lizhi, letter, and telephone interview, September 18, 1995.

245 *moved into heavily guarded compound:* Interviews with Jin Luo and Qin Guo.

245 *preferred to eat his meals in silence:* Interview with Yucon Tsien.

245 *left his wife in hysterics:* People's Daily, March 31–April 10, 1992, part 8.

246 *serve as Chairman Mao's tutor:* Qian Xuesen, "I Shall Remember Chairman Mao's Kind Teachings All My Life," *People's Daily,* September 16, 1976.

246 *"intention was to urge me":* Ibid.

246 *"The present method of education":* Bill Brugger, *China: Radicalism to Revisionism 1962–1979* (Totowa, NJ: Barnes and Noble, 1981), p. 36.

246 *"We need determined people who are young":* Ross Terrill, *Mao: a Biography* (New York: Harper & Row, 1980), p. 315.

246 *"Then came the ten year catastrophe":* Qian Xuesen, "Reading a Letter From Sun Yefang Again," *People's Daily,* June 30, 1984, p. 5.

246 *quietly cultivated a network:* Salisbury, *The New Emporers,* p. 237.

246 *first Red Guards appeared in Beijing:* Jack Chen, *Inside the Cultural Revolution* (New York: Macmillan, 1975), p. 226.

247 *first poster went up in the Seventh Ministry:* Interviews with missile scientists.

247 Details of the Cultural Revolution from Chen, *Inside the Cultural Revolution;* Paul Johnson, *Modern Times: The World from the Twenties to the Eighties* (New York: Harper & Row, 1983); Chang, *Wild Swans;* Salisbury, *The New Emperors.* A brief account of the impact of the revolution on the Seventh Ministry is given in *China Today: Space Industry,* p. 43 passim.

248 *a coup d'état:* Details from author's interview with Ye Zhengguang, July 1, 1993.

250 *violence came to a head:* Information about Yao Tongbing from interviews with missile scientists and Peng Jieqing, *Inside Aeronautics* (Beijing: China Friendship, 1993).

250 *"After the metal expert died"* Interview with missile scientist.

250 *one of only fifty top scientists:* Interview with Yucon Tsien, July 13, 1991.

250 *rank equivalent to general:* Pen Ziqiang; Edward Wu, "Peking Honors Rocketry Expert," *Baltimore Sun,* August 9, 1968.

250 *profound effect on his entire family:* Interview with Yucon Tsien, July 13, 1991; interview with Jin Luo, September 5, 1991.

251 *Ying almost went to the countryside:* Interview with Gao Yun and Guo Daizhao.

251 *death of Lin Biao:* Interviews with former employees of the Seventh Ministry and with former friends of Tsien's.

252 Information about Joseph Charyk's visit comes from the author's telephone interview with Charyk, March 9, 1992.

253 *denounced in 1968 for being "a capitalist roader":* Who's Who in the People's Republic of China, p. 292.

253 *His colleague Guo Yonghuai:* Interview with Qin Guo, July 11, 1993.

254 *Luo Shijin:* Interview with Luo Shijin, July 19, 1991.

254 *"If Premier Zhou Enlai":* Qian, "Recollections and Hopes."

254 Information about Deng Xiaoping and Zhang Aiping from *Who's Who in the People's Republic of China;* Salisbury, *The New Emperors.*

254 *Tsien made speeches:* Interviews with missile scientists.

255 *"Mao not only rescued me from my plight abroad":* Qian, "I Shall Remember Chairman Mao's Kind Teachings." The China News Analysis (no. 1112, March 10, 1978) observed Tsien's "faux pas" in criticizing Deng Xiaoping in his *People's Daily* article but conceded that "to a man of such importance a faux pas may be forgiven."

255 *scrambled to repair the damage:* Chien Hsueh-sen, "Science and Technology: We Must Catch Up with and Surpass World Advanced Levels Within This Century," *Hongqi* [Red flag], no. 7 (1977): 14–18 and *Peking Review*, no. 30 (July 22, 1977); Qian Xuesen and Wu Jiapei, "Zuzhiguanli shehui zhuyi jianshe de jishe—shehui gongcheng" [A technique for the organization and management of socialist construction: Social engineering], *Jingji guanli* [Economic management], no. 1 (1979): 5–9.

255 *"It really didn't make any sense":* Interview with missile scientist.

255 *"things went sour":* Interview with Xu Liangying, June 21, 1993.

256 *"If you had seen him":* Interview with Hua Di, January 13, 1992.

256 *returned to normalcy:* Interview with Jin Luo.

256 *became something of a dilettante:* Interview with Wang Shouyuan. Tsien's involvement in different organizations can be found in numerous *Xinhua* articles and *Who's Who in the People's Republic of China*: methane mentioned in *Xinhua*, June 13, 1979, item no. 061317; UFO research mentioned in *Xinhua*, October 4, 1986, item no. 1004179; waste collection in *Xinhua*, February 9, 1987, item no. 0209032; Chinese women in *Xinhua*, March 6, 1988, item no. 0306115.

256 *"We cannot help saying aloud":* British Broadcasting Corporation, July 25, 1981.

256 *"small bubbles in the sea":* Ibid., October 24, 1989.

256 *"definitely not stupid":* Reuters, September 4, 1982.

257 *boy named Tang Yu:* Xinhua, August 19, 1980, item no. 071810.

257 *Perhaps ESP was linked with qigong:* Qian Xuesen, "Launching Basic Research on the Science of the Human Body," *Nature* (Chinese publication), no. 7 (1981): 483–88 (*Science and Technology*, no. 1 [1981]); *Guangming Daily*, May 12, 1986, p. 1.

257 *"I was frankly astonished":* Interview with Milton van Dyke, January 5, 1992.

257 Information about the pro-democracy movement from Jonathan Spence, *Search for Modern China* (New York: W. W. Norton, 1990); Harrison Salisbury, *Tiananmen Diary* (Boston: Little, Brown, 1989).

259 *"We are shocked":* Kyodo News Service, June 10, 1989.

259 *seen on Beijing television;* Beijing television, June 14, 1989, announcer-read report FE/0484 B2/1, British Broadcasting Corporation, June 17, 1989.

259 *"scum of the nation":* People's Daily, June 28, 1989, p. 5.

259 *won the Williard F. Rockwell, Jr., medal:* Liu Jingzhi, "Qian Xuesen—Pride of the Chinese People," *Guangming Daily*, August 28, 1989, pp. 1–2.

259 Ironically enough, during an August 7, 1989, award ceremony Tsien used the American medal as an excuse to criticize the American government. He retold the story of his detention by the INS and condemned the United States for its reaction to the Tienanmen incident.

259 *"book from Heaven":* Guangming Daily, August 28, 1989, p. 2.

259 *publicly praising senior scientists:* David Sanger, "Beijing Journal: With Democracy in Dust, a Cloud Falls on Science" *New York Times,* September 15, 1989.

260 *"State Scientist of Outstanding Contribution":* Xinhua, October 18, 1991.

260 *People's Daily:* Xinhua, October 17, 1991, item no. 1017056.

260 *awards ceremony:* Xinhua, October 16, 1991, item no. 1016208; British Broadcasting Corporation, October 19, 1991.

260 *"We should follow the example":* Xinhua, October 17, 1991; Agence France Presse, October 17, 1991.

260 *Li and Jiang:* Xinhua, October 16, 1991, item nos. 1016198 and 1016179.

260 *issued a circular:* British Broadcasting Corporation, October 21, 1991.

Index